SOLUBILITY OF GASES AND LIQUIDS
A Graphic Approach

SOLUBILITY OF GASES AND LIQUIDS

A Graphic Approach

Data – Causes – Prediction

W. Gerrard
*The Polytechnic of North London
London, England*

PLENUM PRESS · NEW YORK AND LONDON

Library of Congress Cataloging in Publication Data

Gerrard, William.
    Solubility of gases and liquids.

    Bibliography: p.
    Includes index.
    1. Solubility. 2. Gases. 3. Liquids, I. Title.
    QD543.G45   541'.342   76-10676
    ISBN 0-306-30866-5

© 1976 Plenum Press, New York
A Division of Plenum Publishing Corporation
227 West 17th Street, New York, N.Y. 10011

All rights reserved

No part of this book may be reproduced, stored in a retrieval system, or transmitted,
in any form or by any means, electronic, mechanical, photocopying, microfilming,
recording, or otherwise, without written permission from the Publisher

Printed in the United States of America

Preface

The solubility of gases and liquids in liquids is of great importance in large areas of operations based on chemical concepts. Phenomena have appeared to be so varied that even experts have from time to time remarked on the difficulty of seeing a consistent pattern. Now for the first time the essential pattern of all known gas solubility data is set out in a graphic form for all to see. The continuous merging of the gas–liquid systems and the liquid–liquid systems is also illustrated. The pattern opens the way to rational predictions.

The new data given for the lower alkanes and alkenes, the three methylamines, ammonia, bromomethane, and chloroethane, together with my previously reported data on hydrogen sulfide, dimethyl ether, chloromethane, and sulfur dioxide, have been obtained by a bubbler-manometer procedure which is fully described. Not only are these data of significance in many chemical processes, but they have also been vital to the development of the overall essential pattern covering all gases.

The book is for chemists, chemical engineers, biotechnologists, certain physicists, and teachers and students in these disciplines. It is a book for all those who are concerned with the use and inculcation of the fundamental, even rudimentary, principles of chemistry.

With the kind agreement of The Royal Society, I have given a sketch of Henry's apparatus (1803) based on Henry's original drawing (see Fig. 25). By the kind agreement of The Faraday Society, I have based my Fig. 48 on the data and part of a diagram given by F. H. Campbell (1915). Nine of the figures appearing in Chap. 1 and in the early part of Chap. 2 are based on diagrams in my own papers (1972, 1973) by the kind agreement of the Society of Chemical Industry.

I record the pleasure I have enjoyed in the study of the many papers by J. H. Hildebrand and his co-workers published over a period of some 60 years. In connection with the measurement of the solubility of the hydrogen halides in many organic and certain inorganic compounds, I gratefully acknowledge the collaboration of my former colleagues W. Ahmed, T. Brennan, J. Charalambous, T. M. Cook,

M. J. Frazer, S. V. Kulkarni, E. D. Macklen, R. W. Madden, V. K. Maladkar, A. M. A. Mincer, C. B. Lines, R. G. Luckcock, J. A. Sandbach, P. H. Tolcher, and P. L. Wyvill.

The Polytechnic of North London,   *W. Gerrard*
Holloway, London, N7 8DB, U.K.

Contents

1 The Solubility of Gases in Liquids 1

    1.1. Rationalization of Data 1
    1.2. Measurement of Gas Solubility 3
    1.3. The Bubbler-Tube and Manometer Techniques for the Determination of the Solubilities of Gases in the Upper Half of the bp Range . 3

2 Presentation of Solubility Data on the Reference-Line Diagram 9

    2.1. The Concept of the Reference Line, the R-Line 9
    2.2. Solubilities of Hydrocarbon Gases 11
    2.3. The Basis of the Explanations of the Essential Patterns of Data . . 25

3 Henry's Law and Raoult's Law 29

    3.1. Henry's Law 29
    3.2. Raoult's Law 36

4 Hildebrand's Solubility Parameters 55

    4.1. Intermolecular Forces 55
    4.2. The Solubility Parameter 55
    4.3. Force Constants 56

5 The Hydrogen-Bonding Structure of Water 59

    5.1. The Essential Pattern 59
    5.2. The Situation about 1974 60
    5.3. Significance of the Solubility of Hydrocarbon Gases in Water . . 61

5.4.	Hydrophobic Interactions	63
5.5.	Solubility of Certain Gases in Water and the Influence of the bp/1 atm Factor	64
5.6.	Other Aspects of Water Structure	74

6 Sources and Form of Data 77

6.1.	Review by Markham and Kobe (1941) on "The Solubility of Gases in Liquids"	77
6.2.	Review by Battino and Clever (1966) on "The Solubility of Gases in Liquids"	79
6.3.	Papers by Hildebrand and Colleagues	80
6.4.	Papers by Clever and Battino	80
6.5.	Data Reference Books	81

7 Data for the Noble Gases and Other Gases in the Lower Half of the Boiling Point Range 83

7.1.	References	83
7.2.	The Noble Gases Helium to Radon	83
7.3.	Hydrogen, Nitrogen, Oxygen, Carbon Monoxide	86
7.4.	The Three Factors	87
7.5.	Narcotic Effect of Xenon	87
7.6.	Argon in Dioxane and Water Mixtures	88
7.7.	Other Items	88

8 Carbon Dioxide, Nitrous Oxide, Hydrogen Sulfide, Chlorine, Sulfur Dioxide, and Carbonyl Chloride 89

8.1.	Carbon Dioxide and Nitrous Oxide	89
8.2.	Hydrogen Sulfide, bp $-61°C$	94
8.3.	Chlorine	97
8.4.	Sulfur Dioxide, bp $-10°C/1$ atm	103
8.5.	Carbonyl Chloride, bp $8.2°C/1$ atm	113

9 Hydrogen Halides HCl, HBr, HI 115

9.1.	Acid–Base Function	115
9.2.	Solubility of Hydrogen Halides in Nonaqueous Liquids	120
9.3.	Aspect of Electrolytic Conductance	149
9.4.	Nitrile–Hydrogen Halide Systems	153
9.5.	Solubility in Water	157

10 Ammonia and the Three Methylamines 161

    10.1. Ammonia 161
    10.2. Methylamines 166
    10.3. Appreciation of the Essential Pattern of Data 184
    10.4. Recent Studies on the Interaction of Aliphatic Amines with Water Structure 184

11 Dimethyl Ether 187

12 Halogenoalkanes 191

    12.1. Chloromethane, Chloroethane, and Bromomethane and Non-aqueous Liquids S 191
    12.2. Solubility in Water 202

13 Hydrocarbon Gases 205

    13.1. Saturated Hydrocarbons 205
    13.2. Unsaturated Hydrocarbons 208
    13.3. Selective Absorption of Gaseous Hydrocarbons 209
    13.4. Solubility in Alcohols 216
    13.5. Solubility in Water 217
    13.6. Other Examples 218

14 Examples of Certain Other Gases 223

    14.1. List of Formulas and bp °C/1 atm 223
    14.2. R-Lines 223
    14.3. Data for Certain Gases Containing Fluorine 226
    14.4. Other R-Line Examples 228

15 Effect of Temperature 231

    15.1. Earlier Aspects 231
    15.2. Lannung's Data 232
    15.3. Data by Gerrard and Colleagues 232

16 Prediction 235

    16.1. Approach to the Problem 235
    16.2. Lachowicz's Review 236
    16.3. Other Papers 236

16.4.	New Data on Propene and the Four Butenes	240
16.5.	The R-Line Approach to the Acetylene Systems	244
16.6.	The Carbon Dioxide Systems	246
16.7.	Solubility of Methane, Hydrogen, and Nitrogen in Liquids at High Pressures	251
16.8.	Solubility of Gases in Nitrate Melts	254
16.9.	Solubility of Noble Gases in Molten Fluorides	257

17 Textbook Statements 259

17.1. Introduction 259
17.2. Illustrative Examples of Textbook Statements 259

References 263

Index 273

The Solubility of Gases in Liquids

1.1. Rationalization of Data

1.1.1. Graphical Presentation of Data

The graphical presentation of the way in which solubility varies with pressure has been practiced for more than 100 years, but only in a limited sense, and then with reference to Henry's law or Raoult's law. In this new approach, I disregard these two so-called laws, and also the concept of the ideal model, the ideal solution.

The vapor pressure of a gas A, $p°_A$, over the liquefied gas (or its equivalent) at a stated temperature $t°C$ is a fixed property of the gas A and is quite independent of any property of the liquid S which is selected to absorb the gas. I look upon $p°_A$ as representing the *tendency* of the gas to condense at the stated $t°C$. The bp of A at 1 atm is a rough index of this tendency to condense. In the solubility–pressure diagram, I use a line of reference, the R-line, based exclusively on $p°_A$.

1.1.2. Form of Data

I deem the *mole ratio*, x_A moles of A per mole of S (or M_S moles of S per mole of A), to be the most fundamental form of expressing the mass of gas A absorbed by a given mass of liquid S at a specified temperature $t°C$ and gas pressure p_A. Equilibrium conditions are always understood to prevail. Chemical equations are given in the mole ratio form, and thermodynamic quantities are usually expressed as per mole of a species. For the R-line diagram, the mole fraction form $N_A = x_A/(1+x_A)$ is a more convenient form, *only* because of the peculiar arithmetic way in which N_A changes with x_A (see Fig. 1).

As primary data are given in a variety of ways, I have had to calculate the x_A values in order to convert those data into a form for logical comparison.

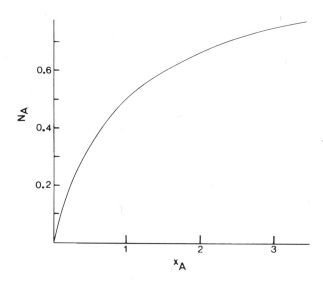

Fig. 1. Plot of N_A versus x_A.

1.1.3. Units

As I trace data and statements to the original sources, I tend to use the older form of units to avoid confusion. It is understood that:

$$1 \text{ cal} = 4.1868 \text{ J}$$
$$1 \text{ mm Hg} = 13.5951 \times 9.80665 \text{ Pa (pascal)}$$
$$1 \text{ atm} = 760 \text{ mm Hg}$$

1.1.4. The Term "Gas"

I use the terms *gas* and *vapor* indiscriminately. For my present purpose, I include as gases all elements and compounds having a bp/1 atm less than about 13°C. If certain complications such as isotope effects are not considered, there are about 140 of them. This decision is arbitrary, but it is based on the operational circumstances that such substances are stored and transported in cylinders or sealed tubes, from which they are usually delivered in the gaseous state. The graphical approach, however, can, without break in form, be used to deal with elements and compounds having bp > 13°C; isopentane, bp 27°C/1 atm, may be treated at 30°C much in the same way as *n*-butane, bp −0.5°C/1 atm, at 3°C. The restriction to 13°C enables me to refer in a simple way to the range of boiling points from that for He to, for example, that for chloroethane, bp 13.1°C/1 atm. I can then refer, for example, to the lower end of the bp range and to the upper end. The bp/1 atm of a gas A is of primary significance in comparing solubility patterns.

The Solubility of Gases in Liquids

1.1.5. The Three Factors

The essential features of all the gas solubility data can be rationalized in terms of three factors:

(a) the *tendency* of the gas A to condense to a liquid at the operational temperature $t°C$
(b) the pattern of intermolecular forces of the liquid S used to absorb gas
(c) the inevitable interaction of A with S.

1.2. Measurement of Gas Solubility

1.2.1. Measurement of Volume of Gas A and Liquid S

The primary observations on gases in the lower half of the bp range, e.g., He, H_2, N_2, Ar, O_2, CO, and CH_4, have involved measurement of volume of gas at a stated $t°C$ and pressure p_A or total pressure $p_A + p_S$. Such volume/volume data have been presented as absorption coefficients. The experimental techniques have been well summarized in the two reviews (by Markham and Kobe[253] and Battino and Clever[17]), discussed in Sections 6.1 and 6.2.

1.2.2. Gravimetric Measurement of Mass of Gas

For gases in the upper half of the bp range, direct measurement of mass of gas becomes convenient. The procedure devised by the present writer is now described.

1.3. The Bubbler-Tube and Manometer Techniques for the Determination of the Solubilities of Gases in the Upper Half of the bp Range

1.3.1. The Bubbler-Tube

The bubbler-tube (B.T.) comprises an absorption vessel of capacity about 20 to 25 cm^3, fitted with two good-quality glass taps, one sealed to a cone B10/19 and the other to a socket (see Fig. 2). Tap T_1 carries the inlet cone and a long inlet tube which passes almost to the bottom of the absorption vessel. Tap T_2 is fitted to the exit socket. The taps are lubricated with a minimum of very low-volatile vaseline which is reapplied after several runs, according to circumstances.

The B.T. is washed out with three small portions of A.R. acetone, then three small portions of sodium-dried ether, and, with the taps open, kept for a few minutes in an oven at about 60°C. With T_1 shut, the B.T. is attached to a water-tap vacuum pump for 2 or 3 min to evacuate the vessel, T_2 being shut before removal from the water-tap assembly. The B.T. is then weighed containing air at about 15 mm Hg

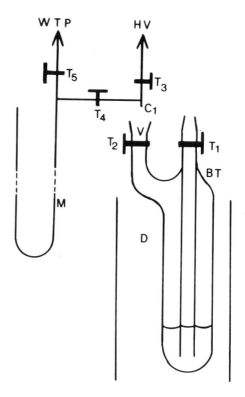

Fig. 2. Diagram of the bubbler-tube (B.T.) and manometer (M) assembly. T_1–T_5 are taps; HV is a high-vacuum pump; WTP is a water-tap pump; C_1 is a cone to fit at V above T_2; D (a dewar) is the constant-temperature bath.

pressure. To get the essential pattern of data, it is unnecessary to evacuate to a pressure of ≈ 0.1 mm Hg by means of a high-vacuum pump. Tap T_2 is opened for a moment and then shut; the difference in weight gives, with sufficient accuracy, the mass of air in the tube at the prevailing atmospheric pressure and temperature, which are measured. From this the capacity of the B.T. in cm^3 is computed. This capacity is not used as such, but it is a helpful background; and the double weighing is always done as a check.

The absorption liquid S is pipetted into the B.T. via the *inlet* cone and T_1, with T_2 open. The amount of liquid S taken depends upon the gas A being examined and the temperature of operation. For example, at 0°C, 760 mm Hg, 1 g of dimethylformamide absorbs about 3.5 g of SO_2, there being a considerable increase of volume of liquid in the tube. Therefore, 1–2 g of S would be taken. Under the same conditions, 1 g of ethylene glycol absorbs about 0.8 g of SO_2, and therefore 3–5 g of S would be taken.

The taps are shut, the inside of the inlet cone is dried by a spool of filter paper, and the exit socket is attached to the water-tap pump. Tap T_2 is carefully opened to

The Solubility of Gases in Liquids

"evacuate" the space above the liquid. Then T_2 is shut, the B.T. weighed, and T_1 carefully opened for a moment and again shut; a second weighing provides information which is used to estimate the mass of gas A above the solution. It is, however, the first weight which is used to obtain the *total* amount of A in the liquid and vapor.

1.3.2. Absorption of Gas at the Barometric Pressure and Measured Temperature $t°C$ Not Less than $-5°C$

Both taps being open, the inlet cone at T_1 is attached via a socket to the source of gas A. The outlet socket at T_2 is attached via a cone to another bubbler-tube containing mercury to a level which just covers the bottom end of the inlet tube of that B.T. This provides a visible flow indicator when the main B.T. is immersed in the constant-temperature bath. For certain gases such as hydrogen sulfide, I use *n*-decane as the liquid in the flow indicator. The effluent gas is chemically absorbed, or passed into the flue of the fume-cupboard.

Temperature control has been manual, and a thermos flask containing water, or ice+water, or acetone cooled by addition of solid carbon dioxide has been used. Usually absorption is quick. For temperatures down to about $-5°C$, the tube is weighed immediately on removal from the bath. The gas flow is shut off, the taps are shut, and the inlet and outlet tubes are disconnected while the B.T. is still in the bath. Gas in the cone and socket is removed by a spool of filter paper. The B.T. is removed, quickly dried with absorbing paper, wiped with a fine cotton cloth, and weighed. The B.T. is then immediately replaced in the bath and attached to the gas supply as before; but a few minutes are allowed to elapse before T_2 and then T_1 are opened, and the gas flow continued. After a few minutes, the weighing process is repeated until the weight is either constant or begins to fall due to entrainment of the vapor of the absorbing liquid S. Significant entrainment occurs only for the more volatile liquids such as ethanol or benzene, and this can be allowed for by carrying out a series of measurements to estimate the rate of loss. In the recent work of the present author, the results of which are described herein, liquids of low volatility were used in order to chart the essential pattern over a wide-enough area by the most expeditious procedure. The pressure recorded for this process is the total pressure. Although attempts to allow for the vapor pressure p_S of the absorbing liquid S are feasible, they are to be made with caution, since much depends on the particular system, especially on the operational temperature. For example, 1 mole of acetone at 10°C and a total pressure of 1 atm absorbs about 1.70 mole of SO_2. At 0°C the x_{SO_2} would be much greater than this. To estimate $p_{acetone}$ for such systems would be a protracted operation; $p_{acetone}$ would be much less than $p°_{acetone}$. In the more recent work of the present writer, the p_A is taken as being essentially the same as the total observed pressure.

For certain gases, such as the hydrogen halides, chemical analysis can be used to avoid the entrainment problem.

1.3.3. Procedures for Temperatures Lower than about $-5°C$

Since the weighing must be done at room temperature, it is hazardous to attempt to weigh the tube after absorption at lower than $-5°C$ because of the development of pressure inside the B.T.

For the earlier work on hydrogen chloride for temperatures down to about $-70°C$, a chemical analysis was made. For HBr down to about $-60°C$, and for HI down to about $-30°C$, a chemical analysis can also be made; but for HBr and HI, a much more direct procedure is available. This is to absorb the evolved gas, as the temperature of the B.T. is gradually raised to about $0°C$, in benzonitrile which avidly absorbs these gases to give crystalline solids. This procedure has been described by Gerrard and Maladkar.[114]

For temperatures down to about $-30°C$, NH_3 can be determined by a chemical procedure, although it is feasible to use an absorbent in a similar way to that used for HBr and HI.

For gases such as chloromethane and dimethyl ether, extrapolation of the $x_A/t°C$ or $N_A/t°C$ curve will give a good-enough value of x_A or N_A for certain purposes. For gases such as propane, the extrapolation would become less reliable as $t°C$ approaches about $-40°C$.

1.3.4. Calculation of the x_A and N_A Values

The estimated weight of the gas A above the solution in the tube is subtracted from the total increase in weight. The x_A and N_A values are then computed as follows:

$$x_A = \frac{\text{Mass of A absorbed}}{\text{Mol. wt. of A}} \bigg/ \frac{\text{Mass of S taken}}{\text{Mol. wt. of S}}$$

$$N_A = x_A/(1+x_A)$$

The molecular weight is always that related to the usual formula weight.

The increase in weight due to absorption can be from 0.6 to 0.15 g for a gas such as propane; but for such a gas as sulfur dioxide, the increase in weight can be as much as 5 g. In estimating the mass of A in the vapor phase, allowance for this increase in liquid volume must be made.

1.3.5. Measurements at Different Pressures

To get x_A/p_A and N_A/p_A data for the revelation of the essential patterns, I have used liquids S of low-enough volatility because I have wished to draw off only gas A from the mixture of A+S by lowering the total pressure.

The primary absorption is done as described, except that the temperature of the bath is about $5°C$ below that of the manometer bath. This is to ensure a first reading

The Solubility of Gases in Liquids

for p_A of about 800 mm Hg. The B.T., with taps shut, is transferred directly from the absorption assembly to the manometer assembly (see Fig. 2). Tap T_3 carries a cone which fits at V on T_2 and controls access to a high-vacuum pump. Tap T_4 controls access to the manometer, and T_5 to a water-tap pump (15 mm Hg). Tap T_1 of the B.T. remains shut throughout. Tap T_2 is initially shut, the socket attached thereto is fitted onto the cone at T_3, and T_3 is opened to evacuate the short lengths of tube between T_2 and T_3 and onto T_4, which is shut. The mercury in the manometer has already been drawn up to the high-vacuum pressure (about 0.1 mm Hg), the end of the other limb being open, and T_4 and T_5 are shut. Tap T_3 is shut. Tap T_2 is gently opened, and then T_4, and the mercury level falls to one registering a p_A of, say, 800 mm Hg. Tap T_4 and T_2 are shut, and T_5 gently opened to draw the mercury up to the 15 mm Hg level; T_5 is then shut, and T_4 and T_3 are opened to bring the level up to the 0.1 mm Hg one. Taps T_3 and T_4 are shut, and T_2 again gently opened, followed by T_4 to get another p_A value. This procedure is incorporated with the detachment of the B.T. for shaking while still immersed in the bath, the B.T. being reattached as before, with T_4 remaining shut. The final reading on the manometer gives the equilibrium total pressure, taken for the purpose as essentially equal to p_A at the $t°C$ of the bath. With T_4, T_3, and T_2 shut, the B.T. is removed, dried as before, and weighed. This gives data for computing the x_A value (and thence the N_A value) for p_A and $t°C$.

The B.T. is shaken gently in the bath to equilibrate the temperature and is then attached to the assembly as described. On opening T_2 and then T_4, the mercury level remains essentially constant. Tap T_5 is now gently opened slightly to withdraw some gas from the solution. The water-tap pump is used to avoid passing more than a very small amount of gas, e.g., ammonia or sulfur dioxide, into the high-vacuum pump assembly. Tap T_5 is shut, and the new pressure p_A measured after attainment of equilibrium achieved by the following drill. Taps T_4 and T_2 are shut. The B.T. is detached, shaken while still in the bath, and reattached by opening T_3 and then closing T_3 and opening T_2 and T_4. Taps T_4 and T_2 are shut, and the B.T. is detached, dried, and weighed. This gives the new x_A (and N_A) for the new corresponding p_A. The process is repeated to get x_A values down to p_A about 25 to 20 mm Hg. Finally, the pressure is reduced to 15 mm Hg to show that the weight of the residual liquid is essentially that of the original liquid S. The refractive index and infrared spectrum have always shown that the remaining liquid is essentially the pure S. Final removal of gas at 0.1 mm Hg has confirmed this conclusion, although a small amount of S can be lost in this final process.

The allowance for the mass of gas A in the vapor phase must be adjusted to comply with the lowering of p_A and the decreasing of the volume of the liquid. The estimation is made according to the particular circumstances.

The values of N_A and the corresponding p_A values give remarkably smooth curves. There is a tendency for plots of the relatively more volatile liquids S such as m-xylene to give curves which tend to drift to the left (N_A lower than it should be) at the low end of the p_A range. This loss is a very small percentage of S, but it appears as a larger one for N_A which by that stage is at the lower end. Although the essential form of the N_A/p_A line is nevertheless quite clear, a correction can be obtained by starting with a fresh sample of S, passing in only a small amount of A, and measuring an isolated value of N_A and p_A by one operation on the manometer assembly.

1.3.6. Accuracy

It is difficult to be precise about accuracy because so much depends upon the system and temperature. For liquids S of low volatility the degree of accuracy is deemed to be within 4%. This affords reliable data for the essential patterns. Time-consuming modifications leading to a precisely specifiable accuracy will be revealed to one who becomes "skilled in the art."

1.3.7. Evaluation of Purity of S and A

To obtain my own data, the best specimens of S were distilled and attested by the infrared spectrum and refractive index. The best commercially procurable specimens of A were used, and certain procedures to ensure freedom from moisture were carried out.

1.3.8. Liquids S of Greater Volatility

I have not yet carried out measurements of N_A for different p_A for liquids S such as ethanol; but certain modifications to the described procedure appear feasible and could lead to a set of data which would give an approximate pattern useful for simple comparisons.

1.3.9. Cleaning the Tube

With both taps open, the residual liquid S is poured out through the socket at T_2, and acetone and then dry ether are added in portions through the cone at T_1, as already described.

2

Presentation of Solubility Data on the Reference-Line Diagram

2.1. The Concept of the Reference Line, the R-line

2.1.1. The Vapor Pressure of n-Butane, bp −0.5°C/1 atm

I pass n-butane gas from a cylinder into the bubbler-tube cooled to about −15°C to collect about 5 g of liquid. I then transfer tube and cooler to the manometer assembly and attach the tube as shown in Fig. 2. The mercury level in the closed limb of the manometer is at its highest, indicating a pressure of about 0.1 mm Hg. Taps T_4 and T_2 are gently opened; the mercury level falls to a certain point on the scale. Tap T_4 is shut; T_5 is gently opened to draw up the mercury to the level corresponding to the water-tap pump pressure. Tap T_5 is shut; T_4 is opened again; and the whole operation is repeated several times to make sure all air has been withdrawn from the tube and nothing but liquid and gaseous n-butane are present. The tube is then shaken at a constant $t°C$, which is recorded, until the level of the mercury in the closed limb is constant. To make sure that equilibrium has been attained, i.e., that gas is leaving and entering the liquid at a constant rate, no matter how long the system is allowed to stand, I shut T_4 and T_2, remove the tube from cone C_1, and shake the tube while it is still immersed in the constant-temperature bath. I attach the tube at cone C_1, open T_3 to evacuate the space V, shut T_3, open T_2 and then T_4. The level of the mercury should remain constant if equilibrium has been attained. The final reading of the manometer gives the pressure $P°_{BuH}$ for the temperature $t°C$. With T_4 and T_2 still open, I can remove n-butane from the tube by gently opening T_5, and after again closing T_5, I can show that $p°_{BuH}$ remains unchanged whatever the amount of liquid remaining in the tube so long as there is liquid.

The temperature is now raised to, for example, −10°C, and the whole procedure repeated to observe $p°_{BuH}$ for the new $t°C$. In this way, the value of $p°_{BuH}$ for each of several $t°C$ is observed, the final $t°C$ being about 1°C for this apparatus. To draw a graph connecting $p°_{BuH}$ with $t°C$ up to 30°C, I look up the data in the literature. It is true that more refined techniques may have been used in the determination of the published data, many man hours being expended; but the procedure now described is discriminating enough for the purpose in hand, i.e., to plot the essential pattern of solubility data. My own measurement of $p°_{BuH}$ serves to assess the reliability of the bubbler-tube–manometer procedure and to attest the purity of the gas.

At a fixed $t°C$, $p°_{BuH}$ is a fixed value and is a property of the gas alone. I look upon $p°_{BuH}$ as the *tendency* of the gas *to condense* at the $t°C$ stated. The bp of a gas at 1 atm is a ready approximate index of its tendency to condense.

2.1.2. The R-line

I draw the "box" as shown in Fig. 3. The two vertical axes are divided to represent the pressure $p°_{BuH}$, and the base line is not divided because it does not need to be. To draw the R-line for 5°C, I place a spot at 940 mm Hg ($p°_{BuH}$ at 5°C); from this on the right vertical axis, I draw a straight line to the left bottom corner. This is what I call the reference line, the R-line, for n-butane at 5°C, and it is quite independent of any other concept. Similarly I can draw the R-lines for other temperatures, as shown in Fig. 4. This diagram gives a visual indication of how the tendency to condense becomes less as the $t°C$ becomes higher.

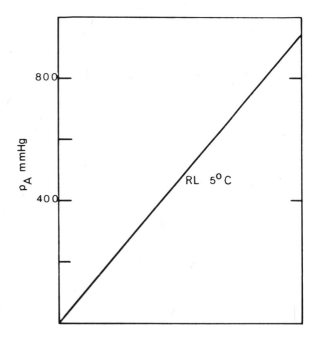

Fig. 3. The reference line RL for n-butane at 5°C. Vapor pressure $p°_A$ (A = n-butane) = 940 mm Hg at 5°C.

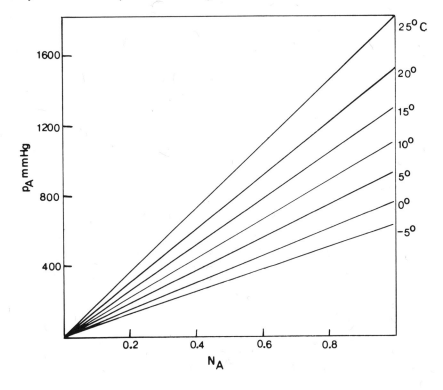

Fig. 4. Reference lines for n-butane at different temperatures $t°C$ with the mole fraction scale N_A added. A = n-butane.

2.2. Solubilities of Hydrocarbon Gases

2.2.1. n-Butane

By means of the bubbler-tube–manometer procedure,[99] I determine the mass of n-butane absorbed by a known mass of a liquid S at $t°C$ and p_{BuH}, the pressure of gaseous n-butane. The apparatus constrains me to measure only the *total* pressure, which is made up of p_{BuH} and p_S. To allow for p_S is a protracted operation, as I shall show later; but to build up the essential pattern of data for my present purpose, I use liquids which have a low-enough p_S to be neglected. Therefore, I take the observed total pressure to be essentially equal to p_{BuH}. In the conventional terminology the expression "partial pressure" is used, i.e., p_{BuH} and p_S would be called "partial pressures." I do not like this, and see no need for it. To me p_{BuH} is the pressure due to n-butane, and p_S that due to S; the measured pressure is $p_{BuH} + p_S$.

Since the usual chemical equations are in the mole *ratio* form, and since thermodynamic quantities are given per mole, I look on the mole *ratio* form as the

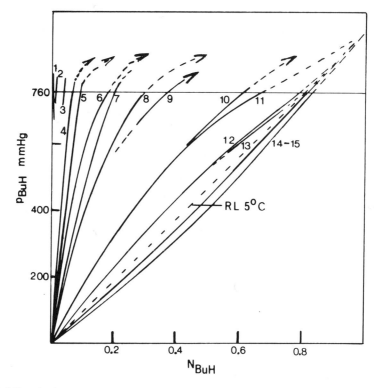

Fig. 5. Solubility of *n*-butane in various liquids at 5°C. Plots of the mole fraction N_{BuH} vs pressure p_{BuH}. The lines are for (1) H_2O; (2) $(CH_2OH)_2CHOH$; (3) $(CH_2OH)_2$; (4) $C_6H_5NH_2$; (5) $HCON(CH_3)_2$; (6) $C_6H_5CH_2OH$; (7) $C_6H_5NO_2$; (8) C_6H_5CN (Cl_3CCH_2OH, o-$NO_2C_6H_4CH_3$, and $C_6H_5NHCH_3$ are just on the left of 8); (9) o-$HOC_6H_4CO_2CH_3$; (10) n-$C_8H_{17}OH$); (11) n-$C_5H_{11}CO_2H$; (12) n-$C_8H_{17}I$; (13) m-$C_6H_4(CH_3)_2$; (14) n-$C_{10}H_{22}$; (15) $(n$-$C_8H_{17})_2O$. The broken line is the reference line RL.

most fundamental one in which to express solubility. I therefore calculate the mole *ratio* x_{BuH}; the number of moles of *n*-butane absorbed by 1 mole of S, is given by

$$\frac{\text{Mass of gas absorbed}}{\text{Mol. wt. of gas}} \bigg/ \frac{\text{Mass of S}}{\text{Mol. wt. of S}}$$

The temperature $t°C$ and p_{BuH} must be specified, and it is always understood that equilibrium conditions prevail. Furthermore, it is understood that the molecular weight refers to the usual molecular formula, e.g., in the present case, n-C_4H_{10}.

For diagram purposes it is convenient to express the solubility as the mole *fraction*, i.e., the number of moles of *n*-butane divided by the total number of molecules, i.e., $x_{BuH}/(1 + x_{BuH})$, and the symbol I use is N_{BuH}.

Using the R-line diagram for 5°C, I divide the base line into numbers from 0 to 1 to represent the mole *fraction* N_{BuH}. I then plot the N_{BuH} values for S = di-*n*-octyl ether against the corresponding p_{BuH} values. The resulting line is concave upward (Fig. 5), on the right side of the R-line. The line must of necessity start from $N_{BuH} = 0$, at the left bottom corner, and finish at $p°_{BuH} = 940$ mm Hg; in effect, $N_{BuH} = 1$. For $N_{BuH} = 0.900$, we have $x_{BuH} = 9.00$, i.e., 9 molecules of n-butane for 1 molecule of the ether. When $N_{BuH} = 0.990$, $x_{BuH} = 99.0$. Therefore, as $p_{BuH} \rightarrow 940$ mm Hg, $N_{BuH} \rightarrow 1$, x_{BuH} increases very rapidly with increase in p_{BuH}; this is shown visually in Fig. 6, which is a plot of x_{BuH} and p_{BuH}. For this reason, the observation of x_{BuH} values between about 800 and 940 mm Hg is troublesome. It is clear, however, that if we start with liquid n-butane at 5°C, the $p_{BuH} = p°_{BuH}$, and add di-n-octyl ether to the extent of $N_S = 0.1$, i.e., $N_{BuH} = 0.9$, $x_{BuH} = 9.0$, p_{BuH} will fall from $p°_{BuH} = 940$ to about 800 mm Hg. Observing the rate at which the change of N_{BuH} with p_{BuH} *changes* in the region near $N_{BuH} = 1$, $p_{BuH} = 940$ mm Hg is a matter for detailed and delicate manipulation. It is easier to follow changes at the lower end of the line, as $p_{BuH} \rightarrow 0$; but even here, delicate manipulation would be required to follow the rate with which the change in N_{BuH} with p_{BuH} changes as we start from pure S and add n-butane in amounts corresponding to $N_{BuH} = 0.001$.

The line for n-decane is also concave upward on the right of the R-line. The line for m-xylene, however, is convex upward, on the left of the R-line. Figure 5 should now be studied patiently. It is evident that the lines for the other liquids named fill the left half of the diagram. Look at the line for aniline, well over on the left of the R-line.

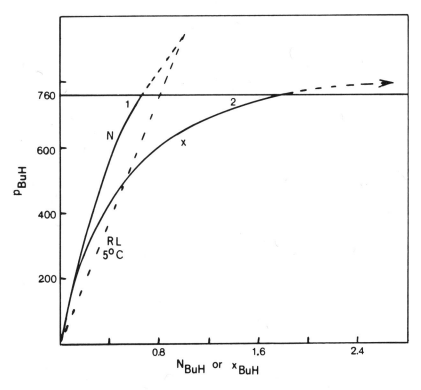

Fig. 6. Solubility of n-butane in n-hexanoic acid at 5°C. Comparison of the plots of pressure p_{BuH} (mm Hg), vs mole fraction N_{BuH} (line 1), and vs mole ratio x_{BuH} (line 2).

Instead of being 5.666, as it is for di-n-octyl ether, x_{BuH} is now 0.074 for p_{BuH} = 760 mm Hg, N_{BuH} being 0.069. The line has to bend round somewhat abruptly and proceed right over to the right in order to reach the point $p°_{BuH}$ = 940 mm Hg on the right vertical axis; and this it must do as p_{BuH} is increased. As p_{BuH} is increased beyond 760 mm Hg, two separate liquids (two liquid phases) appear, both of which contain n-butane and aniline; but there is only one (a common) value of p_{BuH}; and ultimately, as p_{BuH} is continuously increased, the two liquids will merge as p_{BuH} closely approaches $p°_{BuH}$. The actual plot of the data still fits into the R-line diagram, but the N_{BuH} values for these higher pressures refer to the total amount of n-butane added to form the two liquid phases.

Notice the position of ethylene glycol, x_{BuH} = about 0.06, and that of glycerol, x_{BuH} = ~0.01, right over on the extreme left of the diagram. The line for water is registered alongside the left vertical axis. The actual value (see below) is reported to be N_{BuH} = 0.000026 at 1 atm and 20°C.

Figure 7 shows representative lines for 25°C, and also for −5°C. The essential pattern is the same for each temperature, the main factor being the tendency to condense. At −5°C, the system is, in conventional terms, a liquid–liquid one, similar

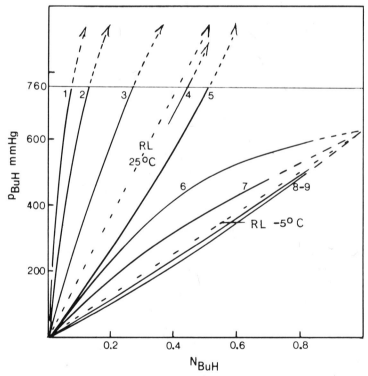

Fig. 7. Solubility of n-butane in several liquids S. Comparison of the pattern for 25°C and for −5°C. The liquids are (1) $C_6H_5CH_2OH$: (2) o-$NO_2C_6H_4CH_3$; (3) n-$C_8H_{17}OH$; (4) n-$C_{10}H_{22}$; (5) $(n$-$C_8H_{17})_2O$; (6) n-$C_8H_{17}OH$; (7) m-$C_6H_4(CH_3)_2$; (8) n-$C_{10}H_{22}$; (9) $(n$-$C_8H_{17})_2O$. The first five curves are for 25°C, and the last four curves are for −5°C. The curve for C_2H_5OH is similar to the curve for $C_6H_5CH_2OH$ (curve 1).

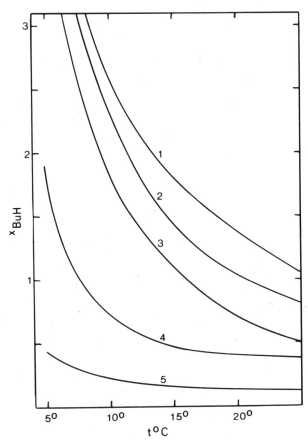

Fig. 8. Change of mole ratio x_{BuH} with temperature ($t°C$) for liquids S: (1) $(n\text{-}C_8H_{17})_2O$; (2) $n\text{-}C_{10}H_{22}$; (3) $m\text{-}C_6H_4(CH_3)_2$; (4) $n\text{-}C_8H_{17}OH$; (5) C_6H_5CN.

to diethyl ether–n-pentane at 25°C; but the essential pattern is the same; in all these examples we observe p_{BuH} and the attendant N_{BuH}.

In Fig. 8 the change in x_{BuH} with $t°C$ is illustrated; the rapid rise in x_{BuH} as the temperature approaches the bp of butane is to be noticed.

2.2.2. Propane

Propane has bp −42.07°C/1 atm, the $p°_{PrH}$ being 7095 mm Hg at 25°C. This means that the tendency to condense is much less than in the example of n-butane at the same $t°C$. After drawing the $p°_{PrH}$ vs $t°C$ line from the published data, I read off $p°_{PrH}$ for, say, 5°C, and draw the R-line. Because of the much smaller N_{PrH} values for the same group of liquids S, I could use a larger scale to indicate N_{PrH} values on the base line. Representative plots are shown in Fig. 9. It is seen that after allowing for the difference in the tendency to condense, the pattern is essentially the same as that of the n-butane one. In other words, the main cause of the difference in N_{PrH} and N_{BuH} values for the same liquid S at the same $t°C$ is due to the lower tendency of propane to condense at that $t°C$.

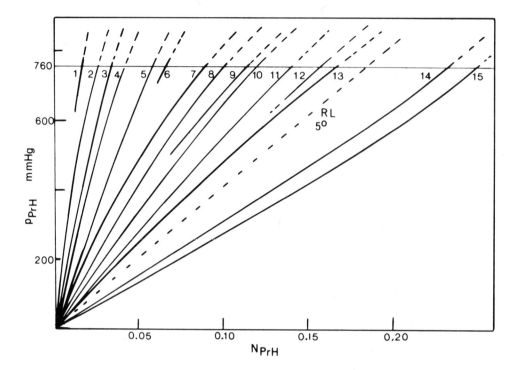

Fig. 9. Solubility of propane at 5°C in various liquids S. Mole fraction N_{PrH} vs pressure plots. The liquids are: (1) $C_6H_5NH_2$; (2) $HCON(CH_3)_2$; (3) $C_6H_5CH_2OH$; (4) $C_6H_5CH_2NH_2$; (5) C_6H_5CN; (6) $C_6H_5N(CH_3)_2$; (7) $C_6H_5OC_2H_5$; (8) $m\text{-}BrC_6H_4CH_3$; (9) $n\text{-}C_5H_{11}CO_2H$; (10) $n\text{-}C_8H_{17}OH$; (11) C_6H_5I; (12) $n\text{-}C_8H_{17}NH_2$; (13) $n\text{-}C_8H_{17}I$; (14) $n\text{-}C_{10}H_{22}$; (15) $(n\text{-}C_8H_{17})_2O$.

On the same scale for x_{PrH} as for the x_{BuH} values in the x vs $t°C$ plots, the slopes of the lines for propane are all less than that for No. 5 line in the n-butane group (Fig. 8). A larger scale is therefore used to show the change of x_{PrH} values with $t°C$ (Fig. 10).

2.2.3. Ethane and Methane

Ethane has bp $-88.63°C/1$ atm, $p°_{EtH}$ being 29,290 mm Hg at 25°C. The critical temperature is 32°C, and so that $p°_{EtH}$ is for vapor over actual liquid. Methane, bp $-161°C$, has a critical temperature of $-82.5°C$, and therefore it cannot be obtained as a liquid at, for example, 20°C. To obtain a value of $p°_{MeH}$ at 20°C, an extrapolation procedure must be followed. To draw the full R-lines for these gases, I have to use a long "box," as shown in Fig. 11.

As most published data are for about 1 atm, and for about 20°C, the operational area on the scale of Fig. 11 is covered by a tiny spot at the left bottom corner of the

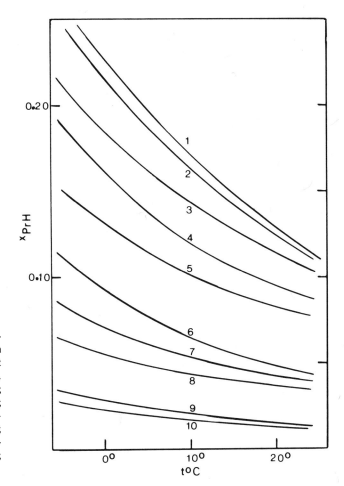

Fig. 10. The mole ratio solubilities x_{PrH} of propane in various liquids S at different temperatures and $p_{PrH} = 1$ atm. The liquids are: (1) $n\text{-}C_8H_{17}I$; (2) $n\text{-}C_8H_{17}NH_2$; (3) C_6H_5I; (4) $n\text{-}C_8H_{17}OH$; (5) $m\text{-}BrC_6H_4CH_3$; (6) $C_6H_5OC_2H_5$; (7) C_6H_5CN; (8) $o\text{-}HOC_6H_4\text{-}CO_2CH_3$; (9) $HCON(CH_3)_2$; (10) $(CH_3CO)_2O$.

diagram. I can show this point a little clearer by increasing the pressure scale, so that I can bring propane and n-butane into the picture, as in Fig. 12. Figures 13–14 show the position of the N_{HC} values.

2.2.4. The Solubility Spectrum

In Figs. 15–20 the R-lines for the noble gases are drawn, and the positions of certain observed N_A values are shown to illustrate the solubility spectrum. For clear delineation, two N_A scales are needed, 0 to 0.02 and 0.02 to 0.10. For gases in the bp range of He to Kr, the N_A values are all absolutely small, but there are relatively large variations. As an example, for 5°C the part of the R-line below the horizontal for 1 atm varies from 1/3500 for helium to nearly the whole of it for n-butane (Fig. 12), a point to be kept in mind when comparing N_A values. Figure 21 shows how nitrogen fits into the spectrum; the R-line for 25°C is immediately on the left of that for argon. Similar diagrams for H_2, O_2, and CO may be drawn (Figs. 22, 23, 24).

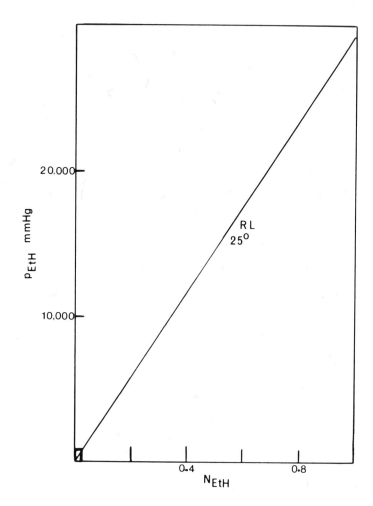

Fig. 11. Complete R-line for ethane at 25°C and $p°_{EtH}$ = 29,290 mm Hg. The frame at the bottom left corner indicates the operational area, N_{EtH} and p_{EtH} values, for 25°C and p_{EtH} up to 1 atm.

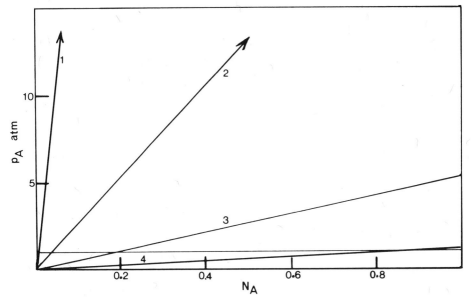

Fig. 12. Indication of the approximate operational areas for the hydrocarbon gases at 5°C. The R-lines are: (1) methane; (2) ethane; (3) propane; (4) n-butane. The frame at the bottom of the diagram shows the areas for p_A up to 1 atm.

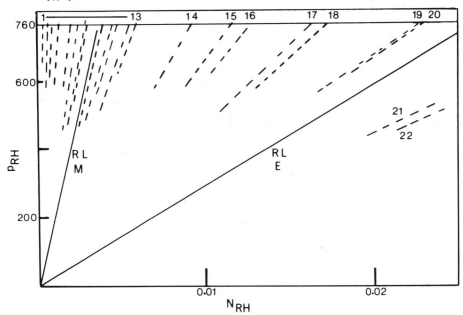

Fig. 13. Solubility of methane (M) and ethane (E) at 25°C and 760 mm Hg pressure, p_{RH}. Registration of the mole fraction N_{RH} values on the R-line diagram. For M, the liquids are: (1) $(CH_2OH)_2$; (2) $C_6H_5NH_2$; (3) CH_3OH; (4) C_2H_5OH; (5) CH_3COCH_3; (6) $n\text{-}C_5H_{11}OH$; (7) $C_8H_{17}OH$; (8) CCl_4; for 9, see below; (10) $(C_2H_5)_2O$; (11) $n\text{-}C_7H_{16}$; (12) $i\text{-}C_8H_{18}$. For E, the liquids are: (9) CH_3OH; (13) C_2H_5OH; (14) $n\text{-}C_3H_7OH$; (15) CH_3COCH_3; (16) $n\text{-}C_5H_{11}OH$; (17) C_6H_5Cl; (18) $n\text{-}C_8H_{17}OH$; (19) $CH_3CO_2C_5H_{11}$; (20) CCl_4; (21) $n\text{-}C_7H_{16}$; (22) $i\text{-}C_8H_{18}$. The pressures p_{RH} are in mm Hg. For M, the curves for C_6H_6, C_6H_5Cl, and $CH_3CO_2CH_3$ are similar to the curve for CH_3COCH_3 (curve 5). For E, the curve for C_6H_6 is similar to the curve for C_6H_5Cl.

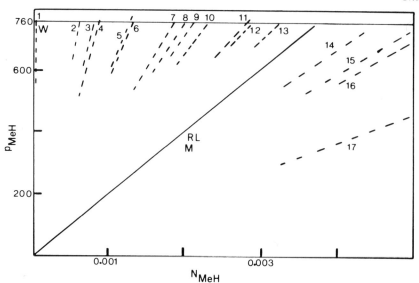

Fig. 14. Solubility of methane in various liquids at 25°C and 760 mm Hg, p_{MeH}. Registration of the mole fraction N_{MeH} values on the R-line diagram on a larger scale than in Fig. 12. The liquids are: (1) H_2O; (2) $C_6H_5NH_2$; (3) $C_6H_5NO_2$; (4) CH_3OH; (5) C_2H_5OH; (6) CS_2; (7) CH_3COCH_3; (8) C_6H_5Cl; (9) C_6H_6; (10) $C_6H_5CH_3$; (11) n-$C_8H_{17}OH$; (12) CCl_4; (13) cyclohexane; (14) $(C_2H_5)_2O$; (15) n-C_6H_{14}; (16) i-C_8H_{18}; (17) C_7F_{16}. Notice the position of water, indicated by W.

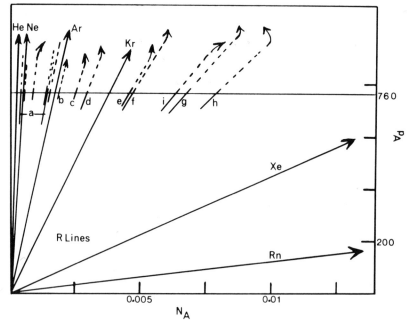

Fig. 15. Reference lines (RL) for the noble gases at 25°C drawn on a common scale. Illustrative registration of a selection of mole fraction N_A values for 760 mm Hg on the R-line diagram. The gas is indicated by A, and the pressure by p_A in mm Hg. For argon: (a) indicates the position of lines for the n-alcohols, left to right, CH_3OH to n-$C_8H_{17}OH$; (b) for cyclo-$C_6H_{11}CH_3$; (c) for n-C_6H_{14}; (d) for i-C_8H_{18}; (e) for cyclo-$C_6F_{11}CF_3$. For krypton: (f) for cyclo-C_6H_{12}; (g) for n-C_7H_{16}; (h) for n-C_6H_{14}. For xenon: (i) for $C_6H_5NO_2$. The pressures p_A are in mm Hg.

Solubility Data and the Reference-Line Diagram

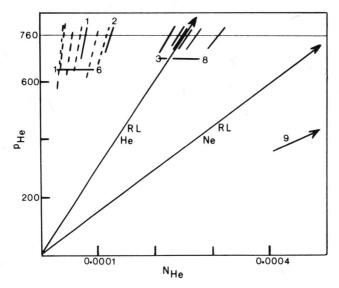

Fig. 16. Solubility of helium in various liquids at 25°C. Registration of the mole fraction N_{He} values on the R-line diagram. Counting from left to right (broken lines): (1) $C_6H_5NO_2$; (2) C_6H_5I; (3) C_6H_5Br; (4) C_6H_5Cl; (5) $C_6H_5CH_3$; (6) C_6H_5F. Full lines: (1) C_6H_6; (2) cyclo-C_6H_{12}; (3) n-dodecane and n-tetradecane; (4) n-C_9H_{20}, n-$C_{10}H_{22}$, and n-C_8H_{18}; (5) n-C_7H_{14} and 2,3-dimethylhexane; (6) n-C_6H_{14}; (7) 2,4-dimethylpentane; (8) 2,2,4-trimethylpentane; (9) C_7F_{16} ($N_{He} = 0.00086$ at 760 mm Hg; see Kobatake and Hildebrand, 1961). The gas pressure p_{He} is in mm Hg.

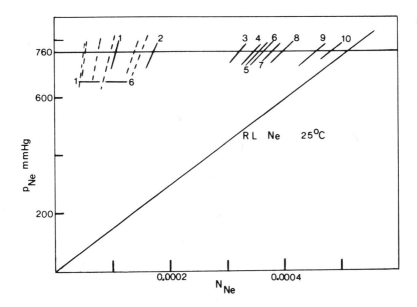

Fig. 17. Solubility of neon in various liquids at 25°C. Registration of the mole fraction N_{Ne} values on the R-line diagram. Counting from left to right, the liquids are (broken lines): (1) $C_6H_5NO_2$; (2) C_6H_5I; (3) C_6H_5Br; (4) C_6H_5Cl; (5) $C_6H_5CH_3$; (6) C_6H_5F. Full lines: (1) C_6H_6; (2) cyclo-C_6H_{12}; (3) n-dodecane and n-tetradecane; (4) n-C_9H_{20} and n-C_7H_{16}; (5) n-$C_{10}H_{22}$ and n-C_8H_{18}; (6) methylheptane and 2,3-dimethylhexane; (7) n-C_6H_{14}; (8) 2,4-dimethylhexane; (9) 2,2,4-trimethylpentane; (10) CCl_2FCClF_2 (from Linford and Hildebrand, 1970).

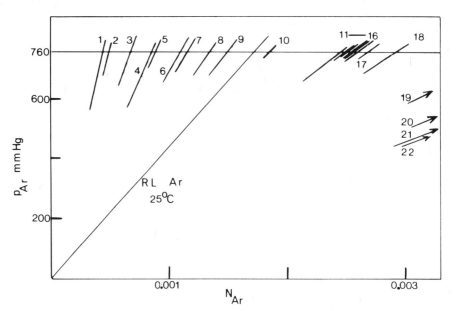

Fig. 18. Solubility of argon in liquids S at 25°C. Registration of mole fraction N_{Ar} values on the R-line diagram. From left to right the liquids S are: (1) $C_6H_5NO_2$; (2) C_6H_5I and CS_2; (3) C_6H_5Br; (4) C_6H_5Cl; (5) C_6H_6; (6) $C_6H_5CH_3$; (7) C_6H_5F; (8) CCl_4; (9) cyclo-C_6H_{12}; (10) cyclo-$C_6H_{11}CH_3$; (11) n-C_8H_{18} and C_9H_{20}; (12) 2,3-dimethylhexane and n-$C_{10}H_{22}$; (13) n-C_7H_{16} and 3-methylheptane; (14) n-C_6H_{14}; (15) n-dodecane; (16) n-tetradecane; (17) 2,4-dimethylhexane; (18) 2,2,4-trimethylpentane; (19) CCl_2FCClF_2; (20) cyclo-$C_6F_{11}CF_3$; (21) $(C_4F_9)_3N$; (22) n-C_7F_{16}. The last four lines simply indicate the lines of extrapolation to the 760 mm Hg horizontal; the N_{Ar} values then are: (19) 0.0039; (20) 0.0046; (21) 0.0050; (22) 0.00530. The values for the fluorine compounds are due to Hildebrand *et al.*

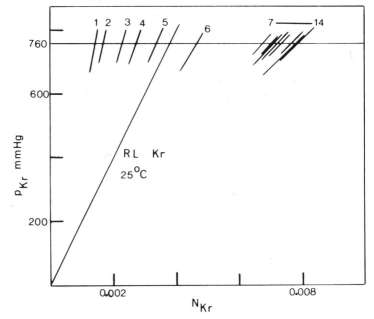

Fig. 19. Solubility of krypton in liquids at 25°C. Registration of the mole fraction N_{Kr} values on the R-line diagram. From left to right the liquids are: (1) $C_6H_5NO_2$; (2) C_6H_5I; (3) C_6H_5Br; (4) C_6H_5Cl and C_6H_6; (5) $C_6H_5CH_3$ and C_6H_5F; (6) cyclo-C_6H_{12}; (7) n-C_6H_{14}; (8) 2,3-dimethylhexane, n-C_8H_{18}, and n-C_9H_{20}; (9) n-C_7H_{16} and 3-methylheptane; (10) n-$C_{10}H_{22}$; (11) 2,4-dimethylhexane; (12) n-dodecane; (13) n-tetradecane; (14) 2,2,4-trimethylpentane.

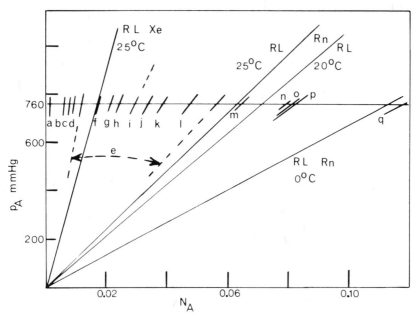

Fig. 20. N_{Xe} and N_{Rn} values at 1 atm. N_{Xe} at 25°C: (b) $C_6H_5NO_2$; (i) n-C_6H_{14}. N_{Rn} at 0°C: (f) $C_6H_5NH_2$; (g) C_2H_5OH; (h) CH_3COCH_3; (k) $CH_3CO_2C_2H_5$; (n) $C_6H_5CH_3$; (o) CS_2; (p) $(C_2H_5)_2O$; (q) n-C_6H_{14}. N_{Rn} at 20°C: (c) $(CH_2OH)_2$; (d) CH_3OH; (j) n-C_4H_9OH; (l) C_6H_6. N_{Rn} at 25°C: (a) HCO_2H; (e) n-RCO_2H [CH_3CO_2H (left) to $C_7H_{15}CO_2H$ (right)]; (m) $CHCl_3$. A = Xe or Rn. Other values are: N_{Xe} (approx. 15.2°C): C_6H_5F (0.01298); C_6H_5Cl (0.01390); C_6H_5Br (0.01222); C_6H_5I (0.00968); $C_6H_5CH_3$ (0.01637); $C_6H_5NO_2$ (0.00627); [C_6H_6 (0.0134); perfluoromethylcyclohexane (0.0188); cyclohexane (0.0237); methylcyclohexane (0.0252); n-hexane (0.0303); isooctane (0.0318); n-dodecane (0.0352); see Clever et al. (1957); Clever (1958); and Saylor and Battino (1958). N_{Xe} at 25°C: CCl_2FCClF_2 (0·0215); see Linford and Hildebrand (1970).]

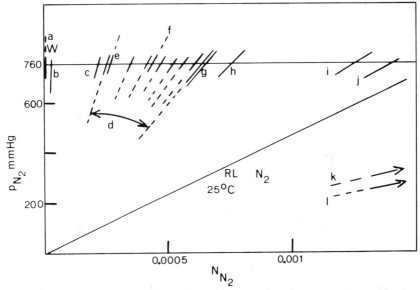

Fig. 21. Solubility of nitrogen in liquids at 25°C. Registration of mole fraction N_{N_2} values on the R-line diagram. The liquids are: (a) H_2O; (b) $(CH_2OH)_2$; (c) CS_2; (d) n-alcohols (CH_3OH, left, to n-$C_8H_{17}OH$, right); (e) cyclo-$C_6H_{11}OH$; (f) C_6H_6; (g) CCl_4; (h) cyclo-C_6H_{12}; (i) $(C_2H_5)_2O$; (j) n-C_6H_{14}; (k) cyclo-$C_6F_{11}CF_3$; (l) n-C_7F_{16}. The last two lines indicate the line of extrapolation to the 760 mm Hg horizontal. The position of the line for water is indicated by W.

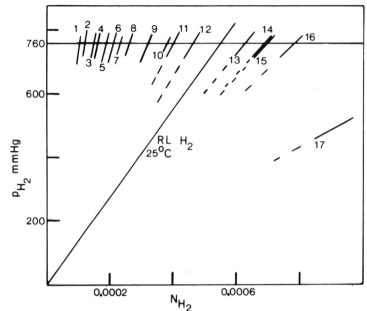

Fig. 22. Solubility of hydrogen in liquids (at 25°C, unless otherwise stated). Registration of mole fraction N_{H_2} values on the R-line diagram. From left to right the liquids are: (1) $C_6H_5NH_2$; (2) $C_6H_5NH_2$ (20°C); (3) $C_6H_5NO_2$ (20°C); (4) CS_2; (5) 1,4-dioxane; (6) $(CH_3)_2CO$; (7) C_6H_6; (8) $CHCl_3$; (9) $C_6H_5CH_3(CCl_4)$; (10) cyclo-C_6H_{12}; (11) m-$C_6H_4(CH_3)_2$ (20°C); (12) $CH_3CO_2C_5H_{11}$; (13) $(C_2H_5)_2O$; (14) n-C_7H_{16}; (15) n-C_9H_{20}; (16) i-C_8H_{18}; (17) n-C_7F_{16}. The line for C_6H_5Cl is near that for C_6H_6 (No. 7). Line 17 is to be extrapolated to the 760 mm Hg horizontal.

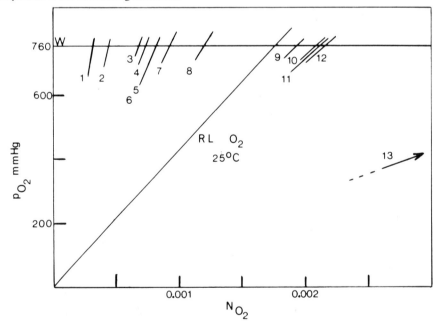

Fig. 23. Solubility of oxygen. Registration of the mole fraction N_{O_2} values for 760 mm Hg on the R-line diagram. The liquids are: (1) $C_6H_5NO_2$ (18°C); (2) CS_2 (25°C); (3) CH_3COCH_3 (19°C); (4) $CHCl_3$ (16°C); (5) C_6H_6 (25°C); (6) near 5, $C_6H_5CH_3$ (18°C); (7) m-$C_6H_4(CH_3)_2$ (16°C); (8) CCl_4 (25°C); (9) $(C_2H_5)_2O$ (25°C); (10) C_8H_{18} (25°C); (11) C_9H_{20} (25°C); (12) C_7H_{16} (25°C); (13) n-C_7F_{16} (25°C). Line 13 indicates the line of extrapolation to the horizontal at 760 mm Hg, corresponding to $N_{O_2} = 0.0054$. The line for water is near the left vertical axis ($N_{O_2} = 0.000023$ at 20°C and $p_{O_2} = 1$ atm) and is indicated by W.

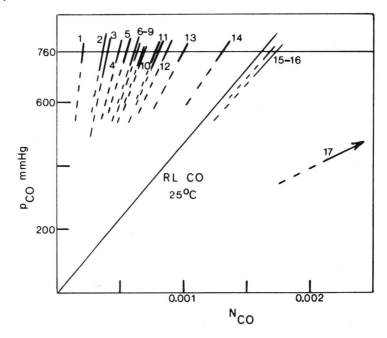

Fig. 24. Solubility of carbon monoxide in liquids (at 25°C, unless otherwise stated). Registration of the mole fraction N_{CO} values on the R-line diagram. From left to right the liquids are: (1) $C_6H_5NH_2$; (2) CS_2; (3) $C_6H_5NO_2$; (4) C_2H_5OH (20°C); (5) n-C_3H_7OH (20°C); (6) i-C_3H_7OH (20°C); (7) C_6H_5Cl; (8) C_6H_6; (9) $CHCl_3$; (10) $(CH_3)_2CO$; (11) $C_6H_5CH_3$; (12) CCl_4; (13) $CH_3CO_2C_2H_5$ (20°C); (14) $CH_3CO_2C_5H_{11}$ (20°C); (15) $(C_2H_5)_2O$; (16) n-C_7H_{16}; (17) n-C_7F_{16} (indicates line of extrapolation to the 760 mm Hg horizontal; $N_{CO} = 0.00386$ at 25°C and $p_{CO} = 760$ mm Hg).

2.3. The Basis of the Explanations of the Essential Patterns of Data

2.3.1. Intramolecular Electron Density Pattern

I now have to direct attention to the area of scientific investigation on the structure of atoms and molecules; for the sake of economy of time and effort, I am obliged to lift from those tremendous achievements the concept of *intra*molecular electron density pattern. The distribution of electron density is in a state of flux; but on a time-averaged basis, there appears to be a mean "structure" or state of "distribution."

The sensitivity of a particular pattern to outside, *inter*molecular influences can vary greatly according to the particular molecule.

I am aware that the concept I am presenting is of universal generality; but unlike so many so-called general statements, it is *true*. It is in general only *qualitative*; but before passing judgment on it because of this, consider to what extent approaches purported to be quantitative have been successful, even for limited and often conveniently chosen examples. This general aspect enables any chemist to appreciate the essential pattern of data and in no way hinders anyone from dealing with individual systems with such finer analytical precision the chosen system allows.

2.3.2. Acid–Base Function

To account for all the data at present known, the concept of electron donor (basic)–electron acceptor (acidic) function is invoked. This function is herein deemed to relate to the change in the *intra*molecular electron-density pattern as molecules approach each other. The facility with which this occurs may be looked upon as polarizability. The acid–base function may vary in intensity from extremely small, e.g., for helium, to decidedly larger, e.g., for n-butane. Steric factors will be involved. The resulting *inter*molecular forces have been given various names; although the expressions "physical interaction" and "chemical interaction" are prevalent, it is far from clear on what criteria a line of demarcation can be drawn. Herein no such distinction is recognized. Each gaseous molecule has a site, or sites, of acid function and of basic function. In hydrogen chloride, there is a highly intensive acid function sited on hydrogen, but the basic function is dissipated by sharing over three lone pairs on chlorine. The noble gases have only lone pairs, but the hydrocarbon gases, CH_4 to n-C_4H_{10}, have no unshared electrons; gases such as methyl chloride, dimethyl ether, sulfur dioxide, and carbonyl chloride have a more complex pattern of sites. Nevertheless, the acid–base function of all gases themselves is herein deemed to determine the tendency to condense, and hence the position of the R-line for a $t°C$ in the spectrum.

The N_A value for a liquid S will then be determined by the mechanisms of the *inevitable* interaction of A with S in the dissolution process. The liquid S itself has its own pattern of acid–base function sites which give what is herein referred to as intermolecular structure. There are no simple rules, but there are clear trends which, by the exercise of intuition backed by chemical knowledge, enable one to appreciate and to rationalize the essential pattern of data.

At, for example, 25°C and $p_A = 1$ atm, the N_A values for all gases having a bp less than, for example, $-100°C$ are absolutely small for all liquids S examined so far; but there is a relatively wide spread of values from He, through N_2, O_2, and on to Rn. The N_A value for water is, for these gases, at the bottom of each list of liquids and is relatively far removed from the respective R-line. The values for methane to n-butane for water are lower than any value for methane for any other liquid. The highly symmetrical and well-organized hydrogen-bonding structure of water is the determining factor. Alcohols offer similar but much less resistance to change, as exemplified in Fig. 21.

Gases in the upper half of the bp range show that the acid function sited on hydrogen in hydrogen sulfide, and in acetylene, is incisive enough to offer effective competition with those functions determining the intermolecular structure of S, although the N_{H_2S} or $N_{C_2H_2}$ value for water is still right at the bottom of each list and well removed from the respective R-line. Sulfur dioxide clearly uses sites of basic function, and dimethyl ether sites of acidic function of S, but again the N_A value for water is at the bottom of each list. The hydrogen halides (HCl, bp $-85°C$; HBr, bp $-67°C$; and HI, bp $-35°C$) each have a highly intensive acidic-function site on hydrogen that can seek out and use basic-function sites of relatively low intensity in

S. This is deemed to be why the N_{HX} values clearly tend to be on the right side of the respective R-lines. Although water is not at the bottom of each list, it is nowhere near the top; and the clustering of values, N_{HCl}, 0.29, N_{HBr}, 0.33, N_{HI}, 0.28, despite the wide differences in the position of the respective R-line, may be considered alongside a similar clustering of the very small N_A values for the hydrocarbon gases CH_4 to n-C_4H_{10}. Ammonia, bp $-33°C$, gives a spread of N_{NH_3} values on each side of the R-line at $0°C$, and that for water is nearly 0.5 (mole ratio x_{NH_3} nearly 1.0) at $0°C$, p_{NH_3} 1 atm. All these points will be brought out more clearly as individual systems are examined.

3

Henry's Law and Raoult's Law

3.1. Henry's Law

3.1.1. The Observations of Henry (1803)

On the 23rd of December 1802, Henry's paper[156] entitled "Experiments on the quantity of gases absorbed by water, at different temperatures, and under different pressures," was communicated to the Royal Society. Henry determined the number of cubic inches of gas (H_2, O_2, N_2, H_2S, CO, CO_2, PH_3, N_2O, and carbureted hydrogen gas) absorbed by 100 in.^{3} of water at 60°F and a pressure p_{gas} of 1 atm. Later in the 19th century, the volume of gas absorbed by one volume of solvent S at a given temperature t°C and p_{gas} was called the *absorption coefficient*, and Henry's values become for *water* 0.0161 (H_2), 0.037 (O_2), 0.0153 (N_2), 0.020 (CO), 0.0214 (PH_3), 0.0140 (carbureted hydrogen gas), 0.47 (N_2O), 0.80 (H_2S), and 0.94 (CO_2). Since chemical equations are expressed as mole ratios, I deem it desirable to convert these volume ratios into *mole ratios*, x_A = number of moles of gas absorbed by *one* mole of solvent S at t°C and 1 atm. The corresponding form of mole ratio is M_S = number of moles of the liquid S required to absorb *one* mole of the gas at t°C and 1 atm.

Henry gave some indication of the effect of temperature; e.g., 100 volumes of water absorbed 86 volumes of hydrogen sulfide at 55°F and 78 at 85°F, and the volumes of nitrous oxide were 50 at 45°F and 44 at 70°F. The outstanding effect was that of pressure, and to discuss this we must first refer to Henry's technique (see Fig. 25). By pouring mercury into the graduated tube B pressures of 1, 2, and 3 atm were obtainable. The results of a series of at least 50 experiments on the five gases CO_2, H_2S, N_2O, O_2 and N_2 showed that water takes up, of gas condensed by 1, 2, or more atm, a quantity which, ordinarily compressed, would be equal to twice, thrice, etc. the volume absorbed under the common pressure of 1 atm.

This is the limit of Henry's statement about the pressure effect; he did not give a mathematical form; but the statement may be expressed as $x_A = Kp_A$, K being the proportionality constant at the stated t°C. It must be emphasized that the relationship is a mole *ratio* one and was based only on water as solvent and for five gases.

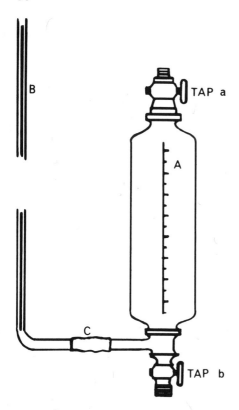

Fig. 25. Henry's apparatus. The calibrated absorption tube A was filled with mercury, "which rose to its corresponding level in the tube B." Tap a is attached to the water supply; tap b is opened to allow mercury to run out and thereby admit a known volume of water at tap a. "A measured quantity of gas is admitted in a similar manner." "Strong agitation is now applied by means of the joint C," which was a tube of rubber covered with leather. Mercury was meanwhile added at the tube B to preserve the pressure; the quantity of mercury so required gave the volume of gas absorbed. By pouring more mercury into the extended tube B, the effect of pressure was measured.

Later, this pressure effect became associated with the term "Henry's law," the term being used in connection with the behavior of *any* gas and *any* liquid. However, it is necessary to point out that although certain chemists continued to look upon Henry's law as a mole *ratio* relationship, others, for reasons not apparent to the present writer, looked upon the law as a mole *fraction* relationship, $N_A = Kp_A$, K being called the Henry's law constant.

3.1.2. The Roscoe and Dittmar (1860)[317] Data for Hydrogen Chloride and Ammonia Absorbed by Water

The data were expressed as "grams of HCl or grams of NH_3 absorbed by 1 g water" at 0°C and several pressures p_{HCl} or p_{NH_3}. These primary data were plotted as shown in Fig. 26. It was concluded that Henry's law was not even approximately followed. These primary data are in the mole *ratio* form; x_A = grams of A/M_A divided by 1/18, A being HCl or NH_3 and M_A being the molecular weight of A. Henry's law was therefore taken to be $x_A = Kp_A$, K being the Henry's law constant. I have calculated these x_A values and plotted them in Fig. 27, which also shows the positions of the N_A values.

Henry's Law and Raoult's Law

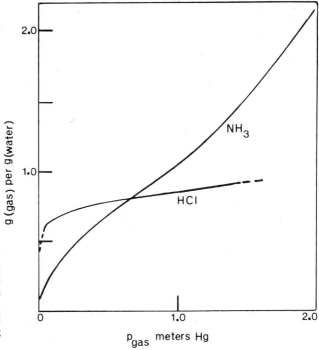

Fig. 26. Solubility of ammonia and hydrogen chloride in water at 0°C and several pressures. Plots of the data of Roscoe and Dittmar (1860). Data were expressed as "grams of gas absorbed by 1 g of water."

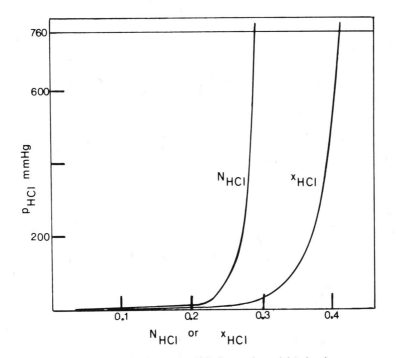

Fig. 27. Solubility of hydrogen chloride in water at 0°C. Comparison of the plots for pressure p_{HCl} vs mole fraction N_{HCl} and p_{HCl} vs mole ratio x_{HCl}.

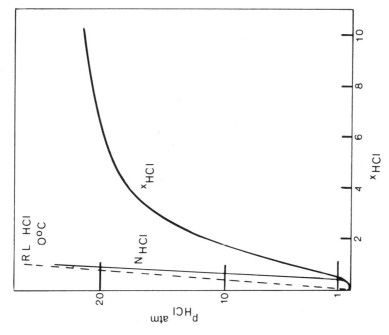

Fig. 29. Solubility of hydrogen chloride in water at 0°C. Plot of the mole ratio x_{HCl} vs p_{HCl} corresponding to the extrapolated mole fraction line from 1 to 26.2 atm. See Fig. 28.

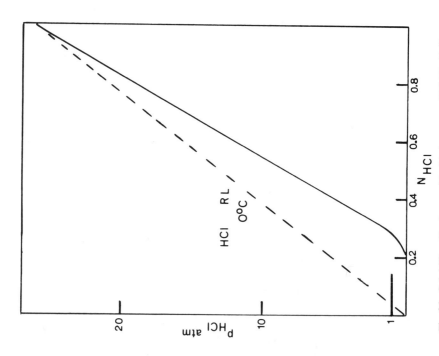

Fig. 28. Solubility of hydrogen chloride in water at 0°C. The R-line of HCl at 0°C and the extrapolation of the mole fraction vs p_{HCl} line beyond 1 atm up to $p°_{HCl} = 26.2$ atm.

Henry's Law and Raoult's Law

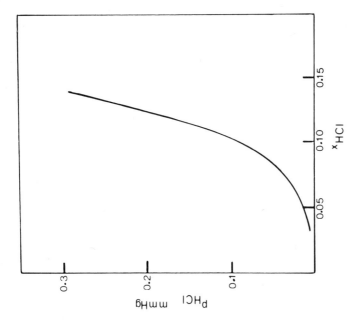

Fig. 31. Solubility of hydrogen chloride in water at 25°C and low pressures p_{HCl} in mm Hg. The mole ratio x_{HCl} vs p_{HCl} plot from the molality data of Bates and Kirschman.[16]

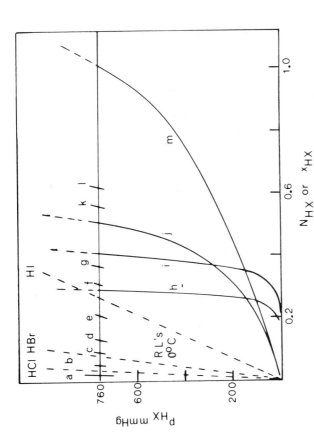

Fig. 30. Solubility of hydrogen halides HX (X=Cl, Br, I) in liquids at 0°C. Registration of mole fraction N_{HX} and mole ratio x_{HX} values on the R-line diagram. (The N_{HX} positions for the named liquids are, for CCl$_4$: (a) HCl; (b) HBr; (e) HI: for m-C$_6$H$_4$(CH$_3$)$_2$: (c) HCl; (d) HBr; (g) HI: for n-C$_{10}$H$_{22}$; (f) HI: for water; (h) HCl (i, for x_{HCl}): for (n-C$_8$H$_{17}$)$_2$O: (j) HCl (m, for x_{HCl}): for n-C$_8$H$_{17}$-OH: (k) HBr; (l) HI.

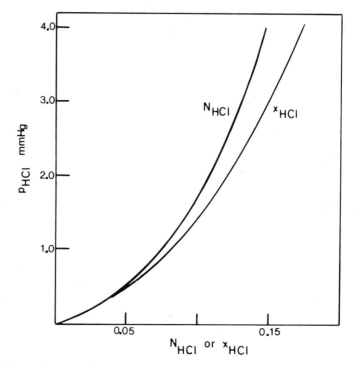

Fig. 32. Solubility of hydrogen chloride in ethanol at 25°C and low pressures. Plots of mole fraction N_{HCl} vs pressure p_{HCl} and mole ratio x_{HCl} vs p_{HCl} from data by Jones, Lapworth, and Lingford (1913)[193] who recorded moles per liter and density.

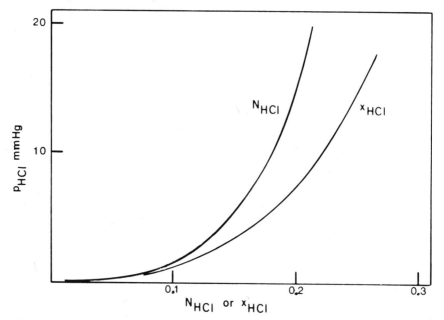

Fig. 33. Solubility of hydrogen chloride in methanol at 25°C and low pressures p_{HCl}. Mole fraction N_{HCl} vs p_{HCl} and mole ratio x_{HCl} vs p_{HCl} from the molality data of Schmid, Maschka, and Sofer (1964).[327] A larger-scale plot of x_{HCl} vs p_{HCl} from the data at the lower end of the given x_{HCl} plot (below $x_{HCl} = 0.03$) is still definitely curved, concave upward.

Table 1. Solubility of HCl

t, °C	Density	g HCl per 100 g solution	g HCl per 100 g water
18	1.2064	42.34	73.41
20	—	—	71.9[a]
23	1.2014	41.54	71.03

[a] Refers to Knight and Hinshelwood's "g per 100 cm^3 saturated solution."

Wynne-Jones (1930)[384] extrapolated the N_{HCl} line to what I now refer to as the $p°_{HCl}$ point (26.2 atm) on the right vertical axis as shown in Fig. 28, which shows also the R-line on the same scale. The x_{HCl} plot is very different, as is shown clearly in Fig. 29. (See also Fig. 30.) The actual shape of the x_{HCl} line below p_{HCl} about 10 mm Hg is not linear, as appears to be indicated on the scale of Fig. 30; Bates and Kirschman (1919)[16] observed the molality data, and these have now been plotted as the corresponding x_{HCl}/p_{HCl} data as shown in Fig. 31. The lines for HBr and HI are similar. If it were in the limits of operational competency, it is more probable that the line would be revealed to be nowhere really straight even at $p_{HCl} = 0.0001$ mm Hg. (See Figs. 32 and 33.)

It is necessary at this stage to refer to a paper by Knight and Hinshelwood (1927)[208] (see Table 1) because it affords an example of how a vital error can occur by the disregard of the discipline of comparing solubility values in a logical way. These workers set out to reveal the condition of hydrogen chloride in water by investigating its partition between water and benzene at 20°C and 760 mm Hg. No solubility data being available for benzene, they determined it and expressed it as "grams of hydrogen chloride per liter of solution." If we assume from want of data that the density of the solution is that of pure benzene, the x_{HCl} value is 0.041, corresponding approximately to 0.045, from the work of Bell.[20]

Since the usually quoted value for water was based on the original determination of Roscoe and Dittmar, Knight and Hinshelwood redetermined the value for 20°C by a different method. The close mean of five determinations was given as 718.8 g hydrogen chloride/liter of *solution*. Roscoe and Dittmar's value was 721 g for 1000 g ($\approx$1 liter) of water at 20°C and 760 mm Hg. The x_{HCl} values for these data are 0.706 (K and H) and 0.361 (R and D), and therefore we see that the K and H value is twice that of the R and D one. As the authors (K and H) did not state the R and D value and did not remark on the very serious discrepancy, it appears that they believed that their 718.8 value (for a liter of *solution*) was essentially the same as the R and D value 721 (for a liter of water). Indeed, in one standard data book, the two forms are so linked in the same table that the unwary could miss the difference (see Table 1).

3.1.3. The Form of Henry's Law

In general, the "law" is "something" = a constant × pressure. Henry's data were in the form "volume of gas/volume of the original liquid taken" (in his example, water). The "vol./vol." data may be converted into mass of gas/mass of liquid; and this form is converted into the x_A form by

$$\frac{\text{Mass of A in grams}}{\text{Mol. wt. of A}} \bigg/ \frac{\text{Mass of liquid S in grams}}{\text{Mol. wt of S}}$$

Therefore, for the same A and S the "vol./vol." and "mass/mass" forms are equivalent to the x_A form which, nevertheless, may still be expressed in two ways:

$$x_A = K' p_A \quad (1) \quad \text{or} \quad p_A = K'' x_A \quad (2)$$

For form (1), "Henry's law constant" $K' = x_A/p_A$ and would be expressed as "mole ratio per unit pressure (mm Hg, atm, etc.)." For form (2), the "constant" would be $K'' = p_A/x_A$, i.e., "pressure per mole ratio." These are very different numbers. If $x_A = 0.10$ at 1000 mm Hg, $K' = 0.0001$ per mm Hg, whereas $K'' = 10,000$ per mole ratio.

A number of writers refer exclusively to the mole *fraction* form, usually corresponding to the expression (2): $p_A = KN_A$, so that K is looked upon as p_A per mole fraction. The mole ratio form and the mole fraction form are mutually exclusive because $N_A = x_A/(1 + x_A)$ and x_A is not a linear function of $(1 + x_A)$. It is true that within the limits of the usual operational competency the difference between N_A and x_A becomes increasingly difficult to discern as x_A becomes successively smaller than 0.015.

3.2. Raoult's Law

3.2.1. Raoult's Work on Determination of Molecular Weights

For the purpose of this monograph, it is desirable to recognize two aspects of Raoult's experimental work, for the expression "Raoult's law" is glibly used altogether too loosely in the original literature and in the textbooks. His experimental work on the depression of the freezing point of a pure liquid by the addition of another substance is described in the papers cited in my list of references which have dates from 1880.[307]

In a long paper on the "general law of freezing of solvents," he gives details of his procedure for the determination of the molecular weight of a substance S by the measurement of the depression of the freezing point of liquid A in which S is dissolved. The lowering of the freezing point caused by "1 g of the substance dissolved in 100 g of the solvent" was called the *coefficient of lowering*. The solvents he used were water, benzene, nitrobenzene, ethylene dibromide, acetic acid, and formic acid. He deemed it advisable to "experiment with very dilute solutions."

A brief report on Raoult's work was published by Cahours, Berthelot, and Debray[43] in which they drew attention to certain historical items such as the observations of Blagden. It should be noticed that when Raoult refers to a solution containing "1 g-molecule in 2 liters (aqueous)" the x_A value is about 0.01.

Henry's Law and Raoult's Law

There were, of course, discussions and counter suggestions. Auwers and Meyer,[9] for example, suggested that glacial acetic acid should always be used as the solvent instead of benzene or water. They believed the Beckmann determinations of the molecular weight of oximes were probably wrong because he had used benzene. It was pointed out that when it was absolutely necessary to use benzene or water, blank experiments with a compound of known molecular weight should be carried out.

3.2.2. Raoult's Experiments on the Depression of the Vapor Pressure of a Liquid

In 1887, Raoult gave a short report on the lowering of the vapor pressure of diethyl ether (herein designated E) caused by the addition of a compound S (S = C_6H_6, $C_{10}H_{16}$, o-$HOC_6H_4CO_2CH_3$, $C_{22}H_{26}N_2O_4$, CNOH, $C_6H_5CO_2H$, Cl_3CCO_2H, C_6H_5CHO, n-$C_8H_{17}OH$, CN_3H_2, $C_6H_5NH_2$, $Hg(C_2H_5)_2$, $SbCl_3$). When the solution contains "1 to 5 moles of the substance (my S) dissolved in 5000 g ether, the difference between the vapor pressure of this solution and that of pure ether is *sensibly* proportional to the weight of the substance dissolved in a constant weight of the solvent." The x_S values for such solutions vary from 0.015 to 0.075; the x_E values vary from about 70 to 15; therefore they represent a large mole fraction value for ether ($N_E = 0.98$ at the upper end).

In Raoult's paper (1888) of 20 pages he described his measurement of the vapor pressures of ethereal solutions of each of five organic liquids S. He had three barometers, one containing only mercury, one with pure ether on the top of the mercury, and the third with the selected solution on the top. "... To prepare solutions with a fixed composition," he writes, "a fixed amount of liquid (S) was weighed in a bottle, 20 cm³ capacity, being equipped with good corks." The ether was then added, and the corked bottle again weighed.

3.2.3. Raoult's Data for Five Liquids S and Diethyl Ether E

Raoult gave his data in the form of number of moles of S per 100 moles of the mixture. I have converted this form into mole fraction N_S by dividing by 100. Raoult gave his *observed* pressure p_E (diethyl ether) in the form $100\, p_E/p°_E$, $p°_E$ being the vapor pressure of ether over pure ether at 16°C (374 mm Hg). I have divided his values by 100 to get simply $p_E/p°_E$ so that I can get the essential value p_E from the quotient multiplied by $p°_E$ (i.e., 374). From $1 - N_S = N_E$, I get the N_E values corresponding to the p_E values; these I have plotted in Fig. 34 for nitrobenzene. His data stopped at the lower end of the full line; but I have myself obtained data which enable me to give the broken-line plot as an extrapolation of the full line.

It is seen that the line is by no means linear, and this is more evident when I add my R-line.

Raoult, however, presented his data very differently. He evidently plotted (although he did not give the plot) $p_E/p°_E$ vs N_S and obtained "a straight line," the

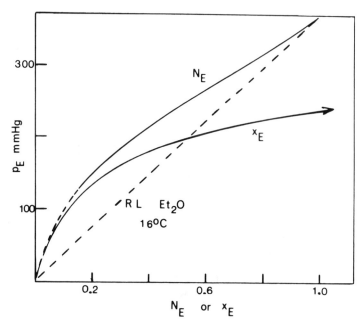

Fig. 34. Raoult's ether–nitrobenzene system at 16°C. Plot of the mole fraction N_E vs p_E and the mole ratio x_E vs p_E on the R-line diagram. E = diethyl ether.

equation being $p_E/p°_E = 1 - kN_S$, where k was called a constant. In Fig. 35, I give this plot for nitrobenzene, for which he gave the "constant" as 0.74, although in a separate table he gave 0.70. The values given to k for the other liquids were: turpentine, 0.90; aniline, 0.90; ethyl benzoate, 0.90; methyl salicylate, 0.82. In none of these examples does $p_E/p°_E = N_E$, i.e., p_E does not equal $N_E \times$ a constant. He emphasized the example in which N_S is very small, which entitled him to put $k = 1$ and arrived at the expression

$$p_E/p°_E = 1 - N_S = N_E$$

a form which found its way into the literature as a "general" law, the so-called Raoult's law.

Now the expression $p_E/p°_E = 1 - N_S$ is equivalent to

$$(p°_E - p_E)/p°_E = N_S$$

referred to as the fractional lowering of the vapor pressure; but it must always be understood that it refers to the condition that k may be taken as unity, i.e., N_S is very small.

In Fig. 36 I have plotted N_E vs p_E values for n-hexanoic acid and 2,2,2-trichloroethanol at 16°C from my own data obtained by the bubbler-tube procedure. The lines are both distinctly concave upwards, on the right of the R-line. In Fig. 37, I show the great difference between the N_E and the x_E plots for methyl salicylate from Raoult's data. This again is illustrated for the example of diethyl ether and sulfuric acid at 30°C from the data of Campbell[44] (Fig. 38).

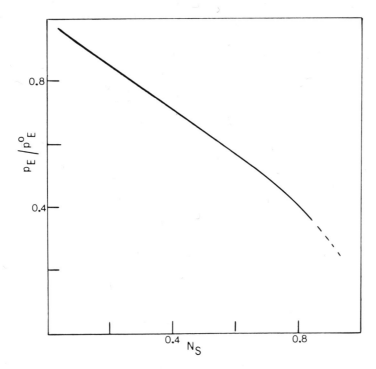

Fig. 35. Raoult's diethyl ether (E) and nitrobenzene (S) system at 16°C. Plot of p_E/p^o_E vs mole fraction of nitrobenzene N_S from Raoult's data; p^o_E is the vapor pressure of ether over the pure liquid at 16°C.

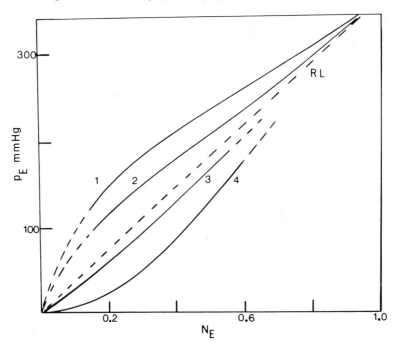

Fig. 36. Mole fraction N_E vs pressure p_E plots for diethyl ether (E) and (1) $C_6H_5NO_2$, (2) o-$HOC_6H_4CO_2CH_3$, (3) n-$C_5H_{11}CO_2H$; and Cl_3CCH_2OH at 16°C. The R-line is for diethyl ether.

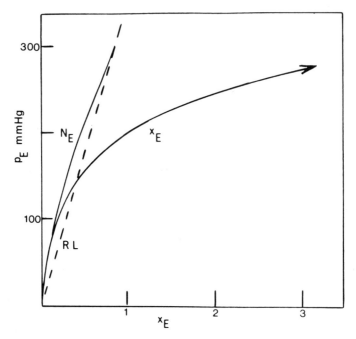

Fig. 37. Raoult's diethyl ether (E) + methyl salicylate (S) system at 14.1°C. Plot of the mole ratio x_E vs pressure p_E on the R-line diagram, which also contains the plot of the mole fraction N_E values for comparison. RL is the reference line for E at 14.1°C.

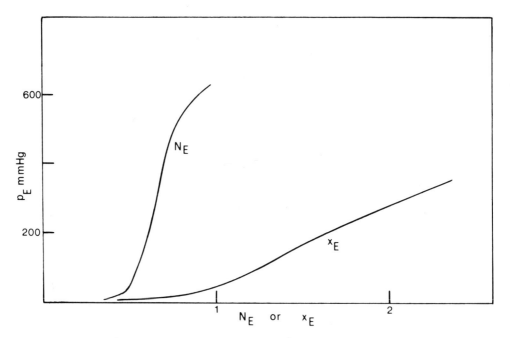

Fig. 38. The diethyl ether (E) + sulfuric acid system at 30°C. The mole fraction N_E vs pressure p_E plot from data of Campbell and the corresponding mole ratio x_E vs p_E plot.

3.2.4. Raoult's 1890 Paper

In Raoult's long paper of 1890, "Sur les tensions de vapeur des dissolutions," he gave a brief history of this subject. For three systems in which A is diethyl ether at 20°C, he gave data covering almost the whole N_S range for S = "essence of terebenthine," ethyl benzoate, and nitrobenzene. I have calculated the p_A and N_A values and plotted these in Fig. 39. I likewise give the corresponding plots (Fig. 40) for the systems $CS_2 + PhCO_2Et$ at 40°C and $EtOH + PhCO_2Et$ at 60°C. All the lines are convex upward on the left of the corresponding R-line.

The plot for his final completely observed system, C_6H_6 + naphthalene at 60°C, is also convex upward on the left of the R-line. In this example, and in the "terebenthine" one, the line appears closely to approach the $p°_A$ point from the right side of the R-line; i.e., the line gently crosses the R-line near $p°_A$, at the top of the R-line.

To avoid confusion, I use my own symbolism in referring to Raoult's data. From the mass of S added, he calculated the N_S, and from the fractional lowering of the vapor pressure $(p°_A - p_A)/p°_A$ he obtained a value Z. The expression Z/N_S then gave him a number which showed a deviation from unity, i.e., a deviation from

$$(p°_A - p_A)/p°_A = N_S$$

In Table 2, I have given examples of his data and my N_A and x_A values.

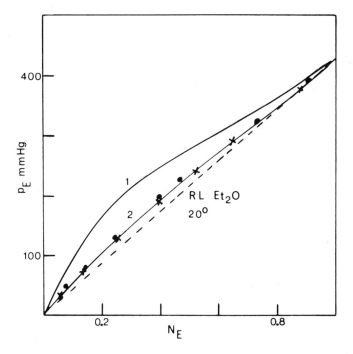

Fig. 39. Diethyl ether (E) + liquid S systems at 20°C. Plots of the mole fraction N_E vs pressure p_E for the liquids S = $C_6H_5NO_2$ (line 1), $C_6H_5CO_2C_2H_5$ (line 2, circles), and "térébenthine" (line 2, crosses). From data by Raoult (1890).

Table 2. Examples of Raoult's Data (1890)

S	N_S	$\dfrac{p^o{}_A - p_A}{p^o{}_A}$ (Z)	Z/N_S	N_A	x_A
For A = CS_2 at 24°C					
Ethyl benzoate	0.0315	0.0296	0.935	0.968	31.2
Ethyl benzoate	0.0773	0.0666	0.862	0.923	11.9
"Terebenthine"	0.0402	0.0435	1.080	0.960	24.0
Naphthalene	0.1171	0.1089	0.930	0.883	7.5
Hexachloroethane	0.0661	0.0644	0.974	0.934	14.1
Camphor	0.102	0.0849	0.834	0.898	8.8
Aniline	0.0912	0.0500	0.548	0.909	10.0
Acide valerianique	0.0959	0.0450	0.469	0.904	9.4
Nitrobenzene	0.0733	0.0394	0.538	0.927	12.7
Thymol	0.0629	0.0331	0.525	0.937	14.9
For A = "amylene" (bp 36.5°C) at 22°C					
"Terebenthine"	0.0866	0.0903	1.042	0.913	10.5
Thymol	0.1364	0.0750	0.549	0.864	6.35
Acide valerianique	0.1039	0.0500	0.481	0.896	8.6
For A = $CHCl_3$ at 23°C					
Ethyl benzoate	0.0756	0.0827	1.094	0.924	11.9
For A = C_2H_5Br at 23°C					
Ethyl benzoate	0.0767	0.0829	1.081	0.923	11.9
Benzoic acid	0.0577	0.0315	0.545	0.942	16.2

It is seen, therefore, that even for small N_S values there is a significant variation in Z/N_S. Those wishing to feel the full impact of his results should study his extensive table in the original.

3.2.5. Planck's 1888 Paper[302]

In the same volume in which Raoult's 1888 paper appeared, M. Planck wrote about the vapor pressure of *dilute* solutions of volatile substances.

3.2.6. Guthrie's Data (1885)[144]

Three years previously, Guthrie described the determination of total vapor pressure of a mixture at different compositions by weight using a barometer procedure. I have calculated the corresponding mole fraction data and plotted them as shown in Fig. 41. The lines for $Et_2O + CS_2$ and $CS_2 + CHCl_3$ are convex upward, whereas the line for $Et_2O + CHCl_3$ is concave upward. Guthrie attributed these effects to the formation of the compounds $C_4H_{10}O$, $CHCl_3$; CS_2, $CHCl_3$; and $C_4H_{10}O$, $2CS_2$.

Henry's Law and Raoult's Law

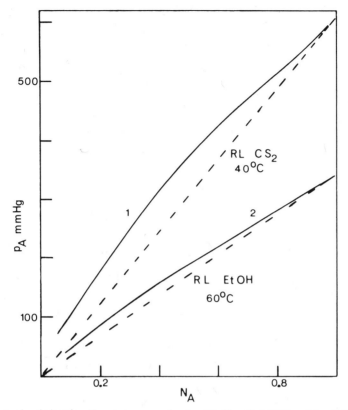

Fig. 40. Plots of mole fraction N_A vs pressure p_A for $A = CS_2$ (line 1) at 40°C, and $A = C_2H_5OH$ (line 2) at 60°C, each for the liquid $S = C_6H_5CO_2C_2H_5$. From the data of Raoult (1890).

3.2.7. Linebarger's Paper (1895)[241]

Linebarger endeavored to break the total pressure into the two partial pressures p_A and p_S by passing an "inert gas" through the liquid mixture and analyzing the entrained vapor by absorbing or decomposing agents such as an alkali to absorb an acid gas or red hot lime for sulfur or halogen compounds. The following mixtures $C_6H_6 + C_6H_5Cl$, $C_6H_6 + C_6H_5Br$, $C_6H_5CH_3 + C_6H_5Cl$, and $C_6H_5CH_3 + C_6H_5Br$ were said to show that the vapor pressure of these mixtures is a linear function of the concentration, given as mole fraction. For the mixtures $C_6H_5CH_3 + CHCl_3$ and $C_6H_6 + CHCl_3$ the vapor pressure "was less than that resulting from the calculation by the rule of mixtures." The mixtures $C_6H_6 + CCl_4$ and $C_6H_5CH_3 + CCl_4$ showed a very different behavior. The accuracy of this technique has been challenged.

3.2.8. Earlier Work on Vapor Pressure

Before proceeding with the paper which has become classical, it is pertinent to glance back at earlier work on the vapor pressure of mixtures. In 1836, Magnus observed that when alcohol was added to ether in the usual barometer assembly, the vapor pressure of both liquids is less than that of the ether alone.

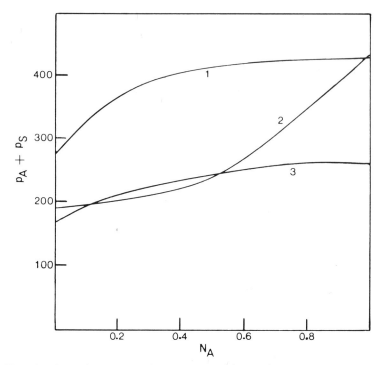

Fig. 41. Plots of mole fraction N_A vs total pressure $p_A + p_S$ (mm Hg) for the systems: (1) A = $(C_2H_5)_2O$, S = CS_2 at 18.9°C; (2) A = $(C_2H_5)_2O$, S = $CHCl_3$ at 19.0°C; (3) A = CS_2, S = $CHCl_3$ at 16.0°C. From the weight % data of Guthrie (1885).

Regnault (1852) aimed at a general idea of the relations between the vapor pressure of mixtures and those of the components. Alcohol + water (1852) and HCN + water (1864) were examined. Duclaux (1878) used a distillation procedure to gather information about the relations between the composition of the liquid mixture to that of the vapor emitted. His mixtures were water + an n-alcohol ROH (R = CH_3 to C_8H_{17}) and water + a carboxylic acid RCO_2H(R = H, CH_3, C_2H_5, C_3H_7); Konowalow (1881) also worked on similar mixtures.

Taylor[353] considered only the acetone–water system, but he made general statements which I must note here. He stated that "the law of Henry is formulated: p_1/c_1 = constant, or p_2/c_2 = constant," p_1 being the partial pressure of the acetone in the vapor state and c_1 the "corresponding concentration" of the acetone in the solution. Correspondingly, p_2 and c_2 refer to the water in the mixture. As it stands, this definition is invalid because it depends on the precise definition of *concentration*.

3.2.9. Zawidski's Diagrams[390]

This paper by Zawidski (1900) has become a classic in this area of chemistry, and one or two of his diagrams (see, e.g., Fig. 42) are reproduced in several textbooks. In the literature a "binary liquid mixture" is one resulting from the mixing of a liquid A with another liquid S, the mole fraction of A and S in the mixture being deemed simply derived from the actual weights of A and S mixed. It may seem strange to labor this point; but the careless disregard of the precise analysis has led to confusion and gross misunderstanding. The term "binary" must imply that A comprises only one species of molecule, e.g., the usual form associated with the usual formula weight. Zawidski considered two forms for acetic acid, CH_3CO_2H and $(CH_3CO_2H)_2$; and if the acetic acid as added to form the mixture is already a mixture of the two forms, then we are in trouble over the term *binary*.

Zawidski examined 13 binary mixtures using the refractive index of a distillate to calculate the partial pressures p_A and p_S, for in his selection both pressures had to be dealt with. His experimental technique is fully described in his paper; likewise, I do not need to consider his "theoretical" arguments; it remains only to draw attention to his diagrams. With the exception of the last-named mixture, the mixtures were examined at one temperature only, that given in parentheses:

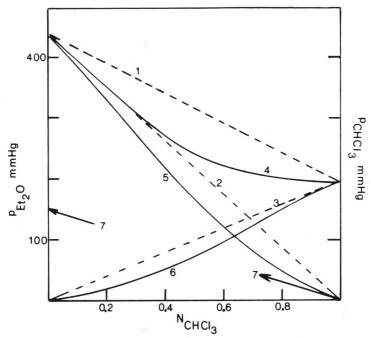

Fig. 42. The Zawidski type diagram. Line 2 corresponds to the R-line for diethyl ether (Et_2O) at 19°C. Line 3 corresponds to the R-line for chloroform at 19°C, and line 1 is the vertical addition of the pressures registered by lines 2 and 3. Line 4 shows the *total* pressure ($p_{Et_2O} + p_{CHCl_3}$) for each value of the mole fraction N_{CHCl_3} of chloroform. Line 5 is the plot of p_{Et_2O} vs N_{CHCl_3} (i.e., $1 - N_{Et_2O}$). Line 6 is the plot of p_{CHCl_3} vs N_{CHCl_3}. The practice has been to declare that for Raoult's law the observed data would give the linear plots, lines 1, 2, and 3. It is a common practice in the textbooks to draw a tangent from the lower end of an "observed" line, such as 5, as shown by line 7, and use the intercept of this, as shown, as a measure of the "Henry's law constant," the assumption being that the plot becomes linear as (in this case) $N_{Et_2O} \to 0$. The purpose and validity of this practice should be investigated.

$C_6H_6 + CCl_4$ (49.99°C)
$C_6H_6 + C_2H_4Cl_2$ (49.99°C)
$CCl_4 + CH_3CO_2C_2H_5$ (49.99°C)
$CCl_4 + C_2H_5I$ (49.99°C)
$C_6H_6 + CH_3CO_2H$ (49.99°C)
$CH_3CO_2H + C_5H_5N$ (80.05°C)
$H_2O + C_5H_5N$ (80.05°C)

$CS_2 + CH_2(OCH_3)_2$ (35.17°C)
$CS_2 + CH_3COCH_3$ (35.17°C)
$CHCl_3 + CH_3COCH_3$ (35.17°C)
$C_2H_4Br_2 + CH_3CHBrCH_2Br$ (85.05°C)
$CH_3CO_2C_2H_5 + C_2H_5I$ (49.99°C)
$CH_3CO_2H + C_6H_5CH_3$ (69.94°C and 80.05°C)

Of all the diagrams, the outstanding one is that for the system 1,2-$C_2H_4Br_2$ (A) + 1,2-$C_3H_6Br_2$ (S) for which the p_A/N_A and p_S/N_S plots are almost straight and, of course, the total-pressure line is almost straight. This example has been quoted again and again as one which "follows Raoult's law." Remarkably enough, Zawidski did not refer to Raoult's law, although he mentioned Raoult's name in a footnote relating to the term mole fraction, citing Raoult's 1890 paper. The system ($C_6H_6 + C_2H_4Cl_2$) gave almost straight lines. The acetic acid–pyridine and chloroform–acetone lines are strongly concave upward; but apart from the two examples of almost straight lines, the remainder of the lines are convex upward.

In the Zawidski diagrams, the two partial-pressure lines were plotted with respect to a base line showing the mole fraction of only one of the members, that of the other being $1 - N$. In my Fig. 43, I have plotted the acetic acid–pyridine data in the R. L. form. In Fig. 44 I show examples of the great difference in the form of the mole ratio and mole fraction plots.

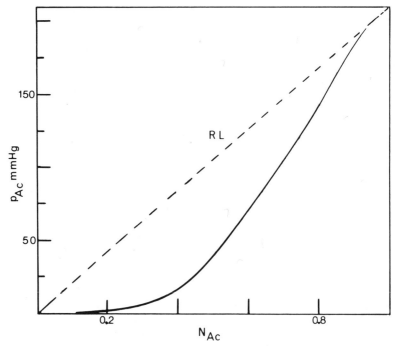

Fig. 43. A simple plot of the Zawidski data on the R-line diagram for acetic acid at 80.05°C. The system is acetic acid (Ac) and pyridine.

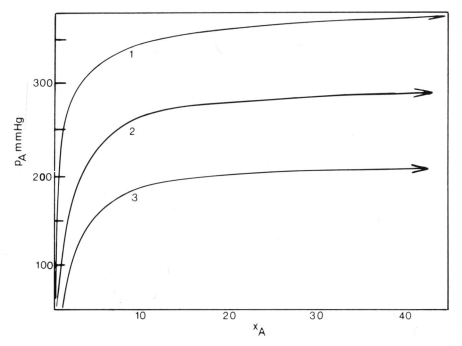

Fig. 44. Plots of mole ratio of A, x_A, vs pressure p_A for the systems: (1) $(C_2H_5)_2O(A) + C_6H_5NO_2$ at 16°C; (2) $CHCl_3(A) + CH_3COCH_3$ at 35.2°C; (3) $CH_3CO_2H(A) + C_5H_5N$ at 80°C.

3.2.10. Dolezalek's Theory[68]

To account for deviation from linearity, Dolezalek assumed certain forms of association and chemical reactivity. He used the data of Zawidski. The chloroform–acetone system was supposed to comprise single molecules of these compounds together with molecules of the compound $CHCl_3,(CH_3)_2CO$. The supposed formation of $(CCl_4)_2$ was deemed to increase $N_{C_6H_6}$ and, therefore, $p_{C_6H_6}$ beyond the "regular" value. An adjustment was made to allow for this and bring the points onto a straight line. To account for the form of the plot for diethyl ether + methyl salicylate, he supposed that the ether remained simple, but the ester existed in solution in the form of double as well as single molecules.

Dolezalek and Schulze (1912)[69] observed that when equal volumes of the vapors of ether and chloroform were mixed at 80°C, so that the total volume remained constant, the total pressure was lowered by 4.8 mm Hg, a result attributed to compound formation.

In his paper of 1910, Dolezalek considered the probable equivalence of liquid–liquid systems and gas–liquid ones. Inglis's data[181] on the solubility of oxygen in liquid nitrogen at −198.4°C (74.7 K) (Figs. 45, 46) were deemed to be explained by assuming the formation of a certain proportion of $(O_2)_2$ molecules. Dolezalek also commented on Blümcke's data (1888)[26] on the solubility of carbon dioxide in liquid sulfur dioxide (see Fig. 47).

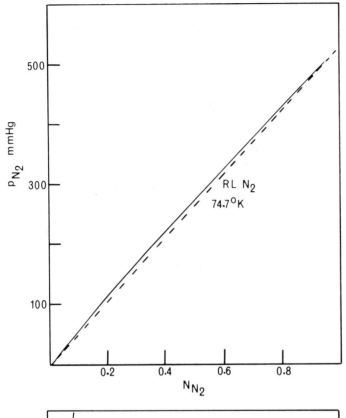

Fig. 45. Solubility of nitrogen in oxygen at 74.7 K. Plot (full line) of the mole fraction N_{N_2} vs pressure p_{N_2} (in mm Hg) on the R-line diagram.

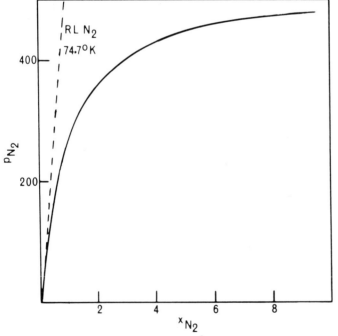

Fig. 46. Solubility of nitrogen in oxygen at 74.7 K. Plot of the mole ratio x_{N_2} vs pressure p_{N_2} (in mm Hg). The broken line is the R-line for N_2 at 74.7 K (see Fig. 45).

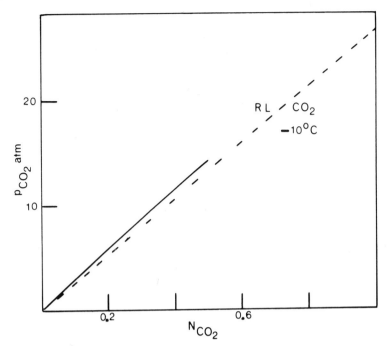

Fig. 47. Solubility of carbon dioxide in liquid sulfur dioxide at −10°C. Plot of the mole fraction N_{CO_2} vs pressure p_{CO_2} on the R-line diagram for carbon dioxide. From data by Blümcke.[26]

Other contemporary workers were thinking in similar terms.

Bein (1909)[19] observed the systems ($C_6H_6 + C_2H_4Cl_2$), ($C_6H_6 + CCl_4$), and ($CHCl_3 + CH_3COCH_3$) and concluded that in the last-named one, the compound $CHCl_3,(CH_3)_2CO$ present in the liquid phase exists to some extent also in the vapor. In the same year, Möller[261] calculated the partial pressures of the components of a binary mixture AB by means of experimental data on AC and BD. From the data on ($C_6H_6 + CCl_4$) and ($C_2H_5I + CH_3CO_2C_2H_5$), the result for the system ($CCl_4 + CH_3CO_2C_2H_5$) was deemed to be excellent. One-third of the molecules of ethyl iodide were supposed to be single, and two-thirds triple. Ethyl acetate was supposed to contain 28% double molecules under the same experimental conditions. Neither of these authors mentioned Raoult's law.

van Laar[362] opposed Dolezalek's theory and held the view that in all but exceptional examples the phenomena could be explained in terms of physical interaction. On the other hand, Dolezalek contended that van Laar's theory was in discord with the principles of thermodynamics.

3.2.11. Work of Young[388,389]

Young looked at the possible connection of the vapor pressure of a mixture of liquids A and S with (a) the heat and volume changes on mixing, (b) the "chemical sameness" of A and S, and (c) the critical pressures of A and S. Young referred to

Speyers' work[339] on bp curves and on the molecular weights of liquids, and he believed that the statement about the application of the equation $(p°_A - p_A)/p°_A = N_S$ was too general because mixtures of liquids of normal molecular weights had been observed (Lehfeldt[231] and Zawidski) to deviate from such behavior.

3.2.12. Campbell's Work[44] (1915)

Campbell plotted N_E vs P $(=p_E/p°_E)$, where E = diethyl ether and the other liquid S was oleic acid at 30°C. The line was concave upward. The corresponding plot for carbon disulfide and oleic acid was concave upward from $N_{CS_2} = 0$ to about 0.6; then it crossed the diagonal and became convex upward up to the top right corner of the diagram, Fig. 48. On the other hand, the aniline–ether line, N_E corresponding to $P(=p_E/p°_E)$, was convex upward from $N_E = 0$ to 0.8, and then it crossed the diagonal and became concave upward. The straight line (diagonal) was called the mixture-law line. In Fig. 49, I show the N_{Ac}/p_{Ac} plot for acetone and S = oleic acid at 30°C. In Fig. 50 I show the corresponding plot for methanol and S = glycerol at 40°C. Acetone, methanol, and glycerol were deemed to be associated, and oleic acid to be "normal."

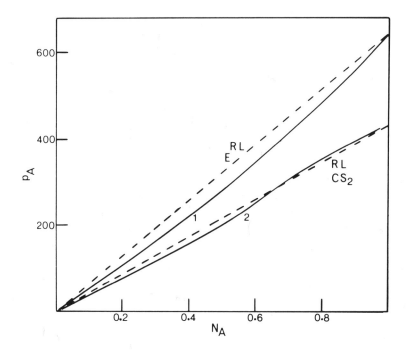

Fig. 48. Line 1: plot of the mole fraction of diethyl ether N_A vs pressure of ether p_A for the oleic acid system at 30°C. Line 2: plot of the mole fraction N_A of carbon disulfide vs pressure of carbon disulfide p_A for the oleic acid system at 30°C. The R-lines are for ether E and CS_2 at 30°C. The pressures p_A are in mm Hg.

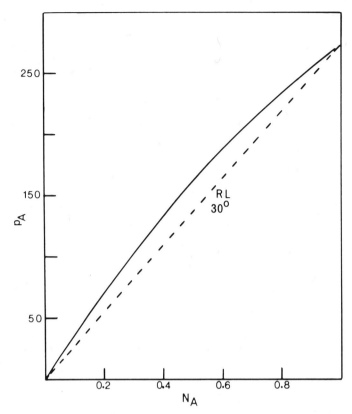

Fig. 49. Plot of the mole fraction N_A of acetone vs p_A (mm Hg) of acetone for the oleic acid system at 30°C.

3.2.13. Lewis' New System of Thermodynamics

At the beginning of this century, G. N. Lewis[235–237] showed that van't Hoff's law of osmotic pressure was seriously at fault, even for those solutions which are usually taken to be dilute, and he was faced with the need to find a new definition of an *ideal solution*. He believed an ideal solution could be defined as one which obeys Raoult's law, or Henry's law, which he took to be "essentially identical" for infinitely dilute solutions.

This aspect appears to relate to what happens during the process of dissolution under the conditions which prevail as p_A and N_A emerge from zero; i.e., the line is near the left bottom corner of the diagram, as in, e.g., Fig. 50. The current opinion appears to be that all gases A (or liquid A) obey Henry's law, $p_A = KN_A$, for small values of N_A, i.e., when N_A is closely approaching zero. The Henry's law "constant" does not need to be $p°_A$, the "constant" of the Raoult's law. I imply from Lewis' statement that as N_A closely approaches zero, there is only one law, essentially only one line. The question is: At what stage, and why, does the Henry's law line begin to

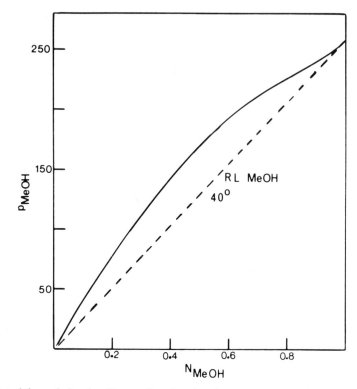

Fig. 50. Plot of the mole fraction N_{MeOH} of methanol vs the pressure p_{MeOH} (mm Hg) for the glycerol system at 40°C.

deviate from the line which *would be* the Raoult's law one? A hydrogen chloride–water system at 0°C and $N_{HCl} = 0.2$ is far from ideal. Why is it significant to look upon a solution having $N_{HCl} = $ (say) 0.0001 as ideal, or nearly ideal? Correspondingly, Raoult's law, $p_A = p°_A N_A$, is deemed to be followed by all systems when N_A closely approaches unity and p_A closely approaches $p°_A$. It should be noticed that I am connecting N_A with p_A (the same A) at each end of the line. Lewis appeared to conclude that Raoult's law (or a modified form of Henry's law) often holds at all concentrations from 0 to 100% of A (called the solvent). The named example was that of benzene + 1,2-dichloroethane.

He introduced the concept of fugacity f and activity a, related as $a = f/RT$; and he defined a perfect solution as one for which $a_A = a°_A N_A$, a_A being the activity of the solvent in the solution, and $a°_A$ being that of the pure solvent. The quantity N_A is, the mole fraction of A. Several decades later, the van't Hoff theory of osmotic pressure was still being propagated. Note Kendall's comment of 1937.[202]

3.2.14. The Elusiveness of Ideal Solutions

The *Discussion* of the Faraday Society of 1937 on the structure and molecular forces in pure liquids and solutions is still worth detailed study, particularly with reference to the currently held distinction between "chemical" and "physical" intermolecular forces, polar–nonpolar molecules, normal–abnormal liquids, and ideality–nonideality. Hildebrand[160] pointed out that despite the time and effort that had been given to the study of solutions, specific examples of systems giving a linear plot for N_A vs p_A were hard to find. Indeed, only one example emerged unscathed from that discussion, and that was the Zawidski system of 1,2-dibromoethane + 1,2-dibromopropane. Kendall[202] recognized that deviations from Raoult's law at that time were "considerable," ideal solutions being almost as rare in practice as infinitely dilute ones. Whereas Guggenheim[142] admitted the existence of such ideal solutions, he acknowledged the scantiness of experimental data. He showed that a "perfect" solution would be formed without heat effect or volume change, but the converse need not be true. He pointed out that up to that time the two examples of Zawidski (benzene + 1,2-dichloroethane and 1,2-dibromoethane + 1,2-dibromopropane) were the only ones for which the laws of perfect solution ($p_A = p°_A N_A$ and $p_S = p°_S N_S$) had been accurately established. However, Glasstone[129] mentioned abnormal results for the dipole moment of 1,2-dichloroethane in benzene, and therefore the near linearity of the Zawidski plot could be attributed to compensation effects. Later (1953), Herington[157] doubted the ideality of this last-named system and, for example, drew attention to Baud's (1915)[18] measurement of the heat of mixing.

In the *Discussion* of 1937, Glockler[131] contended that the division of intermolecular forces into chemical and van der Waals forces was more a matter of convenience than fundamental necessity. To Glasstone,[129] a clear distinction between forces entailing a coordinate link and those of a physical nature due to dipole interaction was somewhat elusive. LeFevre[230] commented on the hazards attending the classification into polar–nonpolar and normal–abnormal liquids. Butler[40] referred to the formation of a complex between the two "polar" molecules ether and chloroform. The difficulty in distinguishing mere dipole interaction and chemical bond formation was indicated by Hildebrand.[160]

In 1965, Longuet-Higgins[245] acknowledged the difficulty in drawing a clear line between the physical and chemical interaction between molecules, and Coulson[60] remarked on the widely different approaches in matters of detail shown by the different contributors. Finally, I refer the reader to the books by Rowlinson[319] and King.[204]

3.2.15. "Positive" and "Negative" Deviations from Raoult's Law

If a line is convex upwards (on the left of the R-line), it means that the p_A is larger than it would be for the same N_A if $p_A = p°_A N_A$; for this reason, the deviation from "Raoult's law" has been deemed positive. On the other hand, if the line is concave upwards (on the right of the R-line), p_A is less than the $p_A = p°_A N_A$ value for the same N_A; therefore the deviation has been said to be negative. It has been assumed that a deviation could not change sign; a line must be entirely concave upward or convex upward.

Metzger and Sauerwald[257] and Högfeldt,[174] however, believed that, although that may be true for a "1:1 compound," all other compositions of the "compound" should give both positive and negative deviations. McGlashan[277] has analyzed the possibilities of a changeover from the concave to convex or vice versa, according to which end of the line one starts from (see also McGlashan and Wingrove[279]). Again, I believe the interaction patterns in real systems are too complex to be treated rigidly as 1:1 or 1:2, etc. compounds. For example, McGlashan and Rastogi[278] obtained an ordinary concave-upward curve for the chloroform–dioxane system although they expected the crossover form because of the possibility of a compound $Cl_3CH \cdots O(C_4H_8)O \cdots HCCl_3$. As always, reality takes its toll. Even a reagent having an intensive acid function such as boron trichloride gives the 1:1 complex which, however, does give the $3BCl_3$:2-dioxane under certain conditions.[94] The x_{HCl} value at 10°C is 1.05 and not 2.0. It is, therefore, unlikely that chloroform would on a time-average basis correspond even to a formal 1:1 compound; the system would appear to be one of dynamic equilibrium, the active making and breaking of hydrogen bonds.

The results of Brown and Smith (1960)[37] on acetone and nitromethane indicated a change of sign of deviation and hydrogen bonding between the oxygen of acetone and hydrogen of the nitromethane in addition to dipole–dipole interaction.

4
Hildebrand's Solubility Parameters

4.1. Intermolecular Forces

During the last 80 years, hundreds of papers have appeared on the attempts to analyze intermolecular forces in liquids. Entailed in these analyses are arguments carrying a bewildering variety of connotations of the terms polar, nonpolar, normal, abnormal, associated, nonassociated, physical interaction, chemical interaction, ideal, and nonideal. I need only refer to the relevant *Discussions* of the Faraday Society held in 1937, 1953, 1965, and 1967. Certain workers pin the distinction of polar from nonpolar to the relative magnitude of the dielectric constant. Others connect these terms with the possession or otherwise of a dipole moment. I can find no systematic correlation between solubility, expressed as a mole ratio, and the dipole moment, or dielectric constant.

Hildebrand[158] considered internal pressure as a basis for comparing the alikeness of molecular forces. Even after his emphasis on his restriction to nonpolar molecules, there remained the uncertainty of relating forces between the molecules of the same species with those between the molecules of different species. He expressed the view that a liquid mixture in which the internal forces of attraction and repulsion do not change with the change in composition of the mixture will follow Raoult's law. This effect, however, requires the following conditions, as Hildebrand expressed them:

(a) The components in the pure liquid state have the same internal pressures.
(b) The different molecules are relatively symmetrical or nonpolar.
(c) The tendency to form chemical compounds is absent.

To allow for the different internal pressures, Hildebrand used his solubility parameter δ.

4.2. The Solubility Parameter

Attempts to account for the mole fraction solubility N_A of a gas A in a liquid S have been dominated by Hildebrand's idea of a solubility parameter δ, defined as the square root of the energy of vaporization per cubic centimeter, and of an ideal solubility N^i_A associated with Raoult's law. In this approach, N^i_A is calculated from $1/p^°_A$, where $p^°_A$ is the vapor pressure in atmospheres of A over liquefied A, or its equivalent, at the operational temperature $t°C$, usually $0°$ to $25°C$. For gases having a critical temperature below the operational temperature, $t°C$, $p^°_A$ must be evaluated by an extrapolation process.

For nonpolar gases and nonpolar liquids, Hildebrand connected N_A with N^i_A, δ_A, and δ_S by the following expression:

$$\log N_A = \log N^i_A - \frac{\bar{V}_A}{2.303RT}(\delta_S - \delta_A)^2 - \left[\log \frac{\bar{V}_A}{V_S} + 0.4343\left(1 - \frac{\bar{V}_A}{V_S}\right)\right]$$

where $\bar{V}_A$ is the partial molal volume of gas "in the saturated solution" and V_S is the molal volume of S.

Based on his results for carbon dioxide and nitrous oxide in a number of organic liquids, Kunerth (1922)[220] concluded that there was no correlation between solubility and the internal pressures of solvent and solute; he expressed disagreement with Hildebrand's views. In his reply, Hildebrand (1923)[159] emphasized that he required the condition of nonpolarity. Of the 21 liquids S used by Kunerth, only 8 had dielectric constants as low as 5. In the many references to the solubility parameter and the parameter equation since that time, there is the inherent difficulty over the terms polar and nonpolar and what constitutes a chemical reaction.

An appreciation of the extent to which the parameter equation responds to actual data can be gained only by a study of all the relevant papers by Hildebrand and his colleagues and of Gjaldbaek and his colleagues. In the light of the solubilities of H_2, CO, N_2, CH_4, C_2H_6, CO_2, and NH_3 in various liquids S, which included liquids deemed to be polar, Hildebrand saw that the polar liquids presented the greatest uncertainty. Carbon dioxide was described as being more polar than the preceding five gases and was described as a *donor* molecule. Ethene was described as an *acceptor* molecule. With reference to the plots of δ_S vs $\log N_A$, Jolley and Hildebrand[192] attributed the divergence of points for C_2H_2 and CO_2 in benzene to chemical effects, i.e., acid–base interaction. Gjaldbaek and Andersen[125] examined the solubility of CO_2, O_2, CO, and N_2 in 12 polar solvents, and the matter of degree of polarity and of association emerged. Toluene, with a dielectric constant of 2.37, was deemed to be weakly polar, whereas nitrobenzene, with a dielectric constant of 34.9, was looked upon as being strongly polar. Of the 12 liquids S, only acetone was deemed to be somewhat associated. With reference to another 12 liquids S,[126] a few were stated to be polar but not associated; Gjaldbaek[124] required the condition that the molecules be nonpolar and not associated.

4.3. Force Constants

As parameters for gases, force constants ε/k were used, ε being the depth in ergs of the minimum of a curve supposed to express the variation of potential between a pair of isolated molecules with distance between their centers, and k the Boltzmann constant. For the gases (He, Ne, N_2, CO, O_2, Ar, CH_4, Kr, C_2H_6) and three liquids (i-C_8H_{18}, CCl_4, C_6H_6) the plots of log N_{gas} vs ε/k were deemed to show excellent correlation, but later[163] this assessment was reduced to "rather good." (Hydrogen was found to deviate from all the other gases in this and other solubility relations.)

The deviation of carbon dioxide in benzene was attributed to donor–acceptor function. A more critical scrutiny with better data showed the inadequacy of the approach via the force constants.

5
The Hydrogen-Bonding Structure of Water

5.1. The Essential Pattern

In a two-page account, Partington (1951)[297] refers to water as "a mixture of polymerized water molecules" and cites about 80 papers. In his inaugural lecture (1963) Ives[185] talked about his "reflections on water," referred to the thermodynamic studies of Frank and his co-workers, and quoted a comment that Rowland made in 1880.[318]

Emerging from detailed studies of the way the various properties of water change with temperature, and from X-ray work (Bernal and Fowler, 1933) and Raman spectra (Rao, 1934), is the essential pattern of hydrogen-bonding structure, herein designated for convenience as $(H_2O)_n$, in which the units "H_2O" are linked in a system of dynamic equilibria. It is of vital importance to distinguish the symbol "H_2O", which should refer precisely only to the molecule

$$H-O\diagdown H$$

in isolation, and the symbol $(H_2O)_n$, which should *always* be used when considering the chemical properties of water under the conditions which usually prevail. Under such conditions, water *never* can react simply as (H_2O), and a critical study of all those reactions which are described so will clearly show how false the treatment is. Again and again one can show clearly that it does matter, and matter vitally.

Lennard-Jones and Pople (1951)[232] suggested that in water there is "a network of bonds extending throughout the whole liquid which is, in a sense, one large molecule." Frank and Wen (1957)[87] accept the "covalent character" of the hydrogen bond in water. They looked upon hydrogen bonding as an acid–base interaction. It was supposed that, on a time basis, there is some "nonbonded" water, but in this, dipole–dipole forces are still very strong and still produce a considerable degree of orientational restriction and powerful binding of an individual molecule to the whole liquid. In three long papers, Némethy and Scheraga (1962)[285] discuss in detail the structure of water and hydrophobic bonding in proteins.

Lennard-Jones and Pople (1951)[232] concluded that very few hydrogen bonds are broken in the liquid near the melting point; but believed they are mostly distorted by bending. Having regard to the "partly covalent character of the hydrogen bond," Frank and Wen were against the concept of large deviations from collinearity. A calculation indicated[285] that about two-thirds of the "molecules" [meaning (H_2O) units] participate at any given instant in the formation of hydrogen-bonded clusters.

Attempts have been made to involve and explain *all* the changes in the properties of water as the temperature changes, or as energy is added. There is disagreement over details. I look upon the "dipole–dipole" attraction of the "unbonded" (not hydrogen-bonded) "H_2O" units as constituting an acid–base function, and I am left with the essential pattern for the present purpose, which may be symbolized simply as $(H_2O)_n$.

But there is still the matter of the so-called hydrogen ion, designated H^+, and of the hydroxyl ion, OH^-. Even the form $H_9O_4^+$, suggested by Eigen and De Maeyer,[73] in which three "H_2O" units are hydrogen-bonded to one central "H_2O" unit, might be deemed to participate, through additional hydrogen bonds, in the formation of clusters in the liquid. The OH^- ion can, in a similar way, participate in hydrogen-bond formation and hence be accommodated in clusters. If in conductivity water there are supposed to be 10^{-7} g-ions of (H^+) per liter of solution, this means that there is one (+) center for every 556,000,000 (H_2O) units. According to Eigen and De Maeyer, the transport of protonic charge entails "structural diffusion" by the formation and disappearance of hydrogen bonds.

5.2. The Situation about 1974

At the Faraday Society discussion of 1967, Frank[85] emphasized the point that he knew of no one who believed there are any free H_2O molecules in liquid water, considered as a mixture—"*free* in the sense in which a molecule is free in the dilute vapor."

Magat[251] referred to the problem of free water molecules and mentioned contradictory results relating to the existence of freely rotating water molecules, which he believed can probably be looked upon as not hydrogen-bonded water.

Holtzer and Emerson[176] have made a detailed analysis of the utility of the concept of water structure in the rationalization of the properties of aqueous solutions of proteins and small molecules and concluded that the arguments put forward by different authors about the *role of water structure* often have led to contradictory conclusions. In many examples, a single argument appeared to them to be sufficiently indeterminate as to be capable of producing several contradictory conclusions. It appeared to them that terminology needs constant vigilance; such expressions as "flickering-cluster model," "making holes," and "icebergs" become less distinct the more closely they are examined. Those writers contended that the position has been little advanced since the work of Frank and Evans[86] and of

The Hydrogen-Bonding Structure

Kauzmann.[201] See papers by Wicke (1966),[375] Luck (1967),[246] Franks,[88] and a book by Eisenberg and Kauzmann (1969)[74] reviewed by Davies.[61] Three symposium publications, refs. 394, 395, and 399, are of interest here.

Scheraga and colleagues have continued work on the structure of water. Shipman and Scheraga (1974)[396] referred to the many empirical intermolecular potentials that have been developed. One by Ben-Naim and Stillinger[398] appears to have had some success, although several defects were stated to have been detected. This potential was based on classical mechanics, whereas it now appears that the intermolecular motions among water molecules are of a quantum-mechanical nature.[396] See also Hagler, Scheraga, and Némethy.[397]

On the recent techniques and arguments in the study of the structure of water, the work of Symons and his colleagues at the University of Leicester must be specially noticed.[401] The compelling conclusion was that the concentration of monomeric water in liquid water under normal conditions is undetectably small.

Of relevance here is the work of Kaulgud and Patil (1974)[400] on the interaction of aliphatic amines with water structure (see Chap. 10).

5.3. Significance of the Solubility of Hydrocarbon Gases in Water

5.3.1. Why Significant

The solubility of hydrocarbon gases in water has attracted attention because of the gas hydrates, hydrophobic effects in protein chemistry, and the reflection of the structure of water.

5.3.2. Free Energy, Enthalpy, and Entropy Changes

The three contributions of special significance here are those of Frank and Evans,[86] Claussen and Polglase,[53] and Némethy and Scherage.[285] The calculation is based on the transfer of *one mole* of hydrocarbon from the pure liquid hydrocarbon to "dilute solution" in water. The *standard states* were taken as "mole fraction unity ($N_{HC} = 1$) in the liquid and the solution, and the pressure 1 atm in the gaseous state":

$$RH_l(N=1, T, p=1 \text{ atm}) \rightarrow RH_{aq}(N_{std}, T, p=1 \text{ atm})$$

Methane, ethane, propane, and *n*-butane are gaseous at temperatures above 0°C, and therefore a gaseous stage must be interposed. Because methane and ethane have critical temperatures (c.t.) below the temperatures considered (up to 70°C, although data were given for 25°C, and the c.t. of ethane is 32°C), their vaporization from "nonpolar" solvents, CCl_4, C_6H_5Cl, C_6H_6, $(CH_3)_2CO$, $CH_3CO_2CH_3$, $(C_2H_5)_2O$, was considered (see Frank and Evans[86]).

Némethy and Scheraga recognized a considerable scatter of the $\Delta F°$, $\Delta H°$, and $\Delta S°$ values, and attributed these to the different polar character of the six liquids just named.

To account for the thermodynamic effects, Claussen and Polglase looked to the structure of the inert gas hydrates, for which Claussen had already suggested a model. Gas hydrates are far from new; they have an attenuated history.[63,340] Interest in them flared up when it was concluded in 1934 that "freezes" in natural gas transmission lines could be caused by hydrocarbon gas hydrates, and not simply by water ice. For methane, for instance, "water molecules" were supposed to occupy the corners of a polyhedron with methane occupying the void space inside. The polyhedral structure of the *iceberg* was supposed to involve the faces of 12 pentagons. The highly ordered state of these icebergs could account for the large negative entropies of solution. Frank and Wen[87] took up the phrase "extra ice-like-ness" but rejected the notion that the effect was due to the formation of cages about solute molecules, similar to those believed to exist in solid clathrate hydrates.

Going back to Claussen and Polglase, we find that ethane and propane were deemed also to fit the "iceberg" model; but *n*-butane was deemed unable to fit into this simple iceberg picture so easily, it being assumed that *n*-butane is too big to remain comfortably in the largest of the voids, the 16-hedron. The iceberg structure surrounding *n*-butane may be the 16-hedron part of the time and still another unknown structure for the remaining time.

Némethy and Scheraga wrote of the probability of finding a cluster being greater in the neighborhood of a hydrocarbon molecule in the solution than at a point in the bulk of pure water. The question to be asked is: How can there be any "bulk of pure water"? At 760 mm, 20°C, and *at equilibrium*, for each individual molecule of methane about 35,000 "H_2O" units are required on a time-average basis. By what argument can it be denied that on a time-average basis *one* molecule of CH_4 has the exclusive *use* of 35,000 molecules of water? The term molecule of water is that used in the papers cited and means the (H_2O) unit. If any "pure water" were present, it would of necessity have the properties of water and absorb more molecules of methane.

An absorption coefficient of 0.04 means that there are about $1/0.04 = 25$ molecules of hydrocarbon gas in 1 ml of gas for each molecule of hydrocarbon in 1 ml of solution. If, at equilibrium, it can be claimed that "ordinary" water is available for this heavy bombardment from the gaseous phase, how can the "ordinary" water discard its true character and resist the acceptance of one more molecule of methane. Of course the whole system is deemed to be in *dynamic* equilibrium. Even ordinary water does not appear to like the hydrocarbon gas; but what the ordinary water is changed into positively repels the molecule of methane.

Similarly for the thermodynamic values. These are give as per mole of methane: $\Delta \bar{H}_{sol}$ at 25°C = -3.052 kcal; $\Delta \bar{S}°$ at 25°C = -31.2 cal; but it requires *more* than 35,000 (H_2O) to produce them. How can one escape from the conclusion that about 35,000 (H_2O) units as "pure water" are changed into a condition which is *not* that of pure water?

Table 3. Solubility of Four Gaseous Hydrocarbons in Water at 19.8°C and 760 mm Hg Pressure

RH	Absorption coefficient	Mole ratio x_{RH}	Mole ratio[a] M_{H_2O}
CH_4	0.0351	0.0000282	35,450
C_2H_6	0.0498	0.0000400	25,080
C_3H_8	0.0394	0.0000317	31,580
$n\text{-}C_4H_{10}$	0.0327	0.0000263	38,060

[a] M_{H_2O} is the number of moles of water (as H_2O unit) per mole of RH.

The same questions apply equally well to the other three hydrocarbons. Look at Table 3. Compared with the steep increase in R-line x_{RH} value for 20°C, 0.0036 for CH_4 to 0.988 for C_4H_{10}, the x_{RH} values for water are all in the 0.000026 to 0.00004 range, and there is no systematic order of change.

5.4. Hydrophobic Interactions

In 1959, Kauzmann[201] indicated the effects of nonpolar groups on the structure of water, the so-called hydrophobic effects in stabilizing protein structures. Wishnia[381] examined the solubility of ethane, propane, and n-butane in aqueous solutions of bovine serum albumin (BSA), hemoglobin (Hb), lysozyme, and sodium lauryl sulfate (SLS).

The increase in solubility in the solution as compared with that in water itself was deemed to be large, except for the example of lysozyme, and this was attributed to a direct interaction between the gas and the other solute. This could be understood as solubility in a "nonpolar material." It was suggested that a study of the solubility values of a graded series of alkanes may lead to an estimate of the number and size of hydrophobic clusters in a given protein.

The solubility of n-butane, expressed in millimoles per 1000 g solution at 1 atm and 10°C, was 2.5 for water and 8.0 for BSA (5% solution). The statement that the mole fraction solubility of butane and propane in BSA is much greater than the ideal solubility needs elucidation in the light of the R-line diagrams I have given for these gases.

In 1970, Wen and Hung[371] cited previous literature (1928–1968) relating to symmetrical tetraalkylammonium salts and the water structural changes induced by large hydrophobic cations. Results of studies by Morrison and Johnstone[265,266] and Ben-Naim[21] were deemed to be rather inconclusive; therefore, a study of the solubilities of methane, ethane, propane, and butane in pure water and in aqueous solutions (0.1 to 1.0 m) of NH_4Br, $(CH_3)_4NBr$, $(C_2H_5)_4NBr$, $(n\text{-}C_3H_7)_4NBr$, $(n\text{-}C_4H_9)_4NBr$, and $(HOC_2H_4)_4NBr$ at 5, 15, 25, and 35°C was undertaken. It must be pointed out that a molality of 1 means a mole ratio x_{R_4NX} of 0.018.

Wen and Hung gave data for pure water expressed as milliliters per 1000 g water, the volume being for 1 atm and 0°C; i.e., the observed volume of gas for t°C and p_{RH} was converted into the volume for 1 atm and 0°C, presumably on the assumption that the "perfect gas law" held and x ($\approx N$) changed linearly with pressure. In Fig. 138, I have plotted the calculated x_{RH} values against the temperature. It will be noticed that the values are all absolutely small, and the relative differences are not large. There is no systematic order. These authors compared their values with those of Claussen and Polglase[53] on the one hand, and those of Winkler[376] and Morrison and Billet[264] on the other.

For 5°C, $x_{CH_4} = 0.00003975$ (W and H) agrees with that by C and P, but not with that of W or M and B (0.00003833); the reverse is so for $x_{C_2H_6}$. It can be time-wasting to argue about relatively small differences in the values submitted by different workers, especially when the details of adjustment to so-called standard conditions of temperature and pressure are not clear. Presumably, in the present instance, the change in density of water with the temperature has been taken care of in the expression "1000 g water."

The original paper must be read to get the details of the argument, but the following statements are relevant to my present purpose. The solubility of the hydrocarbon gas in the tetraalkylammonium salt solution was likewise expressed as milliliters per 1000 g water.

It was stated that large tetraalkylammonium salts such as Pr_4NBr and Bu_4NBr increase the solubility of hydrocarbon gas, whereas NH_4Br and $(EtOH)_4NBr$ decrease it. This, at first sight, appears to mean that here we have a liquid in which a hydrocarbon gas is less soluble than it is in water. However, on a molecular basis, the primary data (volume of gas per 1000 g of "liquid") are not in a valid form for comparison because the average molecular weight of the "liquid" is a material factor. The differences are relatively small, even for 1 m solutions.

The magnitude of gas solubility in water appears to be associated with the number of cavities, according to Eley.[75] Ben-Naim[21] and Namiot[282,283] have used a two-structure model of liquid water. The precise way in which the hydrocarbon molecule is accommodated in the water structure appears to be still a matter of controversy. Namiot (1967) believed that data on the solubility of hydrogen and neon in ice are at variance with current opinions on the reasons for the low solubility of nonpolar gases in water.

5.5. Solubility of Certain Gases in Water and the Influence of the bp/1 atm Factor

With the exception of seven gases (HCl, HBr, HI, NH_3, $MeNH_2$, Me_2NH, Me_3N), the x_A value for each gas A is less for water than for any other liquid reported as having been examined, and the values are well over on the left of the R-line for the particular gas.

In Fig. 51 I show the plot of $x_{C_2H_4}$ vs $p_{C_2H_4}$ for water based on the grams of ethene per 100 g of water data of Bradbury, McNulty, Savage, and McSweeney,[32]

The Hydrogen-Bonding Structure

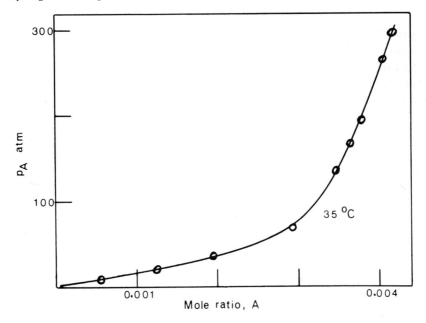

Fig. 51. Solubility of ethene in water at 35°C. Plot of $x_{C_2H_4}$ vs $p_{C_2H_4}$.

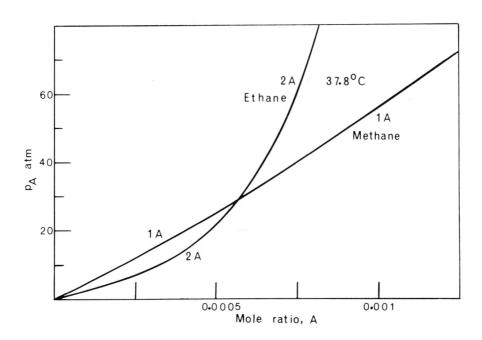

Fig. 52. Solubility of methane (1A) and ethane (2A) in water at 37.8°C. Plots of mole ratio x_A vs p_A.

who gave data for four temperatures and pressures up to as much as 523 atm. The plot is decidedly concave upward; even at 300 atm the $x_{C_2H_4}$ value is only 0.0042. In Fig. 52 I show the plots for methane and water, and for ethane and water from the data of Culberson, Horn, and McKetta, Jr.,[408, 409] who gave mole fraction values for pressures expressed in psia units. The lines are concave upward; at 80 atm, $x_{C_2H_6}$ is only 0.0008 at 37.8°C, and x_{CH_4} is 0.00155 at 25°C.

Brooks and McKetta[34] gave three values for the solubility of 1-butene in water at 37.8°C and at pressures up to about 63 psia, approximately the maximum for the system comprising gas and one liquid phase. In Fig. 53 I show the line for $x_{C_4H_8}$ vs $p_{C_4H_8}$ (atm) corresponding to those data. At $p_{C_4H_8}$, 4.25 atm and 37.8°C, $x_{C_4H_8}$ is only 0.00022. Referring to the work of Reamer, Sage, and Lacey,[308a] Brooks and McKetta mentioned data for n-butane (I have given the plot in Fig. 53) and drew attention to the larger solubility value for the olefin as compared with that for the corresponding paraffin. Their comment that the trend shows a decrease in solubility with increase in molecular weight should be considered with caution; at 1 atm and 25°C the x_A values for n-decane (S) are

$$CH_4 \; (0.0058) < C_2H_6 \; (0.035) < C_3H_8 \; (0.136) < n\text{-}C_4H_{10} \; (0.810)$$

For these gases in n-decane, the bp/1 atm factor is the main one.

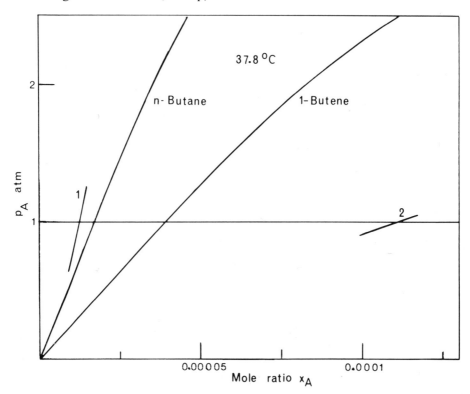

Fig. 53. Solubility of n-butane and 1-butene in water at 37.8°C. Plots of mole ratio x_A vs p_A. Line 1 (slightly concave upward) is for propane at 37.8°C. Line 2 indicates the position of the x_A value for propene at 21°C.

The Hydrogen-Bonding Structure

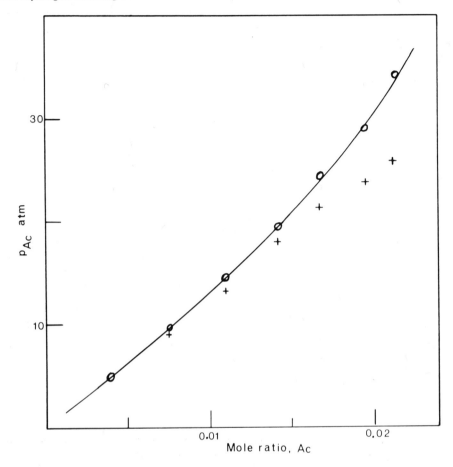

Fig. 54. Solubility of acetylene in water at 20°C. Plot of x_{Ac} vs p_{Ac}. The crosses refer to an adjustment for fugacity (see text).

Hiraoka[170] expressed the solubility of acetylene in water as cubic centimeters C_2H_2/gram H_2O for different pressures $p_{C_2H_2}$ from 5 kg/cm² to 40 kg/cm² and at several $t°C$, mainly 20 and 30°C. The $x_{C_2H_2}$ vs $p_{C_2H_2}$ plot is clearly concave upward, as shown in Fig. 54. Notice that the $x_{C_2H_2}$ value is only 0.02 at 30 atm, 20°C, compared with the value of 4.0 at 25°C for acetylene in dimethylformamide; see Chap. 16.

In Table 4, I give the approximate bp/1 atm and the $p°_A$ pressures in atm for 20°C for gases, which afford further pertinent examples of the influence of the water structure. In Fig. 55 I have shown the R-lines for these gases and have registered the position of the cluster of absolutely small N_A values for water by one line indicated by W. It will be noticed that this cluster is well on the left of the R-line for the gas CH_2F_2 having the lowest bp/1 atm.

The solubility of the gases 9, 10, 11, and 12 in water was expressed as S (wt. % solute) by Carey, Klausutis, and Barduhn (1966),[46] and these workers gave Henry's law as $P = HS$, where P is my p_A, the partial pressure of the gas A. This form is

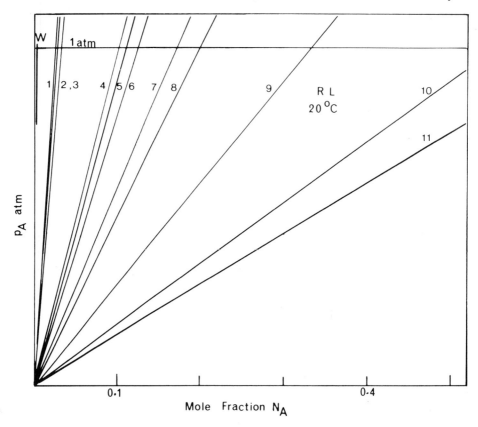

Fig. 55. R-lines for 20°C for the compounds numbered in Table 4. The N_A ($\approx x_A$) values for water are registered as a cluster marked W on the extreme left of the diagram.

Table 4. Bp/1 atm of Gas A and Vapor Pressure of A over Liquid A, $p°_A$, at 20°C

Number for reference in text and diagrams	Formula	bp/1 atm, °C	$p°_A$, atm, 20°C
1	CH_2F_2	−84	36
2	CF_3Cl	−82	31
3	CH_3F	−78	30
4	C_3H_6	−48	9.7
5	$ClCHF_2$	−40.8	9.0
6	C_3H_8	−42	7.8
7	Cl_2CF_2	−30	5.7
8	CH_3Cl	−24	4.7
9	CH_3CClF_2	−9.6	2.9
10	CH_3Br	+3.6	1.8
11	Cl_2CHF	+8.9	1.55
12	$ClCH_2F$	−9.1	2.9

The Hydrogen-Bonding Structure

neither the mole ratio x_A nor the mole fraction N_A form. I have converted the "wt. %" data into x_A ($\approx N_A$) data. In Fig. 56, I have given the x_A ($\approx N_A$) vs p_A plots on a common scale. Whereas the lines for CH_3Br (10) and CH_2ClF (12) may be read as linear over the stated range, the line for CH_3CClF_2 (9) appears to be slightly curved, and that for $CHCl_2F$ (11) distinctly curved convex upward. The "Henry's law constant" based on the slope of the line at the lower end of the x_A values, a conventional notation, would give a seriously erroneous value of x_A for $p_A = 1$ or more atm. The "wt. %" values may not be used for the comparison of "solubilities" on a molecular basis because the molecular weight of A is a material factor in the conversion.

Carey *et al.* stated that their "wt. %" data for $CHCl_2F$ did not "follow Henry's law," and they attributed this to the hydrolysis of the gas:

$$CHCl_2F \underset{}{\overset{H_2O}{\rightleftharpoons}} CClF + Cl^- + H^+$$

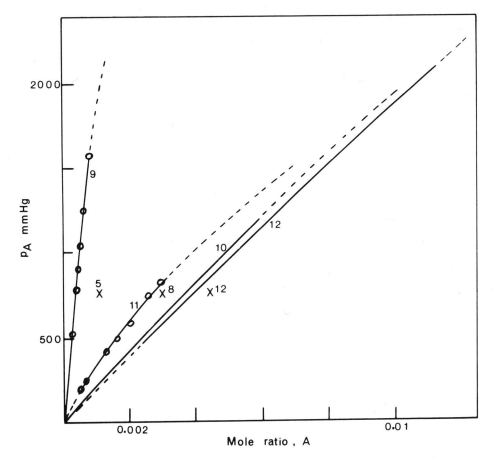

Fig. 56. Solubility of gases A numbered as in Table 4. Plots of x_A vs p_A for the following temperatures: (9) 13.2°C; (10) 14.7°C; (11) 12°C; (12) 18°C. The crosses are for 5, 8, and 12 at 10.2°C.

However, Carey *et al.* invoked a procedure attributed to Vivian and Whitney for allowing for this hydrolysis. Vivian and Whitney[374] were concerned with the chlorine–water system, for which the line is *concave* upward at the lower end, becoming visually straight from about 400 mm Hg to about 1 atm. The concavity is conventionally taken to indicate that chlorine is more soluble in water than it would be if hydrolysis did not occur. It is conventionally believed that when a chemical reaction occurs, the solubility in water is high. It is also believed that when a chemical reaction occurs, Henry's law is not "obeyed"; and when Henry's law is "obeyed," it is concluded that no chemical reaction has occurred. Such generalizations have no scientific content. It will be noticed that the plot for $CHCl_2F$ is decidedly convex upward. To me the hydrolysis depicted by Carey *et al.* in their symbolic terms could not be reversible; the chlorine–water system was assumed to be reversible.

The absolutely small values for water could be due to the resistance of the water to a change in hydrogen-bonding structure. This means that the second and third factors I have mentioned overwhelm the first factor. It remains to be seen if the first factor exerts any influence in deciding the order of the relative differences among the absolutely smaller x_A values. The noble gases give a cluster of x_A values for water,

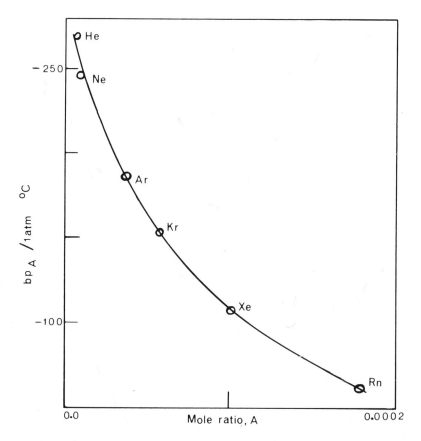

Fig. 57. Bp/1 atm of the noble gases and their solubility in water at 20°C and $p_A = 1$ atm.

The Hydrogen-Bonding Structure

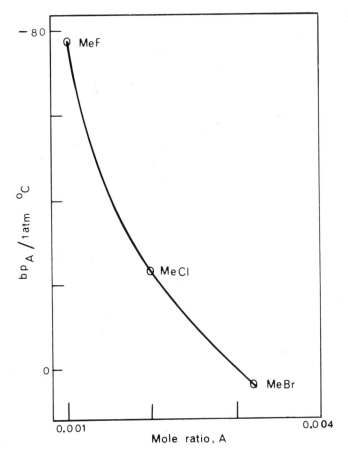

Fig. 58. Bp/1 atm of the halogenomethanes and their solubility in water at 25°C and $p_A = 1$ atm.

and these would be registered well on the left of the R.L. even for He. Nevertheless, the x_A values are spaced out nearly as the bp/1 atm, as my Fig. 57 shows.

The x_A values for CH_3F, CH_3Cl, and CH_3Br (from molarity data of Glew and Moelwyn-Hughes[130]) are absolutely small and form a cluster well on the left side of the R.L. for CH_3F; but within the cluster, there is a clear relationship with the bp/1 atm, as I show in my Fig. 58. Likewise, the x_A values for gases 5, 8, and 12 show a spacing similar to that of the bp/1 atm (Fig. 59). These data are from the "molar concentration" data of Boggs and Buck,[28] and I have registered these x_A values for 10.2°C as crosses on my Fig. 56.

Carey et al.'s comments about lack of conformity among the values given by different workers presents to evaluators cause for concern. Boggs and Buck's data for CH_2ClF were deemed to be 15% lower than the Carey et al. values. Boggs and Buck's data for CH_3Cl showed "considerable disagreement" with those by Glew and Moelwyn-Hughes, and the values for CH_3Br given by the last two workers were deemed to be 20% too high. Other data for CH_3CClF_2 were deemed to be in considerable error.

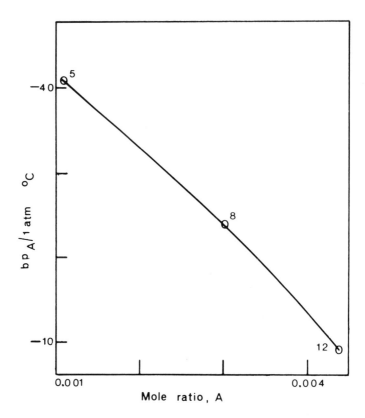

Fig. 59. Bp/1 atm of compounds 5, 8, 12 and their solubility in water at 10.2°C and $p_A = 1$ atm. See Table 4.

A set of data by Boyer and Bircher (1960)[30] enables me to illustrate the invalidity of the Ostwald coefficient L (no matter how defined) in the comparison of "solubilities." The data are for five gases, $A = N_2$, Ar, CH_4, C_2H_4, and C_2H_6, at 25°C and 1 atm and for a series of normal alcohols, methanol to n-octanol.

Whereas the L values for each gas (except ethane) *decrease* from methanol to octanol, the x_A values *increase*, as shown in Fig. 60. For ethane, the L values give a muddled series, whereas the x_A values still increase as with the other gases.

In Fig. 60, I have plotted the x_A values (horizontal axis) vs the bp/1 atm of A for each alcohol. I have purposely excluded the x_A value for ethene for each alcohol, but I have registered the position by a spot and have shown by an arrow the line to which the spot refers. It is clearly discernible that the bp of A/1 atm is the *first* factor to consider, even including ethene in the series of gases. However, excluding ethene for the moment, one can see that the relative spacing of x_A values throughout the series of alcohols is similar for each gas, although the spaces become absolutely larger as the bp of A rises. Furthermore, for each gas, including ethene, the spaces get progressively less as the series of alcohols is ascended, and this brings me to the *second* factor.

Table 5. Mole Ratio Solubilities x_A of Gas A in Liquid S for $p_A = 1$ atm and 25°C

Gas A	Water	CCl$_4$	Hydrocarbon[a]	Ether, etc.[b]
Hydrogen	0.0000135	0.00032	0.0007 (1)	0.00063 (1)
Nitrogen	0.000012	0.00063	0.0014 (2)	0.0013 (1)
Carbon monoxide	0.000019	0.00087	0.00173 (1)	0.00168 (1)
Oxygen	0.000023	0.00120	0.00215 (1)	0.00193 (1)
Carbon dioxide	0.00070	0.0110	(0.0210) (3)	0.028 (2)
Hydrogen sulfide	0.0020	0.042	0.044 (4)	0.10 (3)
Chlorine	0.0016	0.200	0.10 (3)	
Sulfur dioxide	0.032 (18°C)	0.135 (10°C)	0.40 (5)	0.76 (4)

[a] For the hydrocarbons, the numbers in paretheses designate: (1) n-C$_7$H$_{16}$; (2) n-C$_6$H$_{14}$; (3) n-C$_7$F$_{16}$; (4) n-C$_8$H$_{18}$; (5) m-C$_6$H$_4$(CH$_3$)$_2$.
[b] For the ethers, etc., the numbers in parentheses designate: (1) (C$_2$H$_5$)$_2$O; (2) CH$_3$CO$_2$C$_5$H$_{11}$; (3) (n-C$_8$H$_{17}$)$_2$O; (4) (CH$_3$CO)$_2$O.

A significant feature of the structure of the alcohols is hydrogen bonding. It is most prominent in methanol, but the symmetry falls off distinctively in ethanol because the hydrocarbon part is doubled. The *third* factor emerges partly from the interference in the hydrogen-bonding structure caused by the gas, and this part appears to become progressively less as the series of alcohols is ascended. With ethene, however, the acid function sited on hydrogen and the basic function sited at the double bond are incisive enough to cooperate to some extent with the hydrogen bonding; I deem this to give rise to an x_A value out of step, on the more soluble side. It is clearly seen that this deviation decreases as the series of alcohols is ascended.

According to the conventional opinion expressed in textbooks, the absorption coefficient is high if the gas reacts chemically with the solvent. In one book, the following Bunsen coefficients are given for solubility in water at 25°C:

$$\begin{array}{llllll} N_2 & 0.01434 & H_2 & 0.01754 & CO_2 & 0.759 \\ O_2 & 0.02831 & CO & 0.02142 & H_2S & 2.282 \end{array}$$

The sharp rise in the values for CO_2 and H_2S were attributed to "chemical reaction" with the solvent. To illustrate the total invalidity of this concept, attention is drawn to the Table 5.

The increase in x_A value for O_2 to that for CO_2 for water is less than the increase in x_{H_2} from that for water (0.0000135) to that for n-C$_7$H$_{16}$ (0.0007). Is this increase due to chemical reaction? At 25°C and $p_{H_2} = 180$ atm, x_{H_2} for n-C$_8$H$_{18}$ is 70 times the x_{H_2} value for water. In one standard textbook, the author refers to the "high" solubility of hydrogen in water, and this was attributed to a specific interaction with the water depicted as a bridge between two molecules of water:

$$H_2O \cdots H-H \cdots OH_2$$

However, water needs no help from hydrogen in the linking together of H_2O units. I know of only five x_A (25°C and $p_A = 1$ atm) values for water less than that of

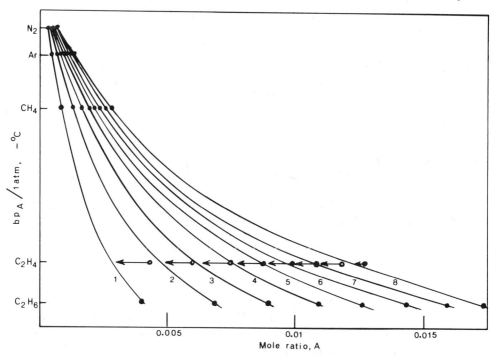

Fig. 60. Solubility of gases A (N_2, Ar, CH_4, C_2H_4, C_2H_6) in the series of n-alcohols $C_nH_{2n+1}OH$, $n = 1$ to 8, at 25°C and $p_A = 1$ atm. Plots of x_A vs bp A/1 atm. See text.

hydrogen. These are: N_2 (0.000012), Ne (0.0000084), He (0.000007), SF_6 (0.00000405), and CF_4 (0.00000362). Under these conditions the concentration of hydrogen in moles per liter is 50 times greater in the gaseous than in the liquid aqueous phase.

5.6. Other Aspects of Water Structure

Attention is drawn to the aspect of water structure relating to the effects of adding water to an organic liquid. Bullock and Tuck[38] explained the nuclear magnetic resonance spectra of a solution of water in tri-n-butyl phosphate by postulating the formation of linear and chain polymers of varying complexity according to the amount of water added:

$$\equiv P \rightarrow O \cdots H-O-H \cdots O \leftarrow P \equiv$$

$$\equiv P \rightarrow O \cdots H-O-H \cdots \underset{\underset{P}{\overset{\overset{|||}{\downarrow}}{O}}}{O} \cdots H-O-H \cdots O \leftarrow P \equiv$$

Johnson, Christian, and Affsprung[190] determined the solubility of water in liquids as a function of the total vapor pressure at 25°C in the low-water-concentration region. The liquids were nitrobenzene, dibutyl phthalate, tributyl phosphate, N,N-dimethylaniline, tributylamine, aniline, N,N-dimethylcyclohexylamine, benzyl alcohol, cyclohexanone, and methyl phenyl ketone. These authors do not refer to the structure of water itself. They assumed that water molecules are present in solution as an equilibrium mixture of monomers and not more than two polymeric species "each of which obeys Henry's law."

It was stated that the Henry's law relationships employed *are strictly valid only in the limit of infinite dilution*. It was pointed out that even in dilute solutions there is no unambiguous way of assessing the relative importance of physical and chemical contributions to deviations from Henry's law. In my opinion there is no line of demarcation, *physical–chemical*. Confusion will prevail so long as the concept "Henry's law at infinite dilution" is uncritically held.

Those authors drew attention to the few quantitative studies on the complexity of water in nonaqueous media, and they cited references. They concluded that water is "approximately 20% associated" in nitrobenzene at "saturation concentration" at 25°C, although Hogfeldt and Bolander stated that water is primarily monomeric in

$$C_6H_5-N\begin{matrix} O\cdots H-O \\ \\ O\cdots H-O \end{matrix}\begin{matrix} H \\ \\ H \end{matrix}$$

nitrobenzene.[175] For p_w near $p°_w = 23.70$ mm Hg at 25°C, N_w appears to be 0.01625; we have to consider the influence of this small amount of water on the intermolecular structure of nitrobenzene.

If, as was suggested, two molecules of water, (H_2O) units, form the hydrogen-bonded complex shown in the diagram, the number of moles of this complex per mole of nitrobenzene originally taken cannot be more than 0.0081 at p_w near $p°_w$. If the remainder of the liquid is deemed to be nitrobenzene as it exists in the original pure liquid taken, why does it not continue to absorb more water and form more of the complex? How can one escape the conclusion that under the conditions of equilibrium specified one molecule, (H_2O) unit, of water is *using* 61 molecules of nitrobenzene on a time-average basis?

6

Sources and Form of Data

6.1. Review by Markham and Kobe (1941) on "The Solubility of Gases in Liquids"[(253)]

6.1.1. Scope of the Review

Numerical solubility data are not given; nevertheless, the review is valuable for the outline of operational procedures and for the list of systems indicating the pressures and temperatures reported, and the 346 references cited. Fifty-five "gases" are named, as follows: helium, neon, argon, krypton, xenon, hydrogen, nitrogen, oxygen, ozone, chlorine, air, methane, ethane, propane, butane, ethylene, propylene, cyclopropane, isobutylene, acetylene, dimethyl ether, methyl chloride, chloroethylene, fluoroethane, fluoroethylene, carbon monoxide, carbon dioxide, carbonyl sulfide, carbonyl chloride, cyanogen, hydrogen cyanide, silane, ammonia, methylamine, dimethylamine, trimethylamine, ethylamine, diethylamine, triethylamine, propylamine, nitrous oxide, nitric oxide, phosphine, methylphosphine, arsine, stibine, hydrogen sulfide, sulfur dioxide, hydrogen selenide, hydrogen chloride, hydrogen bromide, hydrogen iodide, radium emanation, thorium emanation, and actinium emanation.

Attention was drawn to the lack of discipline in the presentation of data up to that time. It had been common practice to observe random pressures near 1 atm and then adjust to 1 atm by means of Henry's law. There was sometimes confusion over partial and total pressure, and there was confusion over the Bunsen and Ostwald coefficients.

6.1.2. Forms of Data

Much data in the literature, especially for earlier dates, are for gases in the lower half of the range of boiling points/1 atm, and so they refer to absolutely small values of $x_A(\approx N_A)$. Operational procedures, therefore, have involved the measurement of a volume of gas A, and this has usually been expressed as an absorption coefficient. The matter of total, p_T, or partial pressure, p_A, must be considered. To equate p_A to

p_T may involve an error which for certain purposes may be neglected. Attempts to allow for the vapor pressure p_S of the original liquid S and thereby arrive at $p_T - p_S = p_A$ should be carefully scrutinized in every case because p_S will be lowered as x_A increases at the different p_A. In the present exposition, p_A is the partial pressure of A, i.e., the pressure due to A. Furthermore, I deem it important to bring into the pattern data for $p_A > 1$ atm, or even 2 atm.

Briefly, the primary data may be recognized as the volume of gas A measured at a stated $t°C$ and p_A (approximately 1 atm). Now this volume may contain the number of moles of A absorbed at that $t°C$ and that p_A, or absorbed at the same or different $t°C$ and a different (could be much greater) p_A which prevailed at the absorption.

The Ostwald coefficient defined as

$$L = \frac{\text{Volume of A at } t°C \text{ and } p_A}{\text{Original volume of liquid S at the same } t°C}$$

is valid. To equate this form to

$$L = \frac{\text{Moles of A per unit volume in liquid}}{\text{Moles of gas per unit volume over liquid}}$$

as is often done, for example, by Burrows and Preece,[39] is in principle invalid. Of course, there are examples for which the difference may be neglected, but there are many examples for which it most certainly may not. Körösy (1937)[213] gave the form

$$M = \frac{\text{Gas dissolved in 1 mole solvent}}{\text{Gas concentration in gas phase}}$$

To convert the volume observed for p_A into that it would have at $p_A = 1$ atm requires *two* decisions. One is about the rate at which the "solubility" is changing with pressure within that area of operation, and the second is about the rate at which the volume of gas changes with pressure, again in the prevailing operational area. These two decisions are often covered by the statements that "Henry's law was assumed," and "the ideal gas law was assumed." There are examples in which both decisions appear to be covered by the "assumption of ideality." It must, however, be kept clearly in mind that there are two quite independent so-called idealities involved. These are the "ideality of the gas" and the "ideality of the liquid solution." The deviation from ideal gas behavior can be a very different matter from the deviation of real solutions from the behavior assumed for ideal solutions.

To convert the volume which has emerged from the adjustment to $p_A = 1$ atm, but still for the observed $t°C$, into the volume at $0°C$ (or 273.16 K) and thus get the Bunsen coefficient, the ideal gas law is usually assumed.

For the comparison of solubilities on a molecular basis, it is invalid to use either the Ostwald or the Bunsen coefficient; this is because the density and molecular weight of the liquid S are always material factors, and the deviation of the gas A from ideality can also be a material factor. Markham and Kobe[254] showed that in the example of carbon dioxide, the assumption of the ideal gas laws could make a

Sources and Form of Data

difference of 0.7% in the results from the "physical" and the "chemical" measurements.

To calculate the number of moles of A in the volume of gas A we have to decide what gram-molecular volume to use. For 0°C and 1 atm, the "ideal" gram-molecular volume is usually taken to be 22.4145 dm^3. The deviation factor may be expressed as follows:

$$\text{Deviation factor} = \frac{\text{Standard density, g/dm}^3 \text{ at 0°C, 1 atm}}{\text{"Ideal density"}}$$

$$= \frac{\text{"Ideal gram-mole volume"}}{\text{Actual gram-mole volume}}$$

In illustration, I mention a few approximate values: CH_4 (≈ 1.000); C_2H_6 (1.01); C_2H_4 (1.0073); C_2H_2 (1.0113); C_3H_8 (1.027); C_3H_6 (1.0203); n-C_4H_{10} (1.0385); H_2S (1.012); HCl (1.0074); HBr (1.0093); CO_2 (1.0070); Cl_2 (1.016); SO_2 (1.024).

Bradbury, McNulty, Savage, and McSweeney[32] referred to Frolich et al.'s conclusion[405] that a gas follows Henry's law over a wide pressure range if it does not form a compound with the solvent, but they believed that this largely depended upon the extent to which the gas "obeys the ideal gas law." There seems to be some confusion here between the two aspects of ideality I have mentioned. I cannot see any obvious correlation between the deviation factor for the gas A and the occurrence of linearity over any part of the x_A vs p_A plot or of the N_A vs p_A plot.

If V_A is the volume of gas A in cubic centimeters at t°C and p_A, M_A is the gram-mole volume of A decided upon, expressed as cubic centimeters, d_S is the density of the liquid S used to absorb the gas, and M_S is the molecular weight of S (taken as the formula weight), then the mole ratio x_A is given by $(V_A/M_A)/(d_S/M_S)$, and N_A then equals $x_A/(1+x_A)$. The x_A value relates to the temperature and pressure of measurement and to the gram-mole volume chosen for these conditions. For arguments on a molecular basis, I can find no useful purpose in the conversion of the Ostwald coefficient into the Bunsen coefficient.

6.2. Review by Battino and Clever (1966)[17] on "The Solubility of Gases in Liquids"

6.2.1. The Scope

These authors continue the literature search from where Markham and Kobe left off, and they cite 686 papers. Apart from a few examples relating to the low bp gases, numerical data are not presented. In the long table under the heading "solubility data" the columns show:

Gas	Solvent	Pressure, atm	Temperature, °C	Measurement value	Reference

Under the column "Measurement value" the number 0, 1, 2 indicate the authors' (B and C) assessment of accuracy or reliability of data. A detailed analysis of methods of determination is given, and this constitutes a valuable contribution to the literature on the subject. Solubility theory and relationship are indicated by detailed reference to relevant papers, but the area covered is mainly that of gases in the lower half of the boiling point range and conventionally called "nonpolar" gases. The mole fraction N_A values are absolutely small, and for $p_A < \sim 2$ atm, the operational area is confined to that near the bottom left corner of the overall R-line diagram.

6.2.2. Numerical Data

The two tables of numerical data are for oxygen in water at 25°C and nitrogen, oxygen, and argon in water at 1 atm and t°C from 0 to 50°C; these tables constitute an indispensable demonstration of the variation of the Bunsen coefficient, for a fixed t°C, observed by a number of workers. These tables were presented to consider the usefulness of the value for oxygen in water as a comparison standard, and the "recommended standard" appears to be $\alpha = 0.02847$. All the α values in the tables show that it decreases as the temperature t°C increases. The calculation of x_A involves the selection of the gram-molecular volume of gas A at t°C and the observed p_A (p_A observed being obliterated in the α value) and also the density of water at t°C. In keeping the procedure free from ambiguity, I deem it to be a help to emphasize that one is selecting, for example, 22,410 cm^3 at 0°C and 760 mm Hg as the volume for the particular purpose, from which the volume at t°C and p_A is computed as $22,410 \times (760/p_A) \times (t\,K/273)$. The density of a gas at t°C and p_A is a real observable property of the gas; to me, it is no longer desirable to refer to so-called ideality.

6.3. Papers by Hildebrand and Colleagues

Since 1916 up to almost the present time, a regular flow of papers on gas solubilities has been due to Hildebrand and his co-workers, but the data relate mainly to his so-called "nonpolar gases" in "nonpolar" liquids.

Others have attempted to extend Hildebrand's theory to "nonpolar gases" in "polar liquids," and there appears to be a propensity to write about "polar associated liquids," and "polar nonassociated liquids."

6.4. Papers by Clever and Battino

Since 1957, a number of papers by Clever and Battino and their co-workers have contained details of work on the gases at the lower end of the bp range. Thermodynamic considerations have been predominant. The review by Wilhelm and Battino[378] appeared in 1973. Again the gases are mainly, almost exclusively, those in the lower half of the bp range.

6.5. Data Reference Books

The standard books are Seidell and Linke,[331] Stephen and Stephen,[344] and Landolt-Bornstein.[224]

7

Data for the Noble Gases and Other Gases in the Lower Half of the Boiling Point Range

7.1. References

Unless there is a special reason, references for individual gases will not be given; they are to be found in the reviews cited, or in the papers cited. There is some doubt about the exact position of the R-line.

7.2. The Noble Gases Helium to Radon

7.2.1. The R-line Diagram and Essential Pattern of Data

In Fig. 15 the R-lines are drawn on a common scale, and examples of N_A values for 760 mm Hg and 25°C are registered on the 760 mm Hg horizontal. In Figs. 16–20 the N_A values are registered on the 760 horizontal for each gas, and the probable pattern of the N_A/p_A plots are indicated by broken lines. It is seen that the base-line scale (N_A) must be reduced stepwise in passing from He to Rn in order to show the relative spread of the N_A values. Figures 16–20 are based on N_A data by Clever, Battino, Saylor, and Gross (1957)[54] for the "h" series (hydrocarbons) (He to Kr), and by Saylor and Battino (1958)[324] for the "b" series (PhX; X = F, Cl, Br, I, CH$_3$) (He to Kr). Two points stand out from the tables in the papers cited. First, the Ostwald coefficient and mole fraction (N_A) values do not correlate in two parallel series because the N_A value depends upon the molecular weight and density of the liquid S and, second, the N_A values for three temperatures ("h" series) and for four temperatures ("b" series) show that for He and Ne the N_A value increases with increase in t°C for every S named. (See Table 6.)

Table 6. Mole Fraction N_A of Argon in Some Organic Compounds

S	N_A		
	15.0°C	25.0°C	41.8°C
Benzene	0.000870	0.000877	0.000906
Cyclohexane	0.00150	0.00149	0.00145
n-Hexane	0.00260	0.00253	0.00237
Remainder of "h" series	N_A decreases with rise in t°C		
C_6H_5F	0.00115	0.00115	0.00116 (40°C)
C_6H_5Cl	0.000864	0.000852	0.00864 (40°C)
Remainder of "b" series	N_A increases with rise in t°C		

With Ar, the effect tends to be borderline. With Kr, all values of N_A ("h" and "b" series) decrease with rise in temperature. Saylor and Battino give N_A values for Xe and the "b" series. These show a marked decrease with rise in t°C.

These workers found that of the several physical properties tried, polarizability of the gases yielded the best fit against the log N_A plot for He to Xe. In Fig. 61, I show that a plot of log N_A (for hexane) and the bp of A (He to Xe) is as good a linear correlation as log N_A vs polarizability of A; an extrapolation to radon fits the N_{Rn} value obtained from another source.

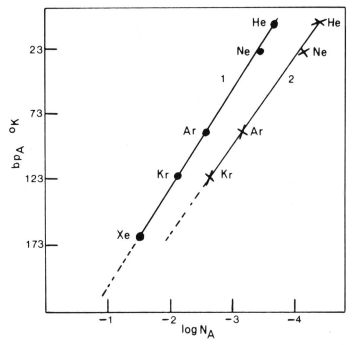

Fig. 61. Solubility of the noble gases (A) in (1) n-C_6H_{14} and (2) C_6H_5Br at 25°C. Plots of the log of the mole fraction, log N_A, vs bp of A in degrees Kelvin. The N_A values are for 1 atm.

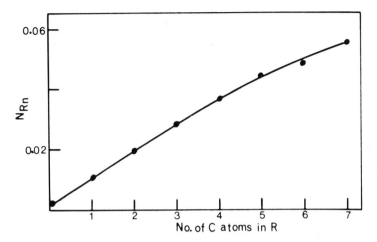

Fig. 62. Solubility of radon in a series of n-carboxylic acids R·CO$_2$H (R = H or CH$_3$ to n-C$_7$H$_{15}$) at 25°C and 1 atm. Plot of the mole fraction N_{Rn} vs the number of carbon atoms in R.

Radon has a bp of −62°C/1 atm and therefore may be expected to show solubility data between those for ethane and for propane. Figure 20 shows the pattern of N_{Rn} data calculated from the absorption coefficients.

Allowing for limitations of the data, it is seen that they fit into the essential pattern for the gaseous hydrocarbons. At 0°C the value for n-hexane is a little higher than the R-line value, whereas the value for aniline is at the low end; still lower is the value for glycerol, and that for water even lower. The series of n-carboxylic acids[286] shows a steady increase in x_{Rn} values, from $x_{Rn} = 0.00162$ for HCOOH at 25°C, p_{Rn} 760 mm up to 0.0617 for C$_7$H$_{15}$COOH, a value approaching the R-line one of 0.0665 (Fig. 62). The plot is almost linear up to C$_5$ and then begins to flatten. The more restricted series of n-alcohols, MeOH to n-BuOH, also show an almost linear plot. See refs. 4, 30, 31, 123, 330.

7.2.2. The Lowest N_A Values Are for Water

For each gas, the N_A value for water is less than for all other examined liquids (see Table 7). The N_A values for the same $t°C$ (about 20–25°C) and $p_A = 760$ mm Hg are all on the left of the R-line for helium. Cady et. al. (1922)[42] gave their data for

Table 7. Solubility of Noble Gases in Water ($p_A = 1$ atm)

	He	Ne	Ar	Kr	Xe	Rn
bp/1 atm	−269°C	−246°	−186°	−153°	−107°	−62°
R-line[a]	0.00025	0.00051	0.0017	0.0038	0.0172	0.063
N_A for water	0.000007	0.0000084	0.000027	0.000047	0.0001	0.00017 (20°C)

[a] Position of R-line at the 760 horizontal, 25°C.

water and helium and commented on the results of other workers. The N_A values are all absolutely small, but there is a relative spread of values which do not form a close cluster as the hydrocarbon gases CH_4 to n-C_4H_{10} do. The $N_A/1$ atm value for each gas *decreases* with rise in temperature.

The outstanding feature is the structure of water and its resistance to change in structure. Whereas the noble gases have no shared electrons, the hydrocarbon gases CH_4 to n-C_4H_{10} have no unshared ones. The N_A (water) values for these hydrocarbon gases form a close cluster well on the left of the R-line for methane.

7.3. Hydrogen, Nitrogen, Oxygen, Carbon Monoxide

7.3.1. The R-line Diagrams

In Figs. 21–24 I have registered the N_A values on the horizontal at 760 mm Hg. The literature data are nearly, if not always, given for $p_A = 1$ atm. The broken lines merely indicate the approximate essential pattern of p_A/N_A data at these extremely low values of N_A. Nevertheless, in principle at least, all these lines must ultimately reach the tip of the R-line as it just approaches the right vertical axis of the full R-line diagram (to be imagined) at $p_A =$ the extrapolated $p°_A$; the form of each line at the much higher pressures is a matter of experimentation.

Although all the N_A values are absolutely small, there are relatively large variations. The list of liquids S is not the same for each gas, and the values of N_A for the same A and S vary somewhat according to the observer. Nevertheless, the essential patterns are remarkably similar, although the scales of the N_A values are different for hydrogen, nitrogen, and the pair oxygen–carbon monoxide (common scale for these two gases); i.e., the positions of the N_A values for a chosen gas with respect to the R-line value for that gas are relatively similar, and predictions can be made about the approximate value of N_A for a liquid S appearing on one list but not on another. Even for a liquid such as bromobenzene, a good guess seems feasible.

Look at Fig. 22 for hydrogen, on the N_{H_2} 0 to 0.001 scale. Aniline appears well over on the left of the R-line, and nitrobenzene appears on the right of the aniline position. Diethyl ether is on the right of the R-line. The named normal hydrocarbons are still further on the right, and the compound n-C_7F_{16} has a value of $N_{H_2} = 0.0014$ at 760 mm Hg, relatively much further on the right and off the scale used.

For nitrogen, it is necessary to lower the scale somewhat. An interesting feature here is the grouped position of the N_{N_2} values for alcohols; the value increases from that for methanol to that for n-octanol, but the group is well over on the left of the R-line. From this group, and from the group MeOH to n-C_4H_9OH for carbon monoxide, one can predict the approximate positions of the normal alcohols for H_2, O_2, CO, as well as for the noble gases. Ethylene glycol occupies a position near the left vertical axis; it may be predicted that glycerol will be between this position and that for water which is right against the left axis. Gjaldbaek and Niemann[128] gave data for argon, nitrogen, and ethane.

Data for Gases in Lower Half of bp Range

The N_A values for nitrogen, bp $-196°C$, carbon monoxide, bp $-191°C$, and oxygen, bp $-183°C$, can all be plotted on a common scale; but I have used a somewhat larger scale for nitrogen to spread the alcohol values. For carbon monoxide, notice the position of aniline, again right over on the left of the R-line, with nitrobenzene somewhat on the right of the aniline position. Diethyl ether and the normal hydrocarbons are on the right of the R-line for carbon monoxide and oxygen, while again the fluorocarbon n-C_7F_{16} is off this scale.

7.3.2. The Lowest N_A Values Are for Water

The values for 760 mm and 25°C are:

H_2	N_2	CO	O_2
0.0000135	0.000012	0.000019	0.000023

These form a cluster well over on the left of the R-line for helium.

7.4. The Three Factors

Three fundamental factors should be kept in mind when attempting to appreciate the causes of these phenomena.

The first, and predominant one here, is the tendency of the gas to condense at an operational temperature of, say, 0 to 25°C. For the gases He, Ne, Ar, Kr, H_2, N_2, CO, and O_2, this is absolutely small, and on this account the N_A values are absolutely small, although, of course, there are relative differences.

The second factor is the pattern of intermolecular forces (the structure) of the liquid S. The third is the inevitable interaction of the gas A with the liquid S.

The intensity of the acid–base function of these gases is very low; therefore the situation is mainly one of interference with the structure of S. Water has an outstandingly symmetrical hydrogen-bonding structure which resists interference. The alcohols have also a decided hydrogen-bonding structure which is not so symmetrical; nevertheless, there is marked resistance. The relative and absolute positions of aniline and nitrobenzene should be noted.

When considering the mathematical approaches to the correlation of the three factors I have mentioned, it is vital to appreciate that the usual operational area on the complete R-line diagram (N_A from 0 to 1, p_A From 0 to $p°_A$) is a tiny area at the left bottom corner of the full, very-small-scale diagram.

7.5. Narcotic Effect of Xenon

A paper by Lawrence and colleagues (1946)[228] has significance here because of comments on the reliability of the published absorption coefficients for water. The narcotic effects of the noble gases He to Xe were outlined, and some relationship with the relative solubility in water and fat was indicated. In connection with decompression effects and aviation medicine, it was necessary to know the solubility of the "inert gas" in water and fat. Those authors stated that for water great discrepancies in the literature were found, especially in the values for neon and krypton.

The reference book (1944) values were found to be in considerable error, being based on the work of Antropoff (1910).[4] It was intimated that several authors (Lannung, 1930[225]; Valentiner, 1930[360]; Van Liempt and Van Wijk, 1937[363]), including von Antropoff himself (1919), have shown these values to be in error. The values for neon given by Lannung and Valentiner differed so much that it appeared difficult to decide which is the more reliable.

7.6. Argon in Dioxane and Water Mixtures

Krestov and Nedel'ko (1970)[216] reported that the addition of a mole fraction of 0.02 of dioxane to water causes an increase of the Ostwald coefficient L of argon as the temperature rises. The quantity L_{Ar} increases from 0.027 at 25°C to a maximum of ~0.030 at 45°C and then remains constant up to 70°C. When a mole fraction of 0.5 of dioxane is added, L_{Ar} increases regularly from 0.085 at 25°C to 0.125 at 70°C. The pattern changes regularly for mole fractions of dioxane between these two values, an effect which was related to the variation in structure of the dioxane–water liquid. Thermodynamic parameters were calculated from the data. It must again be emphasized that these L_{Ar} values represent very small N_{Ar} values (see Fig. 18).

7.7. Other Items

Katayama, Mori, and Nitta (1967)[200] proposed a method for estimating the solubility of nonpolar gases in nonpolar solvents and alcohols by taking benzene as the standard liquid S. The procedure was described to be of "high accuracy," being 5.2% for "nonpolar solvents" and 8.8% for alcohols. Selected alcohols were limited to derivatives of saturated, aliphatic hydrocarbons. The limitation is seen from the reservation that this method cannot be applied to CO_2 and C_2H_2 because of their acid–base interactions with solvents. Noncompliance with fluorocarbon liquids S was also noticed. The expressions "nonpolar" and "polar nonassociating solvents" were used. The extensive significance of solubility data is indicated in refs. 6, 66, 214, and 322. In ref. 6 the gases studied were H_2, He, and Ar, and it was stated that there was a negative deviation from the "ideal behavior" at higher concentrations of hydrogen.

8

Carbon Dioxide, Nitrous Oxide, Hydrogen Sulfide, Chlorine, Sulfur Dioxide, and Carbonyl Chloride

8.1. Carbon Dioxide and Nitrous Oxide

Historically the outstanding paper is that of Kunerth,[220] who cited data (CO_2) by Just.[196] Table 8 shows the N_A values for these gases in a number of liquids, and Fig. 63 shows the R-line diagram for carbon dioxide. That for nitrous oxide would be

Table 8. N_A Values for CO_2 and N_2O and Several Liquids at 20°C/1 atm

Liquids	CO_2	N_2O
$CH_3CO_2C_5H_{11}$	0.0270	0.0321
CH_3COCH_3	0.0209	0.0185
C_5H_5N	0.0129	0.0120
C_6H_5CHO	0.0128	0.0134
$ClCH_2CH_2Cl$	0.0125	
$CHCl_3$	0.0123	0.0182
CH_3CO_2H	0.0121	0.0115
$C_6H_5NO_2$	0.0113	
$CH_3C_6H_5$	0.0107	
$m\text{-}(CH_3)_2C_6H_4$	0.0102	
CCl_4	0.0100	$(0.01815)^a$
C_6H_6	0.0091	$(0.0145)^a$
$i\text{-}C_5H_{11}OH$	0.0082	0.0111
$BrCH_2CH_2Br$	0.0082	0.0100
CH_3OH	0.0071	0.0053
CH_3CH_2OH	0.0070	0.0072
$o\text{-}CH_3C_6H_4NH_2$	0.0066	
$C_6H_5NH_2$	0.0055	0.0056
CS_2	0.0022	$(0.0057)^a$
Water	0.0007	0.0005

^{a}From Yen and McKetta.

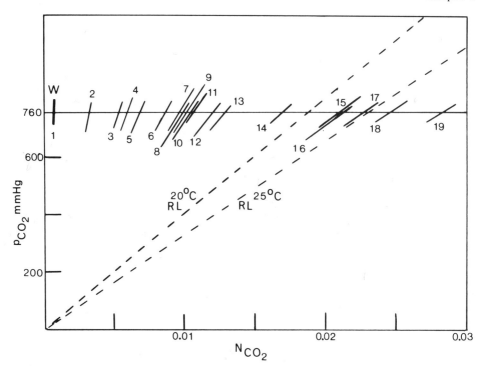

Fig. 63. The mole fraction solubilities N_{CO_2} of carbon dioxide in various liquids at 20 or 25°C and 760 mm Hg pressure, p_{CO_2}, registered on the R-line diagram. The liquids are, for 20°C; (1) water; (3) $C_6H_5NH_2$; (4) CH_3OH; (5) C_2H_5OH; (6) i-$C_5H_{11}OH$; (2) $CHCl_3$; (13) C_5H_5N; (16) CH_3COCH_3; (19) $CH_3CO_2C_5H_{11}$. For 25°C: (2) CS_2; (7) C_6H_6; (8) C_6H_5Cl; (9) $C_6H_5NO_2$; (10) $C_6H_5CH_3$; (11) CCl_4; (14) C_6H_5CN; (15) n-C_7F_{16}; (17) $CH_3CO_2CH_3$; (18) $CH_3CO_2C_3H_7$. The value for water is registered by a thick line near the extreme left (less soluble side) of the diagram. The value for ethylene glycol may be expected to fall on the left, and that for glycerol still farther on the left of the line for methanol; but these values will still be greater than that for water. The position of a line for a stated temperature must be referred to the R-line for that temperature. RL, 20°C is based on $p°_{CO_2}$; RL, 25°C is based on N_{CO_2} (ideal) (ref. 125).

very similar except the R-line would strike the horizontal at 760 mm Hg at 0.0203, with reference to the scale on the base line, instead of 0.0179 for carbon dioxide at 20°C.

Kunerth expressed the solubility as

$$\frac{\text{cm}^3 \text{ of gas under existing barometric pressure and the temperature specified}}{\text{cm}^3 \text{ of solvent under the same conditions}}$$

This form is the Ostwald coefficient, except the p_A is not precisely specified. In every example, the solubility decreased with rise in temperature. Kunerth concluded that Raoult's law rarely, if ever, holds for gases dissolved in liquids. Furthermore, he could see little, if any, correlation between solubility and the difference in internal pressures of the solute and solvent, and therefore he did not accept Hildebrand's conclusion.

Table 9. N_{N_2O} Values for 1 atm

Solvent	10°C	20°C
iso-C_8H_{18}	0.027	0.0229
n-C_7H_{16}	0.0225	0.0196
CCl_4	0.0214	0.01815^a
C_6H_6	0.0165	0.0145
CS_2	0.00625	0.00572

[a] Wrongly given in original table as 0.1815.

Two papers by Yen and McKetta (1962)[385, 386] are relevant here. One is on the solubility of nitrous oxide in some nonpolar solvents (see Table 9). Values of N_A were given for $p_A = 1$ atm and $t°C$ from -10 to $40°C$. The figures interpolated for 25°C were deemed to fit Kobatake and Hildebrand's[209] plot of log N_A vs solubility parameters of the solvents, δ_s. The N_A values were also deemed to fit the linear plot of log N_A vs the Lennard-Jones force constant ε/k. Hildebrand,[163] in his highly significant paper of 1967, appeared to discard the force constant for this purpose.

Yen and McKetta called N_2O weakly polar with a dipole moment of 0.17 Debye units. Their second paper was on a thermodynamic correlation of nonpolar gas solubilities in polar, nonassociated liquids and is a mathematically based approach. The only point to be mentioned here is that of terminology with reference to "polar" and "nonpolar."

Dim and colleagues (1971)[65] reported work on the diffusion of carbon dioxide into primary alcohols and methylcellulose ether solutions. Their solubility data for three alcohols were expressed as g-mole/$cm^3 \times 10^5$, and I now show the effect of the molecular weight and density of S in converting such data into mole ratio x_A and mole fraction N_A. I also give the values for propane. We see further indication of a grouping of normal alcohols on the left of the R-line, with N_A increasing as the series is ascended.

8.1.2. Solubility of Carbon Dioxide in Water

The x_{CO_2} values for CO_2 in water at 1 atm and ambient temperatures is toward the low end of the range, being 0.00129 at 0°C and 0.000950 at 10°C. Again we see a marked lack of accommodation for carbon dioxide; one molecule of CO_2 requires about 1000 molecules of water to hold it in solution at 10°C/1 atm in any form at all. There are a number of papers dealing with the solubility of CO_2 and the conductance of its solution, but for the present purpose the paper by Morgan and Maass[262] needs to be pinpointed. These authors gave their results as % CO_2, referred to the I.C.T. definition of the mole fraction form of Henry's law, and gave Henry's law constant H as

$$H = \frac{p_{CO_2}, \text{mm}}{\text{mole fraction of } CO_2}$$

However, at the low values of x_{CO_2} for p_{CO_2} up to 1 atm, the x_{CO_2} and N_{CO_2} values are almost coincident (see Table 10). The % CO_2/p_{CO_2} plot of these workers does not clearly show curvature, but there is a variation of H values which could be interpreted as a bend around the low end of the p_{CO_2}, i.e., % $CO_2 \to 0$. It was concluded that "practically all the carbon dioxide exists as free carbon dioxide." Olson and Youle[294] stated that the following value for K is accepted by most chemists as the first ionization constant of aqueous carbonic acid:

$$K = \frac{|H^+||HCO_3^-|}{(\text{total } CO_2 \text{ in solution})} = 4 \times 10^{-7}$$

As the amount of H_2CO_3 is deemed to be much less than the amount of carbon dioxide absorbed, i.e., the carbon dioxide absorbed remains as "free" CO_2, K must have a larger value than that just given. A value of 2×10^{-4} was taken to be more realistic, indicating that carbonic acid is about 10 times stronger than acetic acid. Mills and Urey (1940)[259] believed that only about <1% (at a maximum) of the carbon dioxide present "exists in combination with water." Previously, McBain[274] had considered the rate of hydration and rate of dissociation of carbonic acid.

The more one studies the papers cited, the greater must be the conviction that the essential pattern of phenomena must entail the *structure* of water, $[H_2O]_n$, and the expressions *free, uncombined, unhydrated, hydrated, ionized*, must be given much more specific definition in terms of water structure. The expression "highly ionized due to small concentration" is misleading. At 10°C/1 atm the "concentration" referred to is as high as it can be in that particular system. Dilution can only be effected at 10°C by decreasing p_{CO_2}, or by addition of water, which decreases p_{CO_2}. At 10°C/1 atm *more* than 100,000 molecules of water are required to produce one (H^+) and one (HCO_3^-) site. Figure 64 roughly indicates how I picture the situation.

In an 8-page paper, Harned and Davis (1943)[150] discussed the ionization constant of carbonic acid in water and expressed the solubilities as "Henry's law constants," $H = M_{CO_2}/p_{CO_2}$, M_{CO_2} being the molality.

The report on the solubility of carbon dioxide in aqueous solutions of arsenous oxide, arsenic pentoxide, and hydrogen chloride by Robb and Zimmer (1968)[312] affords several points of interest in the present context. That work was done in connection with an attempt to correct to standard states in the combustion

Table 10. Solubility of CO_2 and C_3H_8 in Alcohols at 25°C

Gas A	Liquid S	g-mole A/ cm³ S × 10⁵	x_A	N_A
CO_2	"95% C_2H_5OH"	9.63	0.00563	0.0056
CO_2	n-$C_7H_{15}OH$	7.63	0.0106	0.0105
CO_2	n-$C_8H_{17}OH$	7.42	0.0115	0.0113
C_3H_8	n-$C_7H_{15}OH$	40.8	0.0578	0.0546
C_3H_8	n-$C_8N_{17}OH$	40.3	0.0635	0.0597

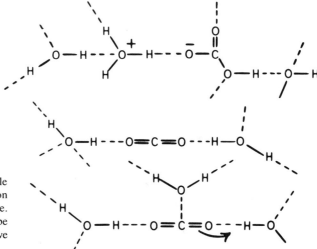

Fig. 64. Indication of the probable modes of interaction of carbon dioxide with the water structure. The system is understood to be three-dimensional and to involve dynamic equilibria.

calorimetry of organic compounds containing chlorine; but that purpose does not call for comment here. It is the aspect of solubility which now calls for attention.

The solubilities were expressed as Bunsen coefficients α for temperatures of 20° and 25°C. I shall deal only with the values for 20°C because of ready access to density data. To adjust for pressure, those workers used Henry's law in the form $m = kp$, m being "the mass of gas." To convert the Bunsen coefficient into values of Henry's law constant expressed as mole · liter^{-1} · atm^{-1}, a gram-mole volume of 22,260 cm^3 was used, allowance thereby being made for deviation from the "ideal gas law" in the example of carbon dioxide.

The α values for the arsenous oxide solutions varied from 0.88 for a molality (As$_2$O$_3$) of 0.0344 to 0.835 for a molality of 0.1113. Taking the value of 0.87 for water itself, the implication is that the solubility of carbon dioxide is less in the aqueous solutions of As$_2$O$_3$ than in pure water. A similar implication emerges from the very similar values of α for the As$_2$O$_5$ solutions. On a molecular basis, the comparison of the α values is invalid, since the conversion of the α value into the x_{CO_2} value involves the density of the liquid and the "molecular weight" of the liquid. For water itself, x_{CO_2} is about 0.0007 for 20°C and $p_{CO_2} = 1$ atm, and for the small molal values for the arsenic oxide solutions, it is questionable if the equivalent x_{CO_2} values would show a distinct decrease in solubility. To deal with the hydrogen chloride systems, I have used the published densities of aqueous solutions of hydrogen chloride for 20°C and have calculated the approximate average molecular weight to use for the computation of the x_{CO_2} value. I show three items as follows:

Molality of HCl	Bunsen coefficient	x_{CO_2}
0.637	0.854	0.00069
10.971	0.938	0.00077
12.754	0.924	0.00079

A molality of 12.754 corresponds to x_{HCl} of 0.197, and p_{HCl} is just beginning to emerge from the low-value range. See Fig. 27.

8.2. Hydrogen Sulfide, bp −61°C

The most recent and extensive data are those of Gerrard (1972).[99] See also Table 11. Figure 65 shows examples of plots on the R-line diagram. The data can be explained in a general way in terms of an acid function sited on hydrogen of sufficient intensity to respond to basic-function sites in a liquid S, which, in turn, are intensive enough. The basic function sited on sulfur is somewhat dispersed over three "lone pairs" of electrons. Whereas hydrogen sulfide does not appear to have decisive-enough acid–base function to compete strongly with the hydrogen-bonding structure of alcohols and carboxylic acids, the lines for $C_6H_5OC_2H_5$, C_6H_5CN, $(CH_3CO)_2O$, o-$C_6H_4(CH_3)_2$, etc. are definitely on the right of the R-line, and those for C_9H_7N and $HCON(CH_3)_2$ are strongly so.

In Fig. 66 are plotted N_{H_2S}/p_{H_2S} and, for comparison, N_{HCl}/p_{HCl} for cyclohexane from data given by Tsiklis and Svetlova (1958).[358] It was stated that the HCl–cyclohexane system "obeys Henry's law," but it is the mole *fraction* form which is meant. Even at these low values of N_{HCl}, there is a slight but discernible curvature to the x_{HCl}/p_{HCl} plot. Therefore, a general declaration that the system "obeys Henry's law" could be grossly misleading if the form of the law and conditions were not defined. The N_{H_2S}/p_{H_2S} plot for cyclohexane has a pronounced curvature, accentuated as the x_{H_2S}/p_{H_2S} plot.

For certain systems the N_A/p_A ratio for low p_A (less than 100 mm) has been quoted as a constant. It is clear from the plot N_{H_2S}/p_{H_2S} that this "constant" would give a grossly false value of N_{H_2S} at pressures approaching 1 atm. It is seen that the N_{H_2S} value at 1 atm is decidedly less than the corresponding R-line value at 10°C. Finally, it will be noticed that the curvature of the N_{H_2S}/p_{H_2S} plot is convex upward.

Another lesson is found in the paper by Bancroft and Belden (1930)[11] who gave their results as m mg of hydrogen sulfide per unit volume of aniline and plotted the m/p values as a straight line, $m = Kp$, where K is taken as the Henry's law constant. This is the mole ratio form. See Fig. 66. The linearity of the plot was taken to mean that no chemical reaction had occurred. They stated that the Henry's law constant is 0.0273 at 22°C; this simple number has been propagated via *Chemical Abstracts* without units and conditions.

Table 11. x_{H_2S} Values Calculated from the Prevailing Primary Form of Data for Liquids S at 1 atm and t°C

Liquids	x_{H_2S}	Temperature	References	Liquids	x_{H_2O}	Temperature	Reference
Octane	0.0460	20°C	Bell[20]	C_2H_5OH	0.031	10°C	Fauser[78]
Cyclohexane	0.050	20°C	Bell[20]	C_2H_5OH	0.019	20°C	Fauser[78]
CCl_4	0.0437	20°C	Bell[20]	$HCON(CH_3)_2$	0.120	25°C	duPont[72]
C_6H_6	0.0597	20°C	Bell[20]	$CHCl_3$	0.115	20°C	Bell[20]
$CH_3C_6H_5$	0.0720	20°C	Bell[20]	$(ClCH_2)_2$	0.0775	20°C	Bell[20]
C_2H_5Br	0.0647	20°C	Bell[20]	C_6H_5Br	0.0391	20°C	Bell[20]
$(C_2H_5)_2O$	0.0269	26°C	Parsons[296]	C_6H_5Cl	0.0404	20°C	Bell[20]
C_2H_5OH	0.046	0°C	Fauser[78]	$(BrCH_2)_2$[a]	0.144	20°C	Bell[20]

[a] Listed as ethyl bromide.

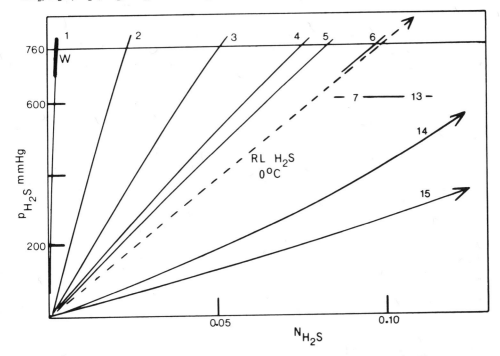

Fig. 65. Solubility of hydrogen sulfide in various liquids at 0°C. Plots of the mole fraction N_{H_2S} vs pressure p_{H_2S} on the R-line diagram. The liquids are: (1) water; (2) $(CH_2OH)_2$; (3) C_2H_5OH; (4) $n\text{-}C_8H_{17}OH$; (5) $n\text{-}C_5H_{11}CO_2H$; (6) C_6H_6. Lines 7 to 13 are not drawn; they appear in the following order in the space shown, and are concave upward: (7) $C_6H_5OC_2H_5$; (8) C_6H_5CN; (9) $n\text{-}C_8H_{17}Br$; (10) $(CH_3CO)_2O$; (11) $o\text{-}C_6H_4(CH_3)_2$; (12) $C_6H_5CH_3$; (13) $o\text{-}HOC_6H_4CO_2CH_3$; (14) C_9H_7N; (15) $HCON(CH_3)_2$. The position of the line for water is shown by a thick line near the extreme left of the diagram.

In Fig. 66 are plotted my calculated x_{H_2S}/p_{H_2S} values, which give a straight line, and the corresponding N_{H_2S}/p_{H_2S} values, which give a clearly discernible curve, concave upward, on the right of the R-line. Certain chemists would take this to indicate an equilibrium system in which the hydrogen-bonding compound, probably $C_6H_5(H_2)N{:}\cdots H_2S$, was existing on a time-average basis. Josien and Saumagne (1956)[194] have reported the infrared frequency shifts indicating the "existence des complexes moléculaires de l'hydrogène sulfuré" with a number of liquids such as benzene, aniline, and diethyl ether. The revelation of acid–base function can be a matter of delicacy of the probe.

In Table 11 are given examples of x_{H_2S} data calculated from the somewhat sparse information in the literature previous to Gerrard's paper. The low bp of hydrogen sulfide as compared with water has long been attributed to the lack of hydrogen bonding, and at one time it was believed that sulfur in hydrogen sulfide, mercaptans, and organic sulfides could not take part in hydrogen bonding. However, the acid–base function appears to be clearly operative in the following systems:

$$Cl_3CH\cdots SH_2 \qquad HCO(CH_3)_2N\cdots H\text{—}S\text{—}H$$
$$C_2H_5Br\cdots H\text{—}S\text{—}H \qquad CH_3C_6H_5\cdots H\text{—}S\text{—}H$$

Table 12. x_{Cl_2} and N_{Cl_2} Values for 1 atm

Liquids	x_{Cl_2} (°C)	N_{Cl_2} (°C)
n-C_7H_{16}	0.370 (0)	0.270 (0)
$SiCl_4$	0.404 (0)	0.288 (0)
CCl_4	0.424 (0)	0.298 (0)
	0.200 (25)	0.167 (25)
CH_2BrCH_2Br	0.208 (25)	0.172 (25)
n-C_7F_{16}		0.164 (0)
		0.110 (20)
		0.097 (25)

The hydrogen-bonding structure of ethanol offers marked resistance to interference by hydrogen sulfide, and water is even more effective in this sense.

Cheung and Zander (1968)[49] refer to the solubility of hydrogen sulfide (and carbon dioxide) in liquid hydrocarbons at cryogenic temperatures.

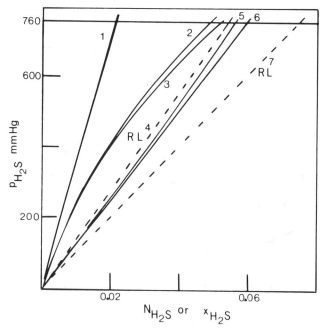

Fig. 66. Solubility of hydrogen sulfide. Line 4 is the R-line for 22°C, and with reference to this, line 5 is the mole fraction N_{H_2S} vs p_{H_2S} plot, and line 6 the mole ratio x_{X_2S} vs p_{H_2S} plot for aniline at 22°C. The N_{H_2S} line is concave upward just on the right of the R-line; the x_{H_2S} plot is almost straight. Line 7 is the R-line for 10°C, and with reference to this, line 2 is the N_{H_2S} plot, and line 3 the corresponding x_{H_2S} plot for cyclohexane at 10°C. Line 1 is the N_{HCl} plot for cyclohexane at 10°C.

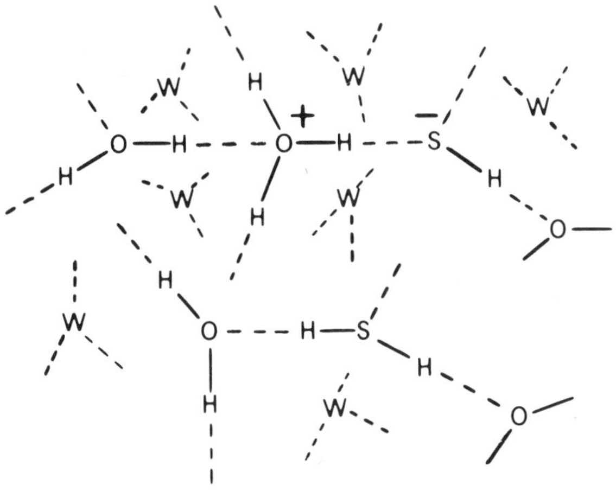

Fig. 67. A representation of the probable modes of interaction of hydrogen sulfide with the water structure. The letters W indicate parts of the water structure which do not contain "H₂S" units at at a given instant of time. It is understood that the system is three-dimensional and that it entails dynamic equilibria.

8.2.1. Hydrogen Sulfide–Water Systems

The $x_{H_2S}(\approx N_{H_2S})$ value for water at 10°C and $p_{H_2S} = 1$ atm is only 0.00267.

Wright and Maass (1932)[383] declared that the reported data on the solubility of hydrogen sulfide in water suffered from the incorporation of Henry's law as an implicit assumption. They expressed their data as moles of hydrogen sulfide per liter of solution and decided that a deviation from Henry's law was evident.

Partington[298] stated that hydrogen sulfide is moderately soluble in water and that Henry's law is followed. Only 0.1% of the dissolved compound was believed to be ionized to "$H^+ + HS^-$"; the second stage to "$H^+ + S^{2-}$" was taken to be very slightly populated.

Wright and Maass[383] tried to sort out the experimental difficulties in the determination of the electrolytic conductance of such solutions and found it necessary to work with the higher values of x_{H_2S} which, however, are still absolutely small. They agreed that the secondary dissociation is "vanishingly small."

The R-line value is about 0.08 at 10°C and $p_{H_2S} = 1$ atm. Water shows a great reluctance to accommodate hydrogen sulfide. To hold one molecule of H₂S, 370 molecules of water (H₂O unit) are required at 10°C and $p_{H_2S} = 1$ atm, and 370,000 molecules of water are required for each (+) and (−) site due to H₂S. In Fig. 67 I indicate how the hydrogen sulfide may be involved in the water structure.

8.3. Chlorine

8.3.1. Nonaqueous Liquids

The solubility of chlorine in nonaqueous liquids is rarely reported presumably because of its propensity to undergo irreversible reactions.

Taylor and Hildebrand (1923)[354] gave data for the four liquids named in Table 12. They calculated the "ideal" solubility, expressed as a mole fraction $N^i_{Cl_2}$ for 0°C

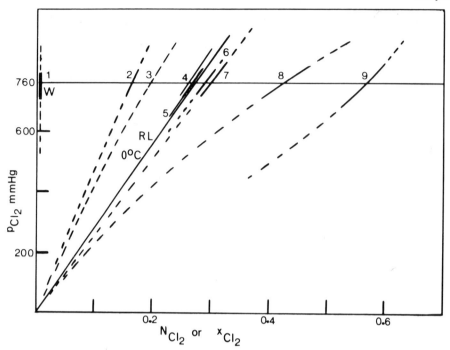

Fig. 68. Solubility of chlorine in various liquids at 0°C. Registration of the mole fraction N_{Cl_2} values for 1 atm on the R-line diagram. The broken lines indicate the probable approximate plots of N_{Cl_2} or x_{Cl_2} vs p_{Cl_2}. The liquids are: (1) water; (2) n-C_7F_{16} (N_{Cl_2}); (3) n-C_7F_{16}(x_{Cl_2}); (4) $CHCl_3$; (5) n-C_7H_{16}; (6) $SiCl_4$; (7) CCl_4; (8) CCl_4(x_{Cl_2}); (9) $HCON(CH_3)_2$. Hildebrand's "fugacity" point is at $N_{Cl_2} = 0.29$ for 1 atm, just on the left of line 6.

and 1 atm, as $1/3.66 = 0.273$; 3.66 atm was taken as $p°_{Cl_2}$, which is the vapor pressure of chlorine over pure liquid chlorine at 0°C. Using fugacity adjustments, they gave the $N^i_{Cl_2}$ as $0.984/3.44 = 0.286$. The actual N_{Cl_2} values were for pressures p_{Cl_2} in mm Hg. The amount of chlorine absorbed was determined for a total pressure of 1 atm. The estimated value of the partial pressure of the liquid S, i.e., p_S, was subtracted from the total pressure to give p_{Cl_2}, corresponding to N_{Cl_2}. They then used Henry's law in the form

$$N_{Cl_2} = N_{Cl_2} \text{ (observed)} \times 760/p_{Cl_2}$$

to compute the N_{Cl_2} value for 1 atm. Thus the mole fraction form of Henry's law was used; the mole ratio form would certainly not be followed in these systems.

In Fig. 68 the line given as the R-line corresponds to that based on $1/p°_{Cl_2} = 0.273$, assumed by Hildebrand to emerge from Raoult's law; the R-line, however, is quite independent of Raoult's law. The fugacity line, based on $f/f° = 0.286$, is just on the right of the R-line. The suggestion is that one should consider deviations from the fugacity value rather than from the R-line value, but I see no advantage in this procedure. Actual N_{Cl_2} values are linked to pressures.

Gjaldbaek and Hildebrand (1950)[127] also invoked the concept of fugacity in reference to the ideal solubility and gave a Bunsen coefficient of 20.1 and $N_{Cl_2} = 0.164$ for n-perfluoroheptane at 0°C and $p_{Cl_2} = 760$ mm Hg. It is seen from Fig. 68 that the x_{Cl_2} vs p_{Cl_2} plot is far from linear. To convert the Bunsen coefficient into the x_{Cl_2} value, the molecular weight and density (taken here to be 1.780) of the perfluoroheptane must be used, and one must also decide what gram-mole volume to take for this purpose. I give the parallel calculations as follows:

$$\frac{20.1}{22{,}400} \times \frac{388.7}{1.780} = x_{Cl_2} = 0.196 \equiv N_{Cl_2} = 0.164$$

$$\frac{20.1 \times 0.00332}{70.918} \times \frac{388.7}{1.780} = x_{Cl_2} = 0.205 \equiv N_{Cl_2} = 0.170$$

Smith (1955)[336] drew attention to the need for more data to test and develop the theories of Hildebrand. He produced data for the $Cl_2 + CCl_4$ system at 40 to 90°C and declared that the two solubility lines (log N_{Cl_2} vs $1/T \times 10^3$) for $Cl_2 + CCl_4$ and $Cl_2 + C_2H_4Br_2$ do not intersect as previously reported.

According to data by Tsiklis and Svetlova (1958)[358] the N_{Cl_2} vs p_{Cl_2} plot for cyclohexane at 10°C is strongly convex upward, as shown in Fig. 69; the x_{Cl_2} plot is also strongly convex upward up to the upper limit of the observed pressure, 500 mm Hg. An extrapolation of the N_{Cl_2} line to 760 mm Hg brings it abruptly across

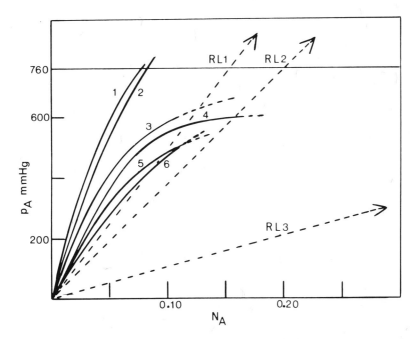

Fig. 69. Solubility of chlorine and nitrosyl chloride in cyclohexane. Mole fraction N_A vs pressure p_A plots on the R-line diagram. (1) Cl_2 at 40°C; (2) NOCl at 40°C; (3) Cl_2 at 20°C; (4) NOCl at 20°C; (5) Cl_2 at 10°C; (6) NOCl at 10°C. Reference lines (RL): (1) Cl_2 at 20°C; (2) Cl_2 at 10°C; (3) NOCl at 10°C.

the R-line to a value of 0.33, and the system therefore needs to be reexamined. The "Henry coefficient" was given as 8240 mm Hg/mole fraction, and this appears to refer to a tangent extrapolation from the N_{Cl_2} values below $p_{Cl_2} = 100$ mm Hg. In Fig. 69 I have plotted their N_{NOCl} values which were given up to 500 mm Hg. A speculative extrapolation gives $N_{NOCl} = \sim 0.225$ at 760 mm Hg, and this is well to the left of the R-line, which refers to a bp of about −6°C, whereas the bp of chlorine is −34.6°C. These authors gave the "Henry coefficient" as 6800 mm Hg/mole fraction for 10°C.

The absorption coefficient of 385 for dimethylformamide[72] at 0°C gives $x_{Cl_2} = 1.32$, $N_{Cl_2} = 0.569$. The position of the N_{Cl_2} value in Fig. 69 clearly shows that the N_{Cl_2} vs p_{Cl_2} plot over the whole range to $p°_{Cl_2}$ must be far from linear and would be concave upward.

In Fig. 70 I have plotted the N_{Cl_2} vs p_{Cl_2} data of Krieve and Mason (1956)[218] for chlorine in titanium tetrachloride at 20°C. The given data were from $N_{Cl_2} = 0.2513$ at $p_{Cl_2} = 1.5$ atm to $N_{Cl_2} = 0.7229$ at $p_{Cl_2} = 4.76$ atm. These data were deemed to show a slight departure from Raoult's law. May one conclude from this that the system is almost ideal? Some workers would assume that Henry's law in the mole fraction form was being followed as the N_{Cl_2} emerged from zero, p_{Cl_2} emerging from zero, and they would give the Henry's law constant as p_{Cl_2}/N_{Cl_2}. An extrapolation of this supposedly straight line to $p_{Cl_2} = 760$ mm Hg or higher would give grossly inaccurate values of N_{Cl_2} at the higher pressures. Furthermore, if one does assume that the mole fraction form of the law holds, this cannot be so for the mole ratio form.

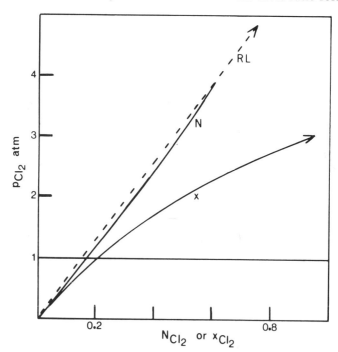

Fig. 70. Solubility of chlorine in titanium tetrachloride at 20°C. Plots of mole fraction N_{Cl_2} vs pressure p_{Cl_2} and mole ratio x_{Cl_2} vs p_{Cl_2} on the R-line diagram for 20°C.

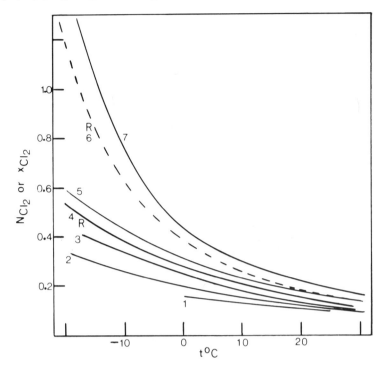

Fig. 71. Solubility of chlorine at different temperatures $t°C$ and 1 atm. Plots of mole fraction N_{Cl_2} vs $t°C$ for the liquids: (1) n-C_7F_{16}; (2) CH_2Cl_2; (3) $CHCl_3$; (5) CCl_4. Line 4 is from R-line N_{Cl} data; line 6 is the corresponding plot from the x_{Cl_2} data. Line 7 is the x_{Cl_2} vs $t°C$ plot for CCl_4.

Krivonos (1958)[219] considered the solubility of chlorine in benzene in connection with the halogenation of benzene. The volume of chlorine absorbed by 1 cm³ of benzene at 20°C appears to be that for $p_{Cl_2} = 300.3$ mm Hg. The multiplication of this volume by $(760/300.3) \times (273/293)$ gave a Bunsen coefficient of 69. To get the moles per liter from this, Krivonos used a gram-mole volume of 22.06, derived from the density of Cl_2 as 3.214 g/liter at 0°C and 1 atm. For 10°C the x_{Cl_2} value is 0.383, and the N_{Cl_2} value is 0.277 at $p_{Cl_2} = 1$ atm. A glance at Fig. 68 indicates that the N_{Cl_2} vs p_{Cl_2} plot would be concave upward. Leonova and Ukshe (1970)[233] have reported on the solubility of chlorine in molten rubidium chloride. Figure 71 gives the solubility of chlorine in various liquids as a function of temperature.

8.3.2. Solubility in Water

The solubility of chlorine in water is entangled with the process referred to as the hydrolysis of chlorine (see Fig. 72). The expressions "concentration of unhydrolyzed chlorine," "concentration of hydrolyzed chlorine," and "concentration of total chlorine" are linked with the equation usually given as

$$Cl_2 + H_2O = Cl^- + H^+ + HClO$$

Work by Whitney and Vivian (1941)[374] is usually referred to. These workers

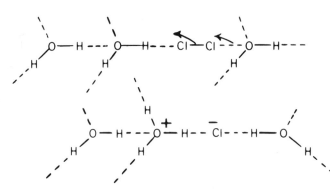

Fig. 72. Representation of the probable modes of interaction of chlorine with the water structure. It is understood that the system is three-dimensional and that a condition of dynamic equilibria prevails.

expressed their data as grams of $Cl_2/100$ g of water, and the corresponding x_{Cl_2} vs p_{Cl_2} plot is given in Fig. 73. The absolutely small value of $x_{Cl_2} = 0.00254$ ($\approx N_{Cl_2}$) for 15°C and $p_{Cl_2} = 1$ atm shows that water is reluctant to find accommodation for chlorine. Partington[298] stated that chlorine is fairly soluble in water and gave the hydrolysis equation as: $Cl_2 + H_2O = HOCl + HCl$. Morris (1946)[263] referred to the extremely rapid rate of hydrolysis of chlorine and suggested that an alternative mechanism could entail interaction with the hydroxyl ion:

$$Cl_2 + OH^- = HClO + Cl^-$$

King (1969)[204] referred to the application of Henry's law in examples involving chemical interaction with the solvent. He gave a detailed analysis of the data of Whitney and Vivian based on the idea that Henry's law in the mole fraction form applies to the "unreacted chlorine." Actually, Whitney and Vivian's data are in the mole ratio form although numerically it does not matter for most purposes because $x_{Cl_2} \approx N_{Cl_2}$ at these pressures.

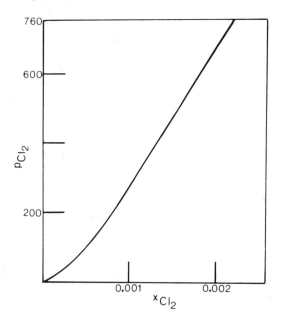

Fig. 73. Solubility of chlorine in water at 15°C. Plot of the mole ratio x_{Cl_2} vs pressure p_{Cl_2} (mm Hg).

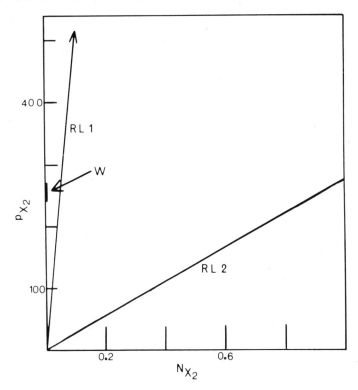

Fig. 74. Solubility of chlorine and bromine in water at 20°C. Registration of the mole fraction values N_{X_2} (X = Cl or Br) on the R-line diagram. RL 2 is the R-line for bromine, drawn from the point $p°_{Br_2}$ = 276 mm Hg at 20°C. The N_{Br_2} value for water at 20°C corresponding to this limiting value of p_{Br_2} in the presence of liquid bromine is about 0.005, and its position on the RL 2 diagram is indicated by the arrow W. At 20°C and p_{Cl_2} = 276 mm Hg, the N_{Cl_2} value for water is about 0.00072 and is toward the extreme left of the R-line for chlorine, RL 1, at 20°C, again indicated by the arrow W. The pressures p_{X_2} are in mm Hg.

In Fig. 74 I have given the R-lines for chlorine and bromine for 20°C on a common scale and have shown the positions of the N_{X_2} values for water. In the example of iodine at 20°C, there is the additional factor in the lattice energy. For the presence of solid iodine in water the p_{I_2} is limited to 0.2 mm Hg. The x_{I_2} value is then 0.00002, i.e., one molecule of iodine for 50,000 molecules (H_2O units) of water.

8.4. Sulfur Dioxide, bp $-10°C/1$ atm

8.4.1. Nonaqueous Liquids

The most recent and extensive data are those of Gerrard (1972).[99] In Fig. 75 are plotted $x_{SO_2}/t°C$ values for chlorobenzene from $N_{SO_2}/t°C$ values given by Horiuti (1931)[177]; agreement with my own results is shown. An outstanding feature is the drastic increase in temperature coefficient between 10 and 0°C, and an even

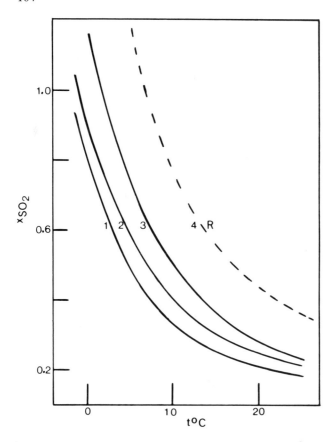

Fig. 75. Solubility of sulfur dioxide in halobenzenes at 1 atm and various temperatures $t°C$. (1) C_6H_5I; (2) C_6H_5Br; (3) C_6H_5Cl. Line 4 (R) is a plot of the mole ratio x_{SO_2} values obtained from the R-lines at different temperatures.

more spectacular increase below 0°C. It is also seen that the x_{SO_2} values at a given temperature are in the order PhCl > PhBr > PhI.

In Fig. 76 are plotted Horiuti's N_{SO_2}/p_{SO_2} values for PhCl at 25°C, together with my calculated values of x_{SO_2}. The N_{SO_2}/p_{SO_2} (25°C) plot shows a slight curvature, convex upward, but the x_{SO_2} values show a more marked curvature, convex upward; those who take Henry's law as the mole fraction form, $p_{SO_2} = KN_{SO_2}$, would declare this plot to be almost linear, indicating that the law is "almost obeyed." However, in terms of the mole ratio form, this conclusion could not be reached.

The x_{SO_2} values for the n-octyl halides (RCl > RBr) are less than the corresponding ones for the aryl halides.

A plot of Horiuti's N_{SO_2} values for benzene at 25°C and a plot of my calculated x_{SO_2} vs p_{SO_2} values would lead the adherents of the mole fraction form of Henry's law to declare that the law is being "obeyed," but the mole ratio supporters would declare that it is not being "obeyed." The $x_{SO_2}/t°C$ plots for benzene from Horiuti's N_{SO_2} data, and for ditolylmethane and dicumylmethane from Gal'perin et al.'s data (1958),[91] are shown in Fig. 77. For carbon tetrachloride at 10°C, the N_{SO_2}/p_{SO_2} plot is slightly convex upward (Horiuti).

Horiuti's N_{SO_2} data for methyl acetate and acetone afford other significant examples of the need to decide the form of Henry's law, whether mole ratio or mole fraction. The x_{SO_2} value for acetone is 1.70 and for methyl acetate, 1.60, at 10°C/1 atm. Whereas the N_{SO_2}/p_{SO_2} plots are decidedly nonlinear, being concave upward, the plots of my calculated values of x_{SO_2}/p_{SO_2} are less curved. An extrapolation of the x_{SO_2}/p_{SO_2} plot from the lower pressures (e.g., total pressure = 300 mm Hg) by Henry's law based on x_{SO_2}/p_{SO_2} would give a rough estimate of the x_{SO_2} value at 1 atm, but the corresponding extrapolation based on the N_{SO_2}/p_{SO_2} form most certainly would not; for acetone this would be about $N_{SO_2} = 1.1$, which is absurd.

The handbooks quote Lloyd's (1918)[244] solubility measurements as grams of sulfur dioxide per liter but do not state whether they are for a liter of solvent (benzene, nitrobenzene, toluene, o-nitrotoluene, acetic anhydride) or solution. Lloyd's data refer to a liter of *solution*, and without the density of the solution they are worthless for comparison purposes. For only one example (acetic anhydride) was the density of the solution ($d_{0°C}, =1.22$) given, and this gives an x_{SO_2} value of 0.22 at 0°C/1 atm, astonishingly small compared with my own value of 2.35. I suspect the decimal place is wrong.

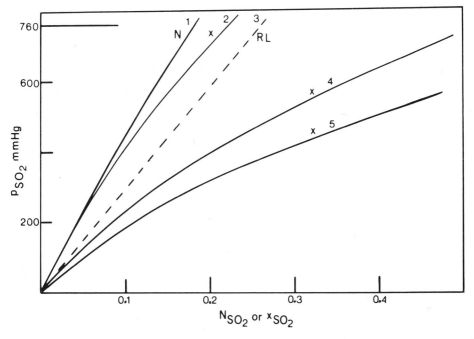

Fig. 76. Solubility of sulfur dioxide in chlorobenzene at different temperatures t°C and pressures p_{SO_2}. Line 1 is the mole fraction N_{SO_2} vs p_{SO_2} plot for 25°C from the data of Horiuti; line 2 is the corresponding mole ratio x_{SO_2} vs p_{SO_2} plot. Line 3 is the R-line for SO_2 at 25°C. Line 4 is the x_{SO_2} vs p_{SO_2} plot for 10°C, and line 5 is the x_{SO_2} vs p_{SO_2} plot for 5°C.

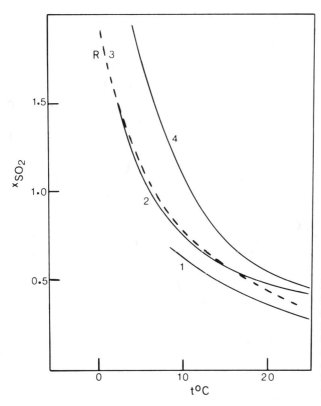

Fig. 77. Solubility of sulfur dioxide in hydrocarbons at 1 atm and different temperatures $t°C$. Mole ratio x_{SO_2} vs $t°C$ plots for: (1) C_6H_6 (from data by Horiuti[177]); (2) dicumylmethane, $C_{19}H_{24}$; (4) ditolylmethane, $C_{15}H_{16}$ (from data by Gal'perin et al.[91] Line 3 (R) is the x_{SO_2} plot based on the R-lines.

Horiuti's data for methyl acetate and acetone also strongly support my conclusion about Lloyd's data and are in line with my own data for acetic anhydride.

Weissenberger and Hadwiger (1927)[370] reported on "the absorption of sulfur dioxide in organic liquids" but mentioned only six compounds such as cyclohexanone and hydrogenated naphthalenes. Thus, 103 cm³ (the molar volume) of cyclohexanone was stated to absorb 2353.87 cm³ SO_2, and an unstable compound

$$(CH_2)_5 \; C\begin{matrix} S-O \\ | \\ O-O \end{matrix}$$

was supposed to form. However, the computed x_{SO_2} value is approximately 0.098, presumably for room temperature and $P_{SO_2} \simeq 1$ atm. The corresponding x_{SO_2} value for methylcyclohexanol is $2082.82/24,000 = 0.087$. In that example, compound formation between the hydroxyl carbon and sulfur was postulated. This work has been checked by the manometer procedure (see Chap. 16).

Figure 78 shows the x_{SO_2} vs $t°C$ plots for methanol based on the data of de Bruyn (1892).[64] The figure also shows my own data for n-octanol and ethylene glycol.

In a previous report[99] I have shown representative plots of my own data. In Fig. 79 the change of the R-line values of x and N (for 1 atm) with temperature is depicted.

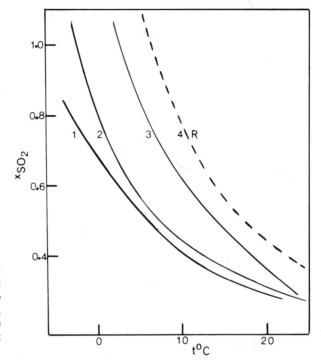

Fig. 78. Solubility of sulfur dioxide is alcohols at 1 atm and different temperatures $t°C$. The mole ratio x_{SO_2} vs $t°C$ are for (1) n-$C_8H_{17}OH$, (2) $HOCH_2CH_2OH$, and (3) CH_3OH. Line 4 (R) is the x_{SO_2} plot based on the R-lines.

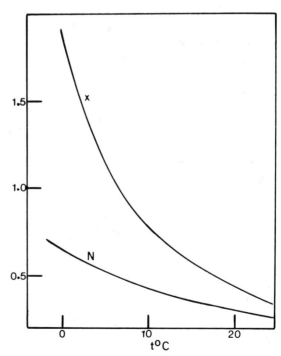

Fig. 79. Change of the R-line value of the mole fraction N_{SO_2} for 1 atm with temperature $t°C$. Contrast of the corresponding mole ratio x_{SO_2} plot. The x_{SO_2} snd N_{SO_2} values are shown on the vertical axis.

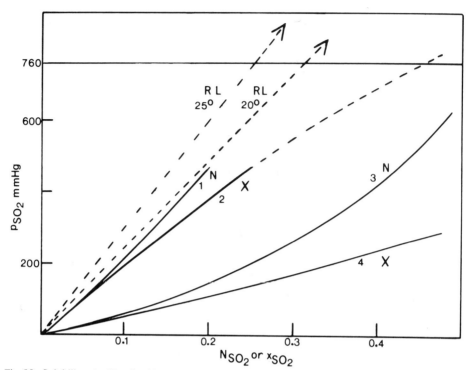

Fig. 80. Solubility of sulfur dioxide at different pressures p_{SO_2}. (1) N_{SO_2} for di-n-butyl ether at 20°C; (2) corresponding x_{SO_2} plot; (3) N_{SO_2} for diethylaniline at 25°C; (4) corresponding x_{SO_2} plot. N_{SO_2} is the mole fraction; x_{SO_2} is the mole ratio.

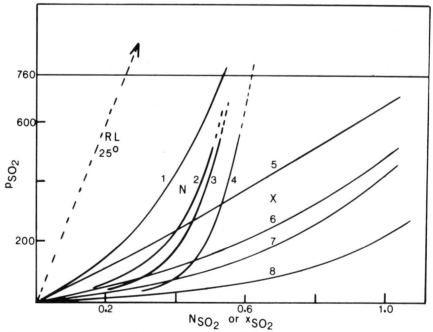

Fig. 81. Solubility of sulfur dioxide in the alkylanilines at 25°C. Plots of the mole fraction N_{SO_2} vs pressure p_{SO_2} and mole ratio x_{SO_2} vs p_{SO_2} on the R-line diagram. The liquids are: (1,N; 5,X) $C_6H_5N(C_2H_5)_2$; (2,N; 6,X) $C_6H_5NHC_2H_5$; (3,N; 7,X) $C_6H_5NHCH_3$; (4,N; 8,X) $C_6H_5N(CH_3)_2$. The pressures p_{SO_2} are in mm Hg.

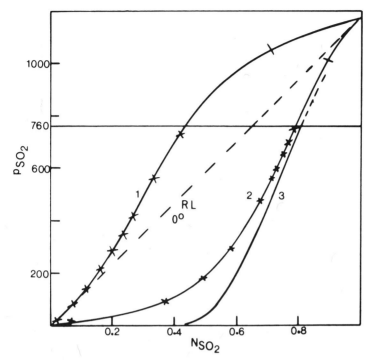

Fig. 82. Solubility of sulfur dioxide at 0°C. Line 1 is the mole fraction N_{SO_2} vs pressure p_{SO_2} plot for ethylene glycol based on published data[83]; the cross bar is the maximum p_{SO_2} observed. Line 2 is a similar plot for diethylaniline. The crosses on lines 1 and 2 are the values obtained by the present writer; line 3 is for dimethylaniline from data likewise obtained by the bubbler procedure. Lines for methylaniline and ethylaniline lie between lines 2 and 3. The pressures p_{SO_2} are in mm Hg.

Hildebrand looked upon sulfur dioxide as an electron-acceptor molecule. The x_{SO_2} data show the decisive acidic function of this gas in the examples of dimethylformamide, pyridine, and benzonitrile.

Hill and Fitzgerald (1935)[167] (see also ref. 166) called di-n-butyl ether a nonreacting solvent because Henry's law was deemed to be followed within fairly narrow limits. The data are for p_{SO_2} up to 472.5 mm Hg, and of the mole ratio form, and therefore it was implied that the law is $x_{SO_2} = Kp_{SO_2}$. Figure 80 shows the plot of x_{SO_2} vs p_{SO_2}, which is convex upward, and the corresponding one of N_{SO_2}, which is slightly concave upward just on the right side of the R-line. Again, to give a Henry's law constant K based on the line as $p_{SO_2} \to 0$ would give a grossly erroneous value of x_{SO_2} or N_{SO_2} at 760 mm Hg. Figure 81 shows plots for x_{SO_2} vs p_{SO_2} for alkylanilines which were deemed to show "positive deviations from Henry's law in all four cases, though the deviation is not great in the case of diethylaniline."

The N_{SO_2} plots now given would be described as showing "strong *negative* deviation from Raoult's law." In Fig. 82, line 2 is the N_{SO_2} plot for diethylaniline, and line 1 for ethylene glycol for 0°C based on data by Foote and Fleischer (1934).[83] The crosses on these lines and the full line for dimethylaniline (similar to methyl- and ethylanilines, new for 0°C) are the present writer's data. In the original work,[83] line

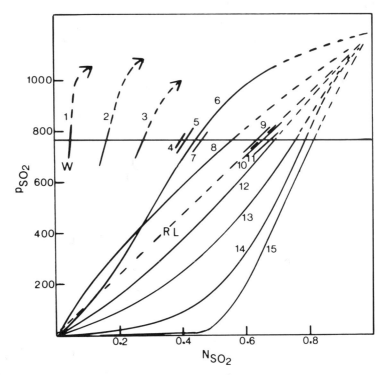

Fig. 83. Solubility of sulfur dioxide in various liquids at 0°C. Registration of the mole fraction N_{SO_2} values and the N_{SO_2} vs pressure p_{SO_2} plots on the R-line diagram. The liquids are: (1) water; (2) CCl_4; (3) $(CH_2OH)_2CHOH$; (4) $n\text{-}C_8H_{17}Br$; (5) Cl_3CCH_2OH; (6) $(CH_2OH)_2$; (7) $n\text{-}C_5H_{11}CO_2H$; (8) C_6H_5Cl; (9) $C_6H_5CH_3$; (10) $C_6H_5NO_2$; (11) $m\text{-}C_6H_4(CH_3)_2$; (12) $o\text{-}HOC_6H_4CO_2CH_3$; (13) $(CH_3CO)_2O$; (14) $C_6H_5N(C_2H_5)_2$; (15) $C_6H_5N(CH_3)_2$. The broken line is the reference line (RL) for SO_2 at 0°C. The pressures p_{SO_2} are in mm Hg.

1 was drawn as a full one beyond the highest observed p_{SO_2} of 1050 mm Hg, and it shows a much more abrupt approach to the point $p_{SO_2} = p°_{SO_2}$ than in the conventional diagrams. Furthermore, there is a slight indication that the line crosses the R-line at the lower end of the p_{SO_2} range. Reading the line as approaching $p_{SO_2} = 0$ along the R-line, we appear to observe an example of a system which "obeys" Raoult's law as $p_A \to 0$ but does not do so as $p_A \to p°_A$, contrary to the conventional notions. In Fig. 83, other N_{SO_2} vs p_{SO_2} plots are compared on the R-line diagram.

8.4.2. Solubility in Water

This system has been extensively examined because of its technological significance in the wood pulp industry. In the current textbooks sulfur dioxide is stated to have a high solubility in water because of the chemical reactions symbolized as

$$SO_2 + 2H_2O \to HSO_3^- + H_3O^+$$
$$SO_2 + 3H_2O \to 2H_3O^+ + SO_3^{2-}$$

Partington (1966)[298] stated that sulfur dioxide is freely soluble in water, and Rawcliffe and Rawson (1969)[308] declared it to be very soluble in water. Such comparisons are unscientifically based and give a wrong impression. The x_{SO_2} value for water for 0°C and $p_{SO_2} = 1$ atm is less than that for any other liquid S reported as having been examined.

Earlier and often-cited measurements were by Sims (1862)[335] and Hudson (1925).[179] Hudson gave his results as grams of sulfur dioxide per 100 g of water at a stated p_{SO_2} and at temperatures from 10 to 30°C. The calculated x_{SO_2} value for 10°C and $p_{SO_2} = 1$ atm is only about 0.04. There was an argument about the use of Henry's law to extrapolate from, for example, 520 to 760 mm Hg, but it was the mole *ratio* form of the law which was considered.

Maass and Maass (1928)[248] set out to measure the conductance of "concentrated" aqueous solutions of sulfur dioxide but had to look more closely at other aspects first. Their plot of % SO_2/total pressure is slightly concave upward, tending to become linear at the higher percent values. Figure 84 shows the x_{SO_2} vs total pressure plot. It must be emphasized that the so-called *concentrated* solutions are not more than $x_{SO_2} = 0.04$, nor can they be less than this for 10°C and $p_{SO_2} = 1$ atm at equilibrium.

According to Smith and Parkhurst (1922),[337] the "concentration" of sulfur dioxide as *sulfurous acid* is proportional to its partial pressure. Campbell and Maass (1930)[45] remarked on the disagreement among the cited values for the dissociation constant. The primary hazard is the value to take for $[H_2SO_3]$, the tendency being to equate "sulfurous acid" to sulfur dioxide absorbed. Tartar and Garretson (1941)[352]

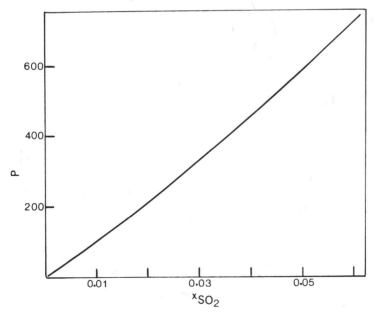

Fig. 84. Solubility of sulfur dioxide in water at 0°C. Plot of the mole ratio x_{SO_2} vs total pressure $P = p_{SO_2} + p_w$ in mm Hg.

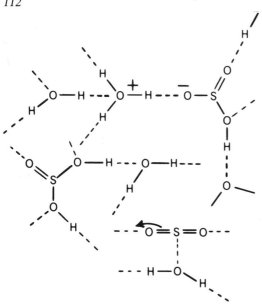

Fig. 85. An indication of the modes of interaction of sulfur dioxide with the water structure. It is understood that the system is three-dimensional and that a condition of dynamic equilibria prevails.

referred to the primary dissociation as due to the following sequence of reactions:

$$SO_2 + H_2O \rightarrow H_2SO_3 \rightarrow H^+ + HSO^-$$

The second dissociation constant $[H^+][SO_3^{2-}]/[HSO_3^-]$ was given the small value of 6.24×10^{-8}. Campbell and Maass[45] declared that this second dissociation may usually be neglected, and although they believed the sulfurous acid to be "almost completely ionized," they asserted that at 10°C, 20% of the total amount of SO_2 absorbed remained "uncombined," and at 22°C the amount was 50%, both estimates relating to what they called a solution of 5% SO_2.

Ostwald[295] and Barth[12] concluded that sulfurous acid acts as a monobasic acid even at "high dilutions."

Johnstone and Leppla (1934)[191] expressed the molality of sulfurous acid as "moles SO_2 per 1000 g H_2O" and referred to "ionized" and "un-ionized" molecules, the latter being present in by far the greater proportion. Sulfurous acid was referred to as a *weak* acid. Henry's law was stated in the mole *ratio* form. Whereas the total SO_2 vs p_{SO_2} gave a slightly curved plot, the "un-ionized" portion was deemed to follow Henry's law.

In his long paper on the dissociation of sulfurous acid, Lindner[240] stated that the expression

$$(\Lambda/\Lambda_\infty)^2/(1-\Lambda/\Lambda_\infty)V$$

could not be made constant by lowering or increasing the value of Λ_∞. Sulfur dioxide in water was deemed to depart from the law of Henry, even at 50°C.

Beuschlein and Simenson (1940)[24] gave a plot of g SO_2/g H_2O vs p_{SO_2} for 20°C, which shows a decided bend at the lower end of the concentration range at about 1 to 1.5 g SO_2/1000 g H_2O. The upper part to $p_{SO_2} = 600$ mm Hg is perceptibly linear.

8.5. Carbonyl Chloride, bp 8.2°C/1 atm

In 1920, Atkinson, Heycock, and Pope[8] published data which I represent by the $x_{COCl_2}/t°C$ plots in Fig. 86. The rapid increase in x_{COCl_2} as $t°C$ approaches the bp, 8.2°C/1 atm, is well illustrated. The vapor pressure data[8] enable me to draw the R-lines for 0 and 20°C (Fig. 87), and from the data of Kireev, Kaplan, and Vasnera (1936)[206] I give the plots of N_{COCl_2}/p_{COCl_2}. All the lines are on the left (less soluble) side of the respective R-line, and the rate of increase of N_{COCl_2} with p_{COCl_2} increases markedly as p_{COCl_2} increases. The system at 0°C is, in conventional terms, a liquid–liquid one, but the form of the diagram is essentially the same as at 20°C. The contrast of the N_{COCl_2} plot and that for x_{COCl_2} is illustrated.

The plot for carbon tetrachloride[206] showed a somewhat abrupt crossing of the R-line at $p_{COCl_2} = 600$ mm Hg, the curve arriving at the 760 mm Hg horizontal at $N_{COCl_2} = 0.70$. This indicated a possible error in the original calculation[206] converting primary mass data into N_{COCl_2} form, and indeed a recalculation showed this to be so. The amended plot fitted into the cluster. Data by Baskerville and Cohen

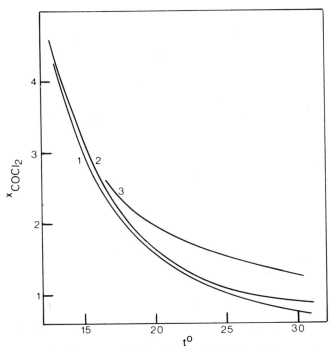

Fig. 86. Change of mole ratio x_{COCl_2} solubility of carbonyl chloride with temperature $t°C$ for $p_{COCl_2} = 1$ atm. The liquids are: (1) xylene; (2) chlorobenzene; (3) 1,2-dichloroethane.

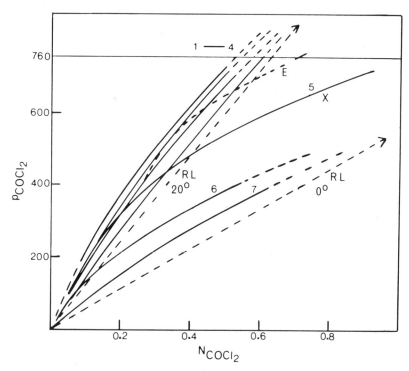

Fig. 87. Solubility of carbonyl chloride. Mole fraction N_{COCl_2} vs pressure p_{COCl_2} plots on the R-line diagram. For 20°C, the liquids are, from left to right: (1–4) xylene, 1,2-dichloroethane, carbon tetrachloride (the broken line E is from the erroneous data; see text), and tetrachloroethane. Line 5 (X) is the mole ratio x_{COCl_2} plot corresponding to line 1 (xylene). For 0°C, the liquids are: (6) 1,2-dichloroethane; (7) xylene (chlorobenzene closely follows line 7). Pressures p_{COCl_2} are in mm Hg.

(1921)[15] for CCl_4 ($N_{COCl_2} = 0.301$), C_6H_6 ($N_{COCl_2} = 0.439$), and $C_6H_5CH_3$ ($N_{COCl_2} = 0.382$) at 20–21°C are seen to be much lower than those given in Fig. 87. The value for "paraffin oil" was stated to be zero,[15] but "kerosene" (59.44 g), bp 180–280°C, d_4^{20} 0.821, absorbed 40.56 g $COCl_2$ at $p_{COCl_2} = 757$ mm Hg.[206]

9

Hydrogen Halides HCl, HBr, HI

9.1. Acid–Base Function

9.1.1. Significance of the Term Function

In this monograph, the concept of acid–base function is based on the inevitable change in *intra*molecular electron density pattern as molecules approach each other, a change giving rise to *inter*molecular forces. It is appreciated that there is a wide range of intensities, but I contend that any attempt to draw a line of demarcation between what certain persons call physical as distinct from chemical interactions is not merely arbitrary but unnecessary and undesirable. Descriptions involving such a distinction carry *ad hoc* reservations and excuses which tend to cancel out on passing from one specialist to another.

The charge-transfer theory of Mulliken (1952)[268] is a quantum-mechanical explanation of electron donor–electron acceptor *function*. In a conference on the properties of liquids (Paris, 1964), I contributed to the discussion on Mulliken's paper[269] by expressing my opinion that it is helpful to look upon electron acceptor or electron donor function as a property of a particular part of a molecule. At one part of a molecule there may be acceptor function, and at another part of the same molecule there may be donor function. A donor part can donate to the acceptor part of another molecule, or to the acceptor part of the same molecule. In his reply, Mulliken emphasized that one should speak rather of electron donor or acceptor *functioning* than of donors and acceptors. I can find no unequivocal argument in the literature against my expression acid–base function in place of electron acceptor (acid)–electron donor (base) function. In a comment on Mulliken's exposition,[268] H. C. Brown referred to the utmost importance of the widespread phenomena relating to acid–base function and pressed for simplicity of expression in order to ensure ready and clear communication of such ideas among all chemists.

There does not appear to be any universally valid quantitative order of intensity of acid–base function.

9.1.2. How Is the Function Diagnosed and Measured?

When it all comes down to details, we are confronted with the same fundamental problem which overshadows the investigation of intermolecular forces. The mass of publications on so-called "acids" and "bases" defies any *generalization* less general than the one I have adopted. Reviews of reviews tend to get further and further away from the reality of the original experimental data and do no more than add to the confusion in terminology. The outstanding causes of confusion result from the emphasis on aqueous systems, and extremely "dilute" ones at that, and, above all, the tendency to look upon the formation of ions as an essential and predominant feature of acid–base function.

With the hydrogen halides, we have acid–base function which I deem to be essentially hydrogen bonding; but there is always a probability (small in many examples) of a so-called "proton transfer" resulting in the formation of ions, which nevertheless still entail hydrogen bonding (Fig. 88). Chemists are brought up on the notion that there are two types of acid, the *"proton"* acids ("proton" donors) and *Lewis* acids (electron acceptors). The "proton" acids came through from the early days, being associated with a sour taste (from the Latin *acidus, acere*, be sour), the evolution of hydrogen by interaction with a metal such as zinc, and with changes in color of so-called indicators. On the other hand, the recognition of the other type of acid, Lewis acids, emerged from the electron pair theory of valence and may be timed about 1916.

The so-called "proton" acids involve a hydrogen atom at the acid site, whereas the so-called Lewis acids are deemed to have acid function because of a tendency to

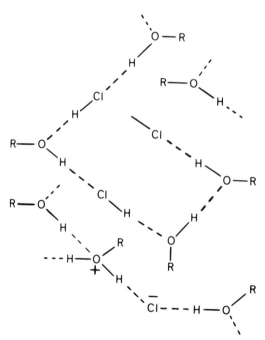

Fig. 88. An indication of the modes of reversible interaction of hydrogen chloride with an alcohol such as *n*-butanol. It is understood that the system is three dimensional, and that a condition of dynamic equilibrium prevails.

accept electrons at a site, by implication, other than at a hydrogen atom, e.g., at the boron atom in boron trichloride. Chemists write and speak of Lewis bases, and this must of necessity imply that there are other kinds of bases.

The special terminology associated with the use of indicators of hydrogen ion concentration, pH and the camp followers pK_a and pK_b, referring in the main to very dilute aqueous solutions, will probably never be ousted from its commercially vested position. This does not matter at all, *provided* we approach the analytical significance of pH in the right way. But I believe that instruction in acid–base function should never now be *initiated* from the properties of dilute aqueous solutions. The use of the terms acid and base to designate classes of compounds should be discontinued; the colloquial convenience is enjoyed at too great a price.

Many compounds clearly have both acidic and basic function, e.g., acetic acid, alcohols, and acetylenes. The careless and indiscriminate use of the expression "strong" and "weak" should be ridiculed out of practice; in the quickly expanding area of nonaqueous systems, the use of the expressions strong acids, weak acids, strong bases, and weak bases leads to muddle.

All references to the "free proton" of hydrogen ion should be avoided. The Lowry–Brönsted theory should be relegated to a historical survey unconnected with the teaching of first principles. In this connection, I recommend a detailed scrutiny of G. N. Lewis' paper of 1938,[239] coming as it does after the establishment of the quantum-wave mechanical theory of valence. In that paper he was able to look back at his book of 1923.[238] He pointed out that in aqueous solutions only a small part of the range of "acid and basic strengths" can be studied. His assessment of Brönsted's conclusions should be noticed.

To refer to H^+ as a general functional entity can only mean that any source will do, a conclusion which is out of harmony with reality.[96,303] Energy requirements *prevent* the formation of the free proton in ordinary systems. Homolytic fission giving the hydrogen atom is much less energy demanding.

Detailed work by Davis and Hetzer (1951–1958)[62] on acid–base systems in nonaqueous liquids such as benzene constitutes an attempt to clarify the position relative to the Brönsted–Lowry and the Lewis concepts of acidity. Even in this work, however, liquids such as benzene were described as "inert" because they do not give rise to "significant quantities of free hydrogen ions." It was asserted that in such solvents the product of the interaction of a base with a Brönsted acid and the product of the interaction of that base with a Lewis acid are essentially of the same type, a "highly polar" addition compound. For organic bases in liquids such as benzene, these authors urged the adoption of *association* rather than dissociation constants: A + B $\rightleftharpoons$ C. They referred to spectroscopic evidence showing that the anion is not completely detached from the proton of the cation. In the system

$$Et_3N + HCl \rightleftharpoons Et_3NH^+ \cdots Cl^-$$

the proton was seen as forming a bridge between the nitrogen and chlorine. What other can this be but hydrogen bonding? These authors mentioned conductance measurements which showed that Et$_3$NHCl and other "Brönsted salts" are extremely weak electrolytes in "inert" solvents. Based on conductance data for the picrates of aniline, dimethylaniline, and pyridine, Witschonke and Kraus (1947)[382] concluded that the dissociation constant constitutes a quantitative measure of the relative strengths of acids or bases in nonaqueous "neutral" solvents. Davis and Hetzer, however, deemed it more logical to express the equilibria in terms of an *association* constant and went on so far as to assert that had such a practice been previously adopted, the analogy between the Brönsted and the Lewis concepts of an acid would by that time have become clearly recognized.

9.1.3. Hydrogen Bonding as a Fundamental Acid–Base Function

The several detailed infrared studies of hydrogen bonding in alcohols and phenols are exemplified by the work of Stuart and Sutherland (1956),[349] Lippincott and Schroeder (1955, 1956),[243] Jakobsen and Brasch (1965),[187] and Lake and Thompson (1966).[223] Investigators write about monomers, dimers, polymers, proton donors, proton acceptors, and free hydroxyl, but without reference to the formation of ions.

Basila (1961)[14] accepted the view that hydrogen bonding has been classified as a special case of general charge-transfer interaction, and it was argued that in this aspect the strength of series of electron donor molecules with respect to their interaction with a given electron acceptor should be related to the ionization potential of the donors. He gave data relating to his experiments on the "interactions" of the surface hydroxyl groups on amorphous silica with two sets of "adsorbents," methylbenzenes and chloromethanes. The shift of the strong sharp band at 3743 cm^{-1}, previously assigned to the fundamental OH stretching vibration, was observed when absorbate was added. The figure given[14] showed the plot of $\Delta \nu$ and ionization potential. Donor electrons were "probably the chlorine lone pair."

Here we see CCl$_4$, so often referred to as an *inert* solvent in acid–base studies, being put forward as having basic function ($\Delta \nu = 40$ cm^{-1}). Furthermore, a change in ionization potential of $11.47 - 11.28 = 0.19$ eV (about $0.2 \times 23 = 4.6$ kcal in a matter of $11.5 \times 23 = 264$ kcal) is "correlated" with a change of $106 - 40 = 66$ cm^{-1} (CCl$_4$ to CH$_3$Cl). The I.P. changes from 9.25 to 8.39 (19.8 kcal in some 210 kcal) in passing from benzene to mesitylene (change in $\Delta \nu$ of 106 to 156 cm^{-1}). It was found that the correlation between the ionization potential and $\Delta \nu_{OH}$ is not the same for the benzene series as for the halogen series; i.e., the data do not fall on a single plot. This effect was attributed to the different functions of the π-electrons of the benzene series on the one hand, and "probably" the chlorine lone pairs on the other hand.

The usual observation is that the NMR spectrum of a "hydrogen-bonded proton" is shifted downfield with respect to that of a "nonhydrogen-bonded proton." Berkeley and Hanna (1963, 1964)[22] refer to this as a "hydrogen-bond shift." They use the terms "proton donor," "proton acceptor," and the unqualified terms "donor" and "acceptor" and so unnecessarily introduce confusion with the terms "electron acceptor" (acid) (but called donor) and "electron donor" (base) (but called acceptor). In one approach, one compound (hydroxylic) acts as electron acceptor and electron donor, and the shift of the hydroxyl proton is observed as the compound is mixed with an increasing amount of a so-called *inert* solvent; the other approach involves the use of a binary mixture of the electron-acceptor compound with an increasing amount of the electron-donor compound.

Berkeley and Hanna gave data for one acid function ($HCCl_3$) and four basic function (C_5H_5N, CH_3CN, etc.) sites. The quantity σ_{AD} is the chemical shift of the *pure complex*, and σ_D is that of the pure proton donor:

$$\Delta_{AD} = \sigma_{AD} - \sigma_D$$

They pointed out that it would be unsound to attempt to correlate these NMR shifts in any more than a general way with the functions of the hydrogen bonds formed in these systems.

Kreglewski (1965)[215] considered the influence of donor–acceptor interactions on the free energy of mixing of chloroform and acetonitrile at 40°C. Based on his own and other workers' papers (see original paper for reference citations), the following comments are made. To represent the energies involved, the following symbols were used: g_{AA}^E refers to the self donor–acceptor interactions, g_{AB}^E refers to interaction between sites in two different molecules, and g_{NS}^E refers to nonspecific interactions: London forces, dipole–dipole, and dipole–induced dipole interactions. It appears that the AA and NS contributions are not strictly additive and in various solvents will vary in an independent way. There appears to be no theory on the interdependence of the AA, AB, and NS effects.

From spectroscopic evidence Schleyer and West (1959)[326] concluded that the relative electron-donor ability of the alkyl halides decreases in the order RI > RBr > RCl > RF, and that electronegativity is not a dominant factor. In their table, di-*n*-butyl sulfide appears to have as strong an electron-donor function as di-*n*-butyl ether in relation to the electron-acceptor function of methanol, and nearly as strong in relation to the electron-acceptor function of phenol. At one time, it was believed that sulfur in sulfides and mercaptans could not take part in the formation of hydrogen bonds. The example given by Zellhoefer *et al.*[392] was diphenyl sulfide, but this is an unfortunate example to give since the electron-donating function of even diphenyl ether is weak.

Hite, Smissman, and West (1960)[172] and West and Kraihanzel (1961)[372] referred to the acetylenes as proton donors in hydrogen-bond formation. They measured the shifts of the acetylene C—H band, R—C≡C—H, upon hydrogen bonding to the reference bases: N, N-dimethylacetamide, N,N-dimethylformamide, and 1,2-dimethoxyethane. The system may be depicted as R—C≡C—H···N≡, and proton transfer giving rise to ions does not appear to be involved. The shifts for R = n-C_4H_9, n-C_5H_{11}, and n-C_6H_{13} in R—C≡C—H are about the same, but the shift for R = C_6H_5 is decidedly larger than the average. When R = $BrCH_2$ and $ClCH_2$, the shift is even larger, showing the usual effect of the electron-attracting function of the halogens in increasing the intensity of the acid function sited on the acetylenic terminal hydrogen. By comparison, when R = $CO_2C_2H_5$ the shift for the dimethylacetamide system is still much larger, $\Delta\nu = 123$ cm^{-1}; when R = CN, for the 1,2-dimethoxyethane system the shift is further increased, $\Delta\nu = 153$ cm^{-1}.

The relative intensity of the *basic* function of the acetylenes was similarly studied by measuring the shifts of the O–H stretching band of phenol upon hydrogen bonding to the acetylenes.

The data show that the proton-donating power of an acetylene is not a fixed value. For n-$C_4H_9C{\equiv}CH$ the shifts are $\Delta\nu = 74$ cm^{-1}(DMAc), $\Delta\nu = 61$ cm^{-1} (DMF), and $\Delta\nu = 36$ cm^{-1} (DM ether of glycol)—hence the need always to consider acid–base *functioning*.

Aksnes and Gramstad (1960)[7] concluded that the P=O group in organophosphorus compounds forms very strong hydrogen bonds with phenol. Gramstad (1961)[141] used other acid-function compounds and suggested the formation of hydrogen bonds between the "proton donor" and the "solvent": —O—H···Cl—CCl$_3$.

Weimer and Prausnitz (1966)[369] referred to the electron-accepting function of many aromatic molecules containing "strongly polar" groups such as nitro, cyano, and carbonyl groups. They believed that aliphatic compounds containing these groups can also act as electron acceptors. They measured the ultraviolet spectra for mixtures of p-xylene, the electron donor, and each of 13 "polar organic solvents," which were deemed to act as electron acceptors. These were acetone, cyclohexanone, triethyl phosphate, methoxyacetone, cyclopentanone, butyrolactone, N-methylpyrrolidone, propionitrile, nitromethane, nitroethane, citraconic anhydride, 2-nitropropane, and 2-nitro-2-methylpropane.

9.2. Solubility of Hydrogen Halides in Nonaqueous Liquids

9.2.1. Significance of Data

An outstanding point which emerged from my early work on the interaction of hydroxy compounds and inorganic nonmetal halides, such as thionyl and phosphorus halides, was the vital function of the hydrogen halide formed in the reaction pattern. The interaction of hydrogen halides with alcohols, esters, and ethers has also operational importance.

Of particular significance is the reactivity of the carbon–oxygen bond, e.g.,

$$\overset{\displaystyle\diagdown\diagup}{\underset{\displaystyle\diagup}{-C}}-\ddot{\overset{\displaystyle}{O}}{\diagup}$$

which can vary considerably according to the atoms or groups attached to the carbon atom and the oxygen atom. The influence of the rest of the molecule on this particular bond can be due to the so-called electronic (polarizability or polarization) effects, or due to a steric hindrance caused by bulky groups which either hinder the fruitful approach of the reactants or hinder the bonds attached to the C—O center from reaching the necessary disposition in space. Indeed, all these effects can operate simultaneously. In any particular molecule, the breaking of the C—O bond appears to depend upon the withdrawal of electron density from oxygen by an external molecule having electron acceptor (acid) function.

The interaction of hydrogen chloride with an alcohol such as *n*-butanol, which I have always looked upon as a convenient standard for reference, appears to entail a reversible interaction followed by an irreversible one resulting in the formation of an alkyl chloride:

$$\begin{bmatrix} R \\ \diagdown \\ O \\ \diagup \\ H \end{bmatrix} + HCl \rightleftharpoons \begin{matrix} R \\ \diagdown \\ O \cdots H-Cl \\ \diagup \\ H \end{matrix} \rightleftharpoons \begin{matrix} R \\ \diagdown \\ \overset{+}{O}-H-\overset{-}{Cl} \\ \diagup \\ H \end{matrix}$$

(Hydrogen bonded)

$$\rightarrow ROH + [[H_2O]_n \cdots HCl] + R-Cl$$

By some mechanism or mechanisms, Cl becomes detached from hydrogen and attached to the carbon atom by a severance of the C—O bond. Little or no success has attended attempts to determine the mechanism by kinetic studies on these systems. The last step is not kinetically reversible. The "water" formed competes with the alcohol for the hydrogen chloride, and therefore the formation of alkyl chloride virtually stops before all the alcohol has been converted into alkyl chloride. This is not because of the reversibility wrongly depicted in the textbooks as

$$ROH + HCl \rightleftharpoons RCl + H_2O$$

It is because there is insufficient hydrogen chloride to keep the alcohol in the necessary hydrogen-bonding condition. One mole of *n*-butanol will quickly absorb about 1 mole of hydrogen chloride at 0°C, but the *maximum* (by no means 100%) production of *n*-chlorobutane is achieved only at the end of several months. At −30°C, 1 mole of *n*-butanol will absorb about 1.5 moles of hydrogen chloride at 1 atm, and when the tube is sealed and allowed to stand at room temperature, the production of alkyl chloride is quicker and reaches almost the 100% amount. Now for an alcohol such as 2,2,2-trichloroethanol, the solubility of hydrogen chloride is very small at 0°C. Consequently, the initial concentration of ROH_2Cl is very small, and one would expect the instantaneous rate to be small, in contrast to that of the *n*-butanol system. But what of the specific rate? For details of such systems, see the papers by Gerrard and colleagues.[56, 57, 93, 95, 101, 102, 107, 119, 120, 122]

In the phosphorus trichloride system, the sequence of reactions may be depicted as follows:

$$\text{ROH} + \text{PCl}_3 \xrightarrow{\text{Very quick}} \text{ROPCl}_2 + \text{HCl}$$

$$\text{ROH} + \text{ROPCl}_2 \xrightarrow{\text{Quick}} (\text{RO})_2\text{PCl} + \text{HCl}$$

$$\text{ROH} + (\text{RO})_2\text{PCl} \xrightarrow[\text{the second step}]{\text{Quick, but slower than}} (\text{RO})_3\text{P}$$

$$(\text{RO})_3\text{P} + \text{HCl} \xrightarrow{\text{Quick}} (\text{RO})_2\text{P(O)H} + \text{RCl}$$

where R is an alkyl group such as *n*-butyl.

The sequence of events is much more complicated than shown and depends on the order and rate of mixing of the reagents. Association of the liberated hydrogen chloride with alcohol will reduce electron density on oxygen, hence the probability of the four-center approach, but the quantitative effect will depend on the nature of R and on experimental conditions such as temperature. The main intervention of hydrogen chloride is the formation of alkyl chloride. See Gerrard and Hudson[102] for details and complete literature citation.

In the silicon tetrachloride system, hydrogen chloride has a profound effect in reducing the rate of the last alkoxylation step[103, 104, 121]:

$$\text{ROH} + \text{SiCl}_4 \xrightarrow{\text{Very quick}} \text{ROSiCl}_3 + \text{HCl}$$

$$\text{ROH} + \text{ROSiCl}_3 \xrightarrow{\text{Quick}} (\text{RO})_2\text{SiCl}_2 + \text{HCl}$$

$$\text{ROH} + (\text{RO})_2\text{SiCl}_2 \xrightarrow{\text{Slower}} (\text{RO})_3\text{SiCl} + \text{HCl}$$

$$\begin{matrix} \text{H} \\ / \\ \text{R}-\text{O} \\ \vdots \\ \text{HCl} \end{matrix} + (\text{RO})_3\text{SiCl} \xrightarrow[\text{slow}]{\text{Extremely}} (\text{RO})_4\text{Si} + 2\text{HCl}$$

Now the liberated hydrogen chloride can become involved in a rather remarkable way, specifically related to its solubility in the reactants and products of each step. Supposing the *n*-butanol is added dropwise *to* the silicon tetrachloride, the formation of the monoalkoxytrichlorosilane is quick:

$$\text{ROH} + \text{SiCl}_4 \rightarrow \text{ROSiCl}_3 + \text{HCl}$$

and as the solubility of hydrogen chloride in the trichlorosilane is low, the hydrogen chloride will tend to leave the system as gas. Indeed, shaking is a considerable help, for it encourages the loss of hydrogen chloride and, by virtue of the heat of evaporation, keeps the temperature of the reaction liquid down. However, as the chlorine atoms are replaced by alkoxyl groups, hydrogen chloride is more likely to be retained, for the solubility in the trialkoxychlorosilane is by no means low. Therefore, during the later stages of addition of n-butanol, the hydrogen-bonded structure can accumulate because the interaction of this with the monochlorosilane is very slow. This effect is even more definite when the silicon tetrachloride is *added to* the alcohol, for there is now a greater chance for the hydrogen chloride to remain in the system.

The thionyl chloride system[93, 95] is again different and may be approximately depicted for n-butanol as follows:

$$ROH + SOCl_2 \rightarrow ROSOCl + HCl$$

If thionyl chloride be added to the alcohol by a properly controlled procedure, hydrogen chloride will be retained as (ROH_2Cl), which does not react with ROSOCl. Furthermore, the dialkyl sulfite $(RO)_2SO$ immediately reacts with HCl to give

$$ROSOCl + (ROH_2Cl \rightleftharpoons ROH_2^+Cl^-)$$

and so the final conditions for $(SOCl_2 + 2ROH)$ will essentially be

$$ROSOCl + (ROH_2Cl \rightleftharpoons ROH_2^+Cl^-)$$

Addition of ROH (1 mole) *to* $SOCl_2$ (1 mole) will tend to give ROSOCl + HCl as the final products. The formation of alkyl chloride by decomposition of the chlorosulfite ROSOCl involves a complicated sequence of events and is very sensitive to certain catalysts, of which HCl may be one. If the R group is such that the C—O bond is much more "reactive" than in n-butanol (e.g., as in t-BuOH), then the formation of RCl may mainly involve a mechanism other than the decomposition of the chlorosulfite.

The boron trichloride system is even more different. See Gerrard[94] for details and complete literature citation. The mode of intervention of hydrogen chloride can vary considerably not only according to the nature of R in ROH, but also according to the nature of the inorganic "center," e.g., S, P, P=O, Si, Ti, B, etc. and the experimental conditions. In the systems involving hydrogen bromide and hydrogen iodide, the sequence of reactions is much more obscure, mainly because of the stronger nucleophilic function of Br^- and still stronger function of I^-.[23]

Whenever hydrogen halide is evolved, and whenever hydrogen halide is used as a reactant or as a "catalyst," the factor relating to the "solubility" of the hydrogen halide is of considerable significance.

9.2.2. Solubility Data

Most of the solubility data for organic liquids and nonaqueous inorganic liquids are due to Gerrard and colleagues.[1, 48, 56, 89, 95, 97, 98, 108–113, 116–118] Tables 13A–D and 14 show representative x_{HCl} values for 1 atm. Later, values for hydrogen bromide and hydrogen iodide (see Table 15) were included, and measurements at different pressures were also made. Fernandes and Sharma (1965)[80] gave data for HCl in normal alcohols with 12–18 carbon atoms.

Table 13A. Solubility Data for HCl in Some Organic Liquids at 10°C, 1 atm

Liquids	x_{HCl}	Liquids	x_{HCl}
Methanol	0.857	2-Bromoethanol	0.490
Ethanol	0.950	2,3-Dibromopropan-1-ol	0.304
Propan-1-ol	0.956	1,3-Dibromopropan-2-ol	0.122
Propan-2-ol	1.030	1-Chloropropan-2-ol	0.553
Butan-1-ol	0.964	2,2,2-Trifluoroethanol	0.063
Butan-2-ol	1.048	Formic acid	0.078
2-Methyl-propan-1-ol (isobutyl alcohol)	0.973	Acetic acid	0.195
Hexan-1-ol	0.970	Propionic acid	0.155
Heptan-1-ol	0.973	n-Butyric acid	0.204
Heptan-4-ol	1.053	Isobutyric acid	0.165
Octan-1-ol	0.975	Isovaleric acid	0.176
Octan-2-ol	1.06/15°C	Monochloroacetic acid	0.029/50°C
Decan-1-ol	0.977	Trichloracetic acid	0.030/50°C
4-Methylpentan-2-ol	1.052	Ethyl formate	0.387
Cyclohexanol	1.030	Methyl acetate	0.612
3,4,5-Trimethylhexan-1-ol	0.979	Ethyl acetate	0.659
Pentan-1-ol	0.967	n-Propyl acetate	0.665
Pentan-3-ol	1.056	n-Butyl acetate	0.674
2,2-Dimethyl-propan-1-ol (neopentyl alcohol)	0.93	Isopropyl acetate	0.723
2-Phenylethanol	0.831	n-Pentyl acetate	0.678
Benzyl alcohol	0.812	Neopentyl acetate	0.930
Allyl alcohol	0.785	Isobutyl acetate	0.712
2-Methylallyl alcohol	1.26	sec-Butyl acetate	0.730
3-Hepten-1-ol	0.952	n-Octyl acetate	0.682
3-Hexen-1-ol	0.940	Ethyl propionate	0.645
2-Propen-1-ol	0.785	Neopentyl propionate	1.08
3-Buten-1-ol	0.877	Ethyl n-butyrate	0.657
3-Penten-1-ol	0.933	Neopentyl n-butyrate	1.05
4-Penten-1-ol	0.935	Ethyl chloroformate	0.118
1-Butyn-3-ol	0.961	n-Propyl chloroformate	0.138
2-Propyn-1-ol	0.347	n-Butyl chloroformate	0.140
3-Butyn-1-ol	0.644	n-Hexyl chloroformate	0.157
2-Methylcyclohexanol	1.044	Phenyl acetate	0.325
Cinnamyl alcohol	1.332	Ethyl chloroacetate	0.273
Hydrocinnamyl alcohol (3-Phenyl propanol)	0.911	Ethyl dichloroacetate	0.167
		2,2,2-Trichloroethylacetate	0.252
2-Chloroethanol	0.500	Allyl acetate	0.552
2,2,2-Trichloroethanol	0.087	Allyl propionate	0.565
		Allyl butyrate	0.573

Table 13A (contd.)

Liquids	x_{HCl}	Liquids	x_{HCl}
Crotyl acetate	0.668	Neopentyl trichloroacetate	0.559
Ethyl crotonate	0.699	Neopentyl trimethylacetate	1.02
Ethyl cyanoacetate	0.304	Ethyl benzoate	0.450
Ethyl phenylacetate	0.492	Methyl benzoate	0.445
Ethyl bromoacetate	0.293	Benzyl benzoate	0.450
Ethyl trichloroacetate	0.109	Phenyl acetate	0.325
Neopentyl chloroacetate	0.622	Ethyl lactate	0.9

Ethers, 10°C

Liquids	x_{HCl}	Liquids	x_{HCl}
Diethyl	0.920	Methyl ethyl	2.93 (−40°C)
Di-n-propyl	0.930		
Di-isopropyl	0.978	Methyl n-propyl	0.896
Di-n-butyl	0.850	Methyl n-butyl	0.907
Di-n-pentyl	0.895	Methyl n-pentyl	0.910
Di-n-hexyl	0.890	Methyl n-hexyl	0.900
Di-isopentyl	0.808	Methyl n-heptyl	0.895
n-Butyl methyl	0.788	Methyl n-octyl	0.905
Di-n-heptyl	0.905	Ethyl n-butyl	0.965
1,1′-Dichloromethyl ether	0.048	n-Propyl n-butyl	1.020
2,2′-Dichloroethyl ether	0.220	n-Pentyl n-hexyl	0.890
3,3′-Dichloropropyl ether	0.274	n-Heptyl n-octyl	0.900
1,2-Dichloroethyl ether	0.220	Di-isopentyl	0.80
Di-n-octyl	0.878	Ethyl s-butyl	1.05
Anisole (PhOMe)	0.160	Benzyl n-butyl ether	0.680
Phenetole (PhOEt)	0.175	1,4-Dioxane	1.031
Di-phenyl ether	0.089	Tetrahydrofuran	1.38
Benzyl methyl ether	0.720	Tetrahydropyran	1.05
Benzyl ethyl ether	0.726	2-Ethoxyethanol	1.25
Di-benzyl ether	0.540	2-Butoxyethanol	1.21
Methyl o-tolyl ether	0.129		

Phenols, 25°C

Liquids	x_{HCl}	Liquids	x_{HCl}
p-Chlorophenol	0.045	Phenol	0.055
o-Cresol	0.085	p-Ethyl phenol	0.105
p-Cresol	0.105	m-Cresol	0.055
o-Bromophenol	0.060	o-Ethylphenyl	0.087
p-Bromophenol	0.050	o-Chlorophenol	0.055

Table 13A (contd.)

Liquids	x_{HCl}	Liquids	x_{HCl}
Hydrocarbons, 20°C		**Aryl and alkyl halides, 10°C**	
Benzene	0.058	Chlorobenzene	0.046
Toluene	0.072	Benzotrichloride	0.062
m-Xylene	0.080	Bromobenzene	0.043
o-Xylene	0.075	1-Bromohexane	0.070
p-Xylene	0.072	1-Chlorohexane	0.073
n-Decane	0.043 (10°C)	1-Bromobutane	0.065
n-Octane	0.03	1-Bromooctane	0.082
Cyclohexane	0.0156	Carbon tetrachloride	0.021
		Ethylene dichloride	0.048/20°C
Nitrobenzene	0.085	Chloroform	0.019
Esters of inorganic acids, 10°C			
Carbonates			
Diethyl	0.523	Di-isobutyl	0.587
Di-n-butyl	0.555	Ethyl 2-chloroethyl	0.378
Borates			
Trimethyl	0.37	n-Butyl o-phenylene	0.174
Tri-n-butyl	0.355	n-Pentyl o-phenylene	0.177
Tri-n-pentyl borate	0.361	Trichloroethyl	0.047
ClCH$_2$CH$_2$OB(Cl)Ph	0.04	n-BuOBCl$_2$	0.0512
Methyl o-phenylene	0.129	n-Butyldimethylene	0.19
Ethyl o-phenylene	0.160	Methoxyboroxole	0.46
n-Propyl o-phenylene	0.168		
Silanes			
Tetramethoxysilane	1.725	Monochlorotri-n-butoxysilane	0.36
Tetraethoxysilane	2.1		
Tetra-n-propoxysilane	1.7	Dichlorodi-sec-butoxysilane	0.12
Tetra-n-butoxysilane	2.0		

Table 13B. Solubility Data for HCl in Sulfur Compounds of the Type XYSO$_2$

X	Y	x_{HCl} 25°C	x_{HCl} 0°C
i-Pr	i-Pr	0.712	1.174
n-Bu	n-Bu	0.627	—
n-Pr	n-Pr	0.622	1.010
—CH$_2$(CH$_2$)$_2$CH$_2$—[a]		0.402	0.736
Et	OBu-n	0.316	0.630
Me	OBu-n	0.255	0.510
p-CH$_3$·C$_6$H$_4$	OBu-n	0.249	0.445
Ph	OBu-n	0.213	0.419
p-Cl·C$_6$H$_4$	OBu-n	0.083	0.210
i-Pr	Cl	—	0.126
Et	Cl	0.056	0.098
Me	Cl	0.044	—
Ph	Cl	0.042	—
Cl	OBu-n	0.057	0.117
Cl	Cl	—	0.032
OH	OH	—	0.020

[a]
$$\begin{array}{c} CH_2-CH_2 \\ | \quad\quad\;\; \backslash \\ \quad\quad\quad SO_2 \\ | \quad\quad\;\; / \\ CH_2-CH_2 \end{array}$$

Table 13C. Solubility Data for Hydrogen Halides in Some Sulfur Compounds at 0°C

	HCl	HBr	HI
R-line value for 760 mm Hg	0.041	0.091	0.370
(n-Decane)	0.049	0.10	0.37
Thiophene	0.034	0.14	([a])
PhSH	0.093	0.18	0.39
n-BuSH	0.125	0.35	0.67
Ph$_2$S	0.144	0.23	0.50
n-Bu$_2$S	0.640	2.53	2.89

[a] Reacts further.

Table 13D. Solubility Data of Ahmed, Gerrard, and Maladkar[1] for Hydrogen Halides

Compound	°C	HCl	HBr	HI
CCl_4	0	0.025	0.055	0.27
	10	—	0.042	0.16
	20	—	0.033	0.12
$CHCl_3$	0	0.034	0.078	0.30
	10	—	0.056	0.20
	20	—	0.045	0.15
CH_2Cl_2	0	0.040	0.060	—
	10	—	0.043	—
	20	—	0.036	—
$ClCH_2CH_2Cl$	0	0.063	0.125	—
	10	0.050	—	—
	20	0.042	—	—
n-C_7H_{16}	0	0.03	0.070	—
	10	0.02	0.056	—
	20	0.015	0.047	—
n-$C_{10}H_{22}$	0	0.032	0.090	0.45
	10	0.028	0.075	0.30
	20	—	0.059	0.25
C_6H_6	0	0.070	0.150	—
	10	0.053	0.119	—
$C_6H_5CH_3$	0	0.09	0.175	0.56
	10	0.07	0.120	0.35
	20	0.05	0.090	0.23
m-$(CH_3)_2C_6H_4$	0	0.110	0.155	0.58
	10	0.085	0.125	0.42
	20	0.071	0.105	0.29
o-$(CH_3)_2C_6H_4$	0	0.103	—	0.54
	10	0.075	—	0.38
p-$(CH_3)_2C_6H_4$	0	0.111	—	0.58
	10	0.080	—	0.40
	20	0.064	—	—
C_6H_5Cl	0	—	0.100	0.39
	10	—	0.077	0.27
	20	0.0325	0.060	0.18
C_6H_5Br	0	0.053	0.115	0.440
	10	0.042	0.095	0.311
	20	0.030	0.078	0.231
C_6H_5I	0	—	0.131	0.465
	10	—	0.110	0.325
	20	—	0.094	0.240

Table 13D (contd.)

Compound	°C	HCl	HBr	HI
C₆H₅NO₂	0	0.140	0.171	—
	10	0.105	0.130	—
	20	0.085	0.115	—
o-CH₃C₆H₄NO₂	0	—	0.218	—
	10	—	0.170	—
	20	—	0.130	—
n-C₈H₁₇Cl	0	0.12	0.24	0.65
	10	—	0.20	0.50
	20	—	0.18	0.375
n-C₈H₁₇Br	0	0.105	0.164	0.68
	10	—	0.145	0.49
	20	—	0.100	0.37
n-C₈H₁₇I	0	0.13	0.25	0.73
	10	0.10	0.19	0.53
	20	0.08	0.16	0.425
H₂O	0	0.410	0.491	0.389
	10	0.379	0.467	—
	20	0.358	0.445	—
n-C₈H₁₇OH	0	1.024	1.28	1.625
	10	0.890	1.17	1.38
	20	0.810	1.09	1.20
(n-C₈H₁₇)₂O	0	1.06	2.32	2.97
	10	0.88	2.01	2.61
	20	0.685	1.53	—
n-C₈H₁₇CO₂CH₃	0	1.02	1.90	—
	10	0.89	1.35	—
	20	0.74	0.90	—
n-C₅H₁₁CO₂H	0	0.27	0.91	1.04
	10	0.20	0.80	0.78
	20	0.19	0.66	0.60
CH₃CO₂H	0	0.29	0.73	0.625
	10	0.21	0.60	0.61
	20	0.15	0.48	—
Cl₂CHCH₂OH	0	—	0.530	—
	10	—	0.435	—
	20	—	0.351	—
Cl₃CCH₂OH	0	—	0.174	—
	10	—	0.156	—
	20	—	0.146	—

Table 14. x_{HCl} Values for BCl_3, PCl_3, $SiCl_4$, $TiCl_4$, and $SnCl_4$ at 1 atm and Different Temperatures, °C

Liquid	−60°	−50°	−40°	−30°	−20°	−10°	0°	10°	20°
BCl_3		0.043	0.0345	0.0275	0.022	0.017			
PCl_3	0.24	0.13	0.09	0.063	0.048	0.033	0.027		
$SiCl_4$				0.086	0.068	0.0525	0.040	0.032	0.027
$TiCl_4$				0.105	0.080	0.064	0.054	0.047	0.040
$SnCl_4$				0.115	0.092	0.076	0.065	0.058	0.053

Table 15. x_{HX} Values for Halogenoalkanes at 760 mm Hg (Total Pressure) and 0°C

Hydrogen bromide

Ethyl bromide	0.139	n-Propyl iodide	0.161
n-Propyl bromide	0.153	1,2-Dibromoethane	0.142
n-Butyl bromide	0.180	1,3-Dibromopropane	0.170
n-Hexyl bromide	0.205	1,4-Dibromobutane	0.199
Methyl iodide	0.111	1,6-Dibromohexane	0.245

Hydrogen iodide

Methyl iodide	0.450	n-Butyl chloride	0.424
Ethyl iodide	0.540	n-Propyl bromide	0.47
n-Propyl iodide	0.573	n-Butyl bromide	0.492
n-Propyl chloride	0.378		

Table 16. R-line Data for HX

HX	bp., °C at 1 atm	$p°_{HX}$ at 0°C, atm	N_{HX} (x_{HX}) from R-line at 760 mm Hg		
			0°C	10°C	20°C
HCl	−84.9	25.46	0.0393 (0.0409)	0.0305 (0.0315)	0.02405 (0.0246)
HBr	−67.0	12.3	0.0813 (0.0885)	0.0625 (0.0667)	0.0485 (0.0510)
HI	−35.4	3.7	0.270 (0.370)	0.200 (0.250)	0.150 (0.177)

9.2.3. The R-lines

Table 16 shows the bp/1 atm, $p°_{HX}$ in atmospheres, and the N_{HX} (x_{HX}) value corresponding to the position of the R-line at the 760 mm Hg horizontal. Figure 89 shows the R-lines for 0, 10°, and 20°C, and Fig. 90 shows how the R-lines for HBr and HI for 10°C fit into the R-line pattern for a number of gases. These data relate to the tendency of the hydrogen halide to condense at the operational temperature and pressure. The R-line value (1 atm) for N_{HCl} at 0°C is about 0.04—relatively small and corresponding to the low bp of −84.9°C/1 atm. The value $N_{HBr} = 0.081$ for hydrogen bromide is not much greater in proportion, but the value of $N_{HI} = 0.270$ is about seven times that for hydrogen chloride.

The observed N_{HX} values for each gas must be considered in relation to the R-line values. In Fig. 30 I show the general disposition of the observed N_{HX} values on a common diagram. The N_{HX} values for organic liquids S *tend* to be on the right of the respective R-line, and the N_{HX} vs p_{HX} plots *tend* to be concave upward.

The operation of the three factors—tendency to condense, intermolecular structure of the liquid S, and the modes of interaction of HX with S—is illustrated briefly in Table 17. The acid function sited on hydrogen in the hydrogen halides is so incisive that it will seek out sites of basic function of even low intensity. Furthermore, the hydrogen bonding in alcohols and carboxylic acids can offer little effective opposition to the function of the hydrogen halides.

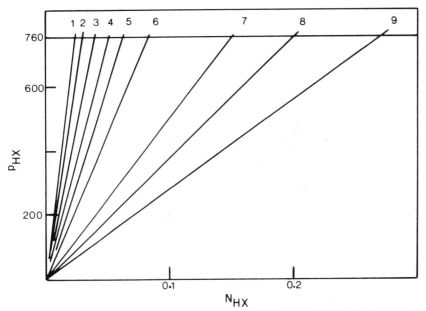

Fig. 89. Reference lines (R-lines) for the hydrogen halides. Hydrogen chloride, bp, 1 atm = −84.9°C, $p°_{HCl} = 25.46$ atm at 0°C; R-lines are: (1) 20°C; (2) 10°C; (3) 0°C. Hydrogen bromide, bp, 1 atm = −67.0°C, $p°_{HBr} = 12.3$ atm at 0°C; R-lines are: (4) 20°C; (5) 10°C; (6) 0°C. Hydrogen iodide, bp, 1 atm = −35.4°C, $p°_{HI} = 3.7$ atm at 0°C; R-lines are: (7) 20°C; (8) 10°C; (9) 0°C. The base line is divided into mole fraction N_{HX} for the reception of solubility data. The pressures p_{HX} are in mm Hg.

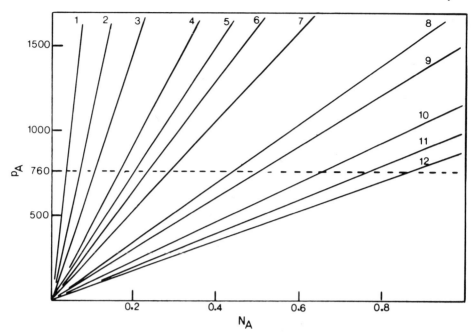

Fig. 90. Reference lines (RL) for a number of higher-bp gases A at 0 or 10°C. The line for HBr gives a tie with the hydrogen halide lines. (1) C_2H_2 (10°C); (2) HBr (10°C); (3) H_2S (0°C); (4) NH_3 (10°C); (5) HI (10°C); (6) NH_3 (0°C); (7) CH_3Cl (10°C); (8) SO_2 (10°C); (9) CH_3NH_2 (10°C); (10) SO_2 (0°C); (11) $(CH_3)_3N$ (10°C); (12) $(CH_3)_2NH$ (10°C). The base line is divided into mole fraction N_A values for the reception of solubility data. The pressures p_A are in mm Hg.

Table 17. x_{HX} Values at 0°C, 760 mm Hg

S	HCl	HBr	HI
CCl_4	0.025	0.055	0.27
$CHCl_3$	0.034	0.078	0.30
$C_6H_5CH_3$	0.090	0.175	0.56
C_6H_5Br	0.053	0.115	0.44
n-$C_8H_{17}Br$	0.105	0.164	0.68
n-$C_8H_{17}OH$	1.024	1.28	1.625
$(n$-$C_8H_{17})_2O$	1.06	2.32	2.97
n-$C_5H_{11}CO_2H$	0.27	0.91	1.04
n-$C_{10}H_{22}$	0.032	0.090	0.45
$[H_2O]_n$	0.410	0.491	0.389

9.2.4. The Essential Pattern of Data

The pattern of acid–base function of alcohols varies according to the nature of the groups attached to the "alcoholic" carbon atom,

$$\diagdown_{\diagup}\!C\!-\!O\diagdown_{H}$$

Earlier descriptions of solids such as $CH_3OH\text{–}HCl$ (formed at low temperatures, about $-90°C$) by Maass and McIntosh (1912, 1913)[250] have tended to dominate attitudes, as indicated in Mellor's treatise (1956).[256] "Oxonium" compounds are frequently referred to; Kohn (1932)[210] believed that his approximate value of $x_{HCl} = 1$ at $4°C$ for R—OH (R = Me, Et, n-Pr, n-Bu, n-Pe) would be general irrespective of the nature of R. Gerrard and colleagues have shown a remarkable correlation between the x_{HCl} values and the constitution of alcohols.

Successive replacement of hydrogen by chlorine at the β-carbon atom materially reduces the x_{HCl} value and the reactivity of the C—O bond to nucleophilic attack. The formation of the corresponding chloride, e.g., $ClCH_2CH_2Cl$, Cl_2CHCH_2Cl, Cl_3CCH_2Cl, becomes slower in the series CH_3CH_2OH, $ClCH_2CH_2OH$, Cl_2CHCH_2OH, Cl_3CCH_2OH, being extremely slow for the last member. On the other hand, the electron-acceptor function at the hydrogen site of the —O—H group increases in that order. Generalizations can be spurious, however, for although in such an alcohol as neopentanol the solubility of HCl is little different from that for n-butanol, the C—O bond is much more resistant to nucleophilic attack than in n-butanol, because this attack is normally restricted to the end-on approach—the so-called S_{N^2} approach—and this is sterically hindered. There is no such hindrance to the basic function of the alcoholic oxygen atom. The reactivity of the C—O bond cannot be simply correlated with the classification as primary, secondary, and tertiary alcohols, as is so often done. 1-Phenylethanol (a secondary alcohol) is every bit as reactive as t-butanol, whereas the tertiary alcohol chloretone, $(CH_3)_2(CCl_3)COH$, has a remarkably stable C—O linkage compared with n-butanol.

Although a double bond

$$\diagdown_{\diagup}\!C\!=\!C\diagup^{\diagdown}$$

can have a basic function itself, the solubility of HCl in 2-propen-1-ol is definitely less than in n-propanol. The value for 2-propyn-1-ol is still less. Replacing the β-hydrogen in 2-propen-1-ol by methyl appears to upset the delocalization effect and actually increases the x_{HCl} value to 1.3 (Cook and Gerrard).[56]

As the double bond is successively moved away from the O—H group, the x_{HCl} value increases:

CH₃CH₂CH₂OH   CH₂=CH—CH₂OH   CH₂=CH—CH₂CH₂OH
$x_{HCl}=0.956$ $x_{HCl}=0.785$ $x_{HCl}=0.877$

CH₃CH=CH—CH₂CH₂OH   CH₂=CH—CH₂CH₂CH₂OH
$x_{HCl}=0.935$ $x_{HCl}=0.933$

Whenever a phenyl group is attached to oxygen in a compound PhO—X, where X could be H, alkyl

$$\diagdown P \diagup \qquad \diagdown Si \diagup \qquad -B-$$

systems, etc., the electron density on the oxygen is considerably reduced from the value for the corresponding n-BuO—X.

Bartlett and Dauben (1940)[13] declared that phenols form complexes with hydrogen chloride in which the phenol hydrogen is attached to chlorine: C₆H₅OH⋯Cl—H. The alternative, and more probable, form C₆H₅(H)O⋯HCl appeared to be ignored (Fig. 91). In glycols the solubility can relate in the main to the basic function of only one oxygen atom, the other being involved in internal hydrogen bonding. Thus, in the unsymmetrically substituted glycol, the α-form is preferred, a result in accordance with infrared data.[111] The second oxygen behaves more independently (increase in solubility) as the distance between the atoms of oxygen increases.[111] See Table 18. For the esters of carboxylic acids, x_{HCl} increases with the length of the hydrocarbon chain in R′. Changing the alkyl group R exerts a smaller effect, and substitution by halogen exerts a profound effect.

Later, Jain and Rivest[186] observed a similar change in the shift of the infrared bands of the C=O and C—O stretching frequency on the formation of the complex (ester–TiCl₄), where ester = CH₃CO₂Et, CH₂ClCO₂Et, CHCl₂CO₂Et, CCl₃CO₂Et, CH₃CO₂CH₂CH₂Cl. On the basis of the x_{HCl} value, acetic acid has a more intense basic function than trichloroethanol but one much weaker than that of n-butanol. The introduction of one chlorine atom in the R group of RCO₂H causes a considerable drop in the x_{HCl} value. For the dialkyl ethers R′OR, there is some untidiness in the order of solubility at 10°C when based on the conventionally accepted order of inductive effect, and this could be due to a steric factor.

H₂C—CH₂   H—C(H)—C(H)—CH₃   H—C(H)—C(H)—CH₃
 | | | | | |
 O   O   O   O   O   O
 / \ /   / \ /   / \ / \
H   H HCl   H   H   HCl   ClH   H   H
                      α-form β-form
$x_{HCl}=0.914$ Calc. 0.996   Calc. 0.924

Found: 0.985

Kapoor, Luckcock, and Sandbach (1971)[199] have reported x_{HCl} values for a number of ethers at temperatures from about 20 to $-70°C$.

Although the cyclic ether tetrahydrofuran has $x_{HCl} = 1.38$ at 10°C and 1 atm, the dioxygen ether 1,4-dioxane has x_{HCl} no more than 1.03 at 10°C and 1 atm, as though only one oxygen site was functioning on a time-average basis. The influence of the benzene ring is seen from the following (see Gerrard[95]):

0.092   0.184

0.187   0.300

Bell (1931)[20] determined the mole fraction, N_{HCl}, values at 20°C and 1 atm for a number of liquids S, which can now be seen to have relatively small values compared with those for many other liquids (see Table 19). Bell sought a connection between the solubility of "dipole molecules" and the electrical energy of dipole fields.

I have registered his N_{HCl} values (for 20°C, 1 atm) on the 760 mm Hg horizontal in Fig. 92. Several values cluster on each side of the R-line, and those for benzene and toluene indicate that the N_{HCl} vs p_{HCl} plot would be concave upward. The value for ethyl bromide, $N_{HCl} = 0.102$, gives a line over on the right, near the PhOMe line (see later). Hamai (1935)[147] looked for a correlation between the solubility of HCl in four halogeno-carbon compounds and the polarity, as measured by the dielectric constant, molecular symmetry, electric moment, and Eötvös' constant (see Table 20).

The $C_2H_4Br_2$ was deemed to fit into the series relating to the *total bond energy*. ("Total bond energy" and "solubility"):

$$C_2H_4Cl_2 > C_2H_4Br_2 > C_2H_2Cl_4 > CCl_4$$

Williams (1936)[380] observed the infrared absorption of hydrogen chloride in C_6H_6, $C_6H_5NO_2$, and m-$NO_2C_6H_4CH_3$, labeled as "nonionizing solvents." He claimed a correlation between the shift of the 3.45-μ band of HCl and the dipole moment of the "solvent." He admitted that there was some difficulty in measuring the HCl absorption in benzene and chlorobenzene owing to the proximity of solvent bands.

Table 18. x_{HCl} for 10°C/1 atm

Liquid	x_{HCl}
CH$_2$CH$_2$(OH)(OH)	0.914
CH$_2$CH$_2$CH$_2$ with OH on C1 and C2	0.985
CH$_2$CH$_2$CH$_2$CH$_2$ with OH on C1 and C3	1.507
CH$_2$CH$_2$CH$_2$CH$_2$CH$_2$ with OH on C1 and C4	1.680

R·C(=O)OEt

	x_{HCl}
R = H	0.387
CH$_3$	0.633
C$_2$H$_5$	0.645
n-C$_3$H$_7$	0.657

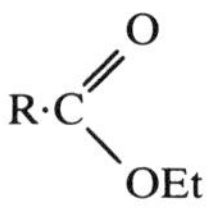

CH$_3$C(=O)OR'

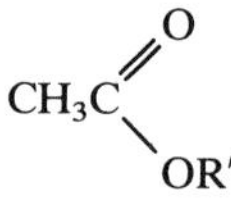	x_{HCl}
R' = H	0.195
Me	0.612
Et	0.659
n-Pr	0.665
n-Bu	0.674
n-Oct	0.682
i-Pr	0.723
s-Bu	0.712
ClCH$_2$·CO$_2$Et	0.273
Cl$_2$CH·CO$_2$Et	0.167
Cl$_3$C·CO$_2$Et	0.109
H·CO$_2$Et	0.387
Cl·CO$_2$Et	0.085
CH$_3$CO$_2$CH$_2$CH$_2$Cl	0.380
CH$_3$CO$_2$CH$_2$CCl$_3$	0.252

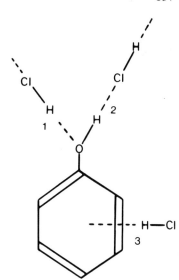

Fig. 91. The main site of attachment of HCl is probably on the oxygen.

Table 19. Solubility of Hydrogen Chloride in Various Liquids S at 20°C and $p_{HCl} = 1$ atm[a]

S	x_{HCl}	N_{HCl}
Hexane	0.0201	0.0197
Octane	0.0305	0.0296
Dodecane	0.0324	0.0314
Cetane	0.0277	0.0270
Cyclohexane	0.0156	0.0154
Carbon tetrachloride	0.0184	0.0181
Benzene	0.0444	0.0425
Toluene	0.0543	0.0507
Chloroform	0.0465	0.0444
Bromobenzene	0.0315	0.0305
Chlorobenzene	0.0325	0.0315
Benzyl chloride	0.0469	0.0448
Benzotrichloride	0.0283	0.0275
Ethyl bromide	0.1134	0.1019
Ethylene dichloride	0.0479	0.0457
"Tetrachloroethylene"	0.0166	0.0163
"Trichloroethylene"	0.0210	0.0206
Pentachloroethane	0.0219	0.0214
Ethylene bromide (given as ethyl bromide)	0.0361	0.0348
Bromoform	0.0316	0.0306
s-Tetrabromoethane	0.0242	0.0236
s-Tetrachloroethane	0.0272	0.0265

[a] The N_{HCl} values are from Bell's work; the x_{HCl} are the corresponding values.

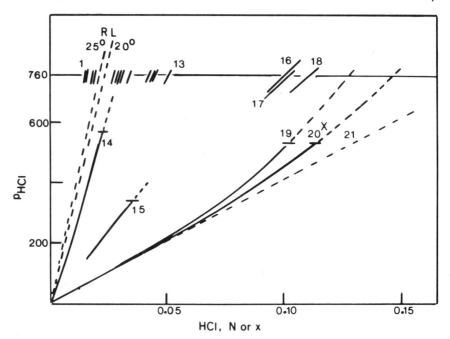

Fig. 92. Solubility of hydrogen chloride in various liquids (at 20°C, unless otherwise stated). Registration of mole fraction N_{HCl} data on the R-line diagram. Lines 1 to 13 and 17 are from Bell's data (see Table 19). Line 16 is for $C_6H_5OCH_3$, and 18 is the corresponding mole ratio x_{HCl} position. Line 15 is for $o\text{-}NO_2C_6H_4CH_3$ at 25°C, and line 14 is for C_6H_6 at 30°C from Saylor's data. Line 19 is for $(ClCH_2CH_2)_2O$, and 20 (X) is the corresponding mole ratio x_{HCl} plot; line 21 is the linear extrapolation of the tangent to the lower end of line 20 (X). The crossbars on the long full lines indicate the limits of observed data. The pressures p_{HCl} are in mm Hg.

The results were attributed to the combined effects of dipole interaction and the formation of complexes, but the nature and site of the complexing function were not discussed. Because of the low solubility of the gas, he had to use "saturated" solutions. But there was no indication of the "concentration" of HCl in these "saturated solutions"; the x_{HCl} values at 25°C are relatively small.

Gordy et al.[133-140] and O'Brien et al.[287-292] published a series of papers between 1937–1949 relating to the correlation of the solubility of hydrogen chloride and the basic function of a solvent. Gordy et al. were mainly concerned with the shift of the 3.42-μ (2890-cm^{-1}) band of HCl, and O'Brien et al. with observation of partial pressure of HCl from the relevant solutions. These papers are linked by an argument about the thermodynamic quantities ΔH and ΔS.

Without mention of the previous work of Gerrard et al., Ionin et al. (1963, 1965)[183, 184] gave data corresponding to $x_{HCl} = 0.802$ (Et$_2$O), 0.799 (i-Pr$_2$O), and 0.808 (n-Bu$_2$O) at 25°C, which were believed to indicate that "complex R$_2$O·HCl," deemed to be "quite stable," is unreactive toward calcium hydride. They expressed the solubility of HCl in alcohols (n-Bu to n-decyl) at 25°C as moles ROH per mole of HCl. Because the ratio was found to be 1:1.2, it was inferred that a complex of the

Table 20. Solubility and the Dielectric Constant and Dipole Moment

	ε 20°C	$\mu \times 10^{18}$	x_{HCl}	N_{HCl}, 20°C/1 atm
Cl$_2$CHCHCl$_2$	8.2	1.6	0.0282	0.02744
CCl$_4$	2.24	0	0.0157	0.01550
"Ethylene chloride"	10.8	1.8	0.0416	0.03993
"Ethylene bromide"	6.3	1.4	0.0356	0.03441

composition ROH$_2$Cl predominates. The solubility of HCl in ROH was deemed to have a constant value.

Howland et al.[178] measured the solubility of HCl and HBr in CCl$_4$ and CHCl$_3$, gave the "Henry's law constants," and believed the results of Bell and Hamai to be inaccurate.

O'Brien and colleagues (1939–1949) approached the study of acid–base function by comparing "Henry's law constants," and the arguments were tied up with infrared studies of other workers. The papers are worth analyzing in some detail because they afford still further pertinent examples of the drastic need to standardize the modes of presentation of solubility data and terminology. By the method of Saylor,[323] which requires many hours for the attainment of equilibrium, O'Brien and colleagues (1939) determined the solubility of hydrogen chloride for a number of liquids S at one or several temperatures and for pressures which, in certain examples, are not greater than about 100 mm Hg and in no example greater than about 540 mm Hg.

Data were given as molality, moles of HCl per 1000 g S, and Henry's law was given as $p = km$, where p is the partial pressure of HCl (i.e., p_{HCl}), m is the molality, and k is Henry's law constant. Now this is a mole ratio form and is equivalent to $p_{HCl} = k \times x_{HCl}$. I have calculated the x_{HCl} values, and I show the x_{HCl} vs p_{HCl} and N_{HCl} vs p_{HCl} plots in Fig. 93. The maximum p_{HCl} was 322 mm Hg; I show the speculative extrapolation to the horizontal at 1 atm. For ethylene glycol, it was agreed that Henry's law does not hold, but many writers would state that Henry's law holds for $N_{HCl} \to 0$. It is clear that such a statement has no useful meaning. In the same Fig. 93, I have registered the positions of the lines for benzene and nitrobenzene and indicate the upper end of the p_{HCl} range by the horizontal bars. Benzene and nitrobenzene were deemed to follow Henry's law, and the acid–base function was designated as

$$HCl + S = S \to HCl$$

Ethylene glycol was supposed to entail the reaction

$$HCl + S = SH^+ + Cl^-$$

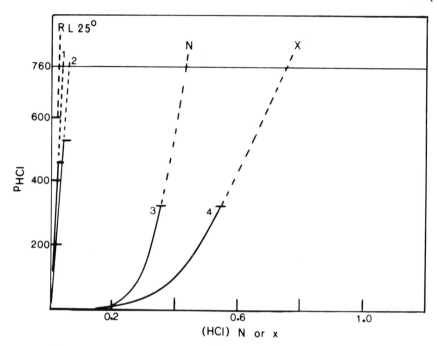

Fig. 93. Solubility of hydrogen chloride in ethylene glycol at 25°C. Mole fraction N_{HCl} vs pressure p_{HCl} and mole ratio x_{HCl} vs p_{HCl} plots on the R-line diagram. Line 3 is the N_{HCl} plot, and line 4 (X) is the corresponding x_{HCl} plot. By contrast, line 1 is the N_{HCl} plot for C_6H_6, and line 2 is that for $C_6H_5NO_2$. The crossbar at the end of each full line indicates the upper limit of the observed data. The pressures p_{HCl} are in mm Hg.

It was deemed possible to determine the extent of such acid–base reactions by assuming that "unreacted molecular hydrogen chloride" in the solutions follows Raoult's law, $f = f_0 N_u$ (f was called the fugacity of hydrogen chloride in solution, f_0 the fugacity of pure liquid hydrogen chloride, and N_u the mole fraction of unreacted hydrogen chloride). Those authors' plot of "log p vs log m" shows the futility of such procedures in presenting the essential pattern of data. Their invocation of fugacity, $f°_{HCl}$, which they calculate to be 30.3 atm at 25°C, does not appear to serve any useful purpose because their m values are given for observed *pressure* values, p_{HCl}. In another paper (1940) "positive" and "negative" deviations from Raoult's law values are deemed to be too uncertain to detect when the deviation is small because of the uncertainty over the value to take as the fugacity of pure hydrogen chloride.

Look at Fig. 92; instead of my R-line based on the observed $p°_{HCl}$ at 25°C, a "fugacity" line would be drawn, crossing the horizontal at 760 mm Hg at the point $1/30.3 = 0.033$ instead of 0.0214 on the scale of the base line. On the scale which would have to be used for a large number of organic liquids S, this substitution has no meaning. Benzene and nitrobenzene would still appear on the right of that new line. The uncertainties involved in the examples of N_{HCl} values falling near the R-line are indicated in the paper by O'Brien and Byrne (1940).[289] Do the monohalogenobenzenes show a positive or a negative deviation from Raoult's law? Bell's solubility values were taken to show a "rather large positive deviation." The shift of the 3.46-μ absorption band of hydrogen chloride shows that chlorobenzene has "a proton attracting tendency" similar to that of benzene, and therefore it was argued that the extent of the negative deviation should be about the same.

O'Brien and Byrne concluded that their results over the pressure range examined showed that Henry's law was followed, but that there was a positive deviation from Raoult's law. This observation seemed to reveal an apparent discrepancy between the solubility data and the infrared shifts in the matter of indicating the intensity of the basic function of these halobenzenes.

It was further pointed out that for these aryl halides, benzene and hexane, there was no "apparent" correlation with internal pressure or dipole moment and the deviation from Raoult's law. All these particular "deviations" are relatively very small, not rather large as stated, and compensating effects could frustrate attempts to analyze in detail the mechanisms of the dissolution process. Again the R-line diagram provides a realistic basis for evaluating the extent of the deviations so that expressions such as "rather large" may be brought into proper focus. O'Brien (1941) referred to the "large differences in solubility" among the members of the list given in Table 21, although in reality the differences are very small against the complete background of solubility pattern. O'Brien used plots of $-\log N_{HCl}$ vs $1/T \times 10^3$ (T in degrees Kelvin) for chlorobenzene, benzene, and nitrobenzene as a basis for the relationship $-\log N_{HCl} = (A/T) + B$, where A and B are constants which were identified with $\Delta H/2.3R$ and $-\Delta S/2.3R$, respectively, ΔH and ΔS being the differential heat and entropy of solution of gaseous hydrogen chloride. Hamai's[147] data for heat of solution of HCl in halogenated hydrocarbons are included, but O'Brien calculated the corresponding ΔS values. O'Brien deemed that the value of $-\Delta H = 25,540$ for CCl_4 was in error; the "correct" value was indicated to be about 15,000 J-mole^{-1}.

Table 21. The Heat and Entropy of Solution of Hydrogen Chloride

Solvent	$-\Delta H$, J	$-\Delta S$, J/deg
Chlorobenzene	16,750	86.7
Benzene	18,000	87.5
Nitrobenzene	16,750	79.1
s-Tetrachloroethane	13,810	77.0
Carbon tetrachloride	25,540	122.3[a]
Ethylene chloride	14,650	77.0
Ethylene bromide	13,400	72.8
Pentachloroethane	9,210	62.8
Trichloroethane	15,060	80.4

[a]Believed in error (see ref. 289).

O'Brien described these solutions as neither ideal nor regular, and he looked upon them as providing exceptions to Hildebrand's rules relating to the heats of solution and deviations from Raoult's law. From data available to him, O'Brien calculated N_{HCl} for diethyl ether to be 0.440 ($x_{HCl} = 0.786$) at 15°C, the ΔH being $-16{,}070$ J-mole^{-1}. I fail to see how this thermodynamic approach contributes to an understanding of the mechanisms of the dissolution process, especially when one considers the uncertainties of the extrapolations from the low p_{HCl} values as 200 or 300 mm Hg. In Fig. 92 I show the plot of p_{HCl} vs x_{HCl} for $(ClCH_2CH_2)_2O$ at 20°C from O'Brien's (1942) molality data up to his maximum p_{HCl} of 533 mm Hg. I have extrapolated this as a broken line and have also plotted the corresponding p_{HCl} vs N_{HCl} line. Now O'Brien concludes that the system follows "Henry's law" in the mole ratio form of $p_{HCl} = km$, where p_{HCl} is in atmospheres and m is molality. By the calculation $N = 1/(1 + km_S)$, where m_S is the number of moles of S in 1000 g, he obtains the value $N_{HCl} = 0.149$ for 1 atm and 20°C, and this is equivalent to $x_{HCl} = 0.175$. O'Brien took a mean value of 0.82 ± 0.03 from his range of k values which vary from 0.75 at 23.7 mm Hg to 0.88 at 533 mm Hg. A plot of x_{HCl} values, calculated from his m values, vs p_{HCl} is not straight, and an extrapolation to the 1-atm horizontal gives a value of x_{HCl} about 0.146, i.e., $N_{HCl} = 0.127$. The linear extrapolation from the actual values of x_{HCl} at the p_{HCl} values approaching zero gives a value of x_{HCl} of about 0.175, which corresponds to $N_{HCl} = 0.149$ given by O'Brien as the solubility at 1 atm. Even if one did accept this linear plot in terms of the mole ratio form of Henry's law, the corresponding N_{HCl} plot would be far from linear. For a value of N_{HCl} of 0.100 for anisole at 20°C and 1 atm, $x_{HCl} = 0.111$, and even in this example the N_{HCl} plot would not be linear. It must be clearly understood that the N_{HCl} line must of necessity reach the point $p°_{HCl} = 41.58$ atm at the right vertical axis corresponding to $N_{HCl} =$ unity.

O'Brien and King (1949) gave the N_{HCl} values at 1 atm for three more ethers but did not give the p_{HCl} and m data; therefore, the x_{HCl} vs p_{HCl} plots cannot be drawn from actual data. However, looking at Fig. 92, it appears highly probable that the p_{HCl} vs x_{HCl} plot would not be linear; and certainly the N_{HCl} plot would not be linear if it were assumed that the x_{HCl} plot were linear. For phenetole at 10°C, it is extremely unlikely that the x_{HCl} vs p_{HCl} plot would be linear, and the N_{HCl} plot would be clearly separated, as for the chloroethyl ether shown in Fig. 92. The lines for n-butyl phenyl ether would also be clearly distinctive.

O'Brien and Bobalek (1940)[290] gave m and p_{HBr} (up to 358 mm Hg) data for hydrogen bromide, benzene, toluene, and o-nitrotoluene at 25°C. I have plotted the x_{HBr} and N_{HBr} values up to the upper limits of p_{HBr} in Fig. 94 and have indicated the probable extrapolations. These authors plotted "log p vs log N" and thereby obliterated the essential pattern. To get the "theoretical line" for that plot, they took the fugacity as 18.3 atm for 25°C, but again they used pressures observed as p_{HBr} for their solubility measurements. In terms of $p_{HBr} = km$, these authors concluded that Henry's law was followed, but there was a negative deviation from Raoult's law.

Brown and Brady (1952)[35] discussed the significance of their measurements of

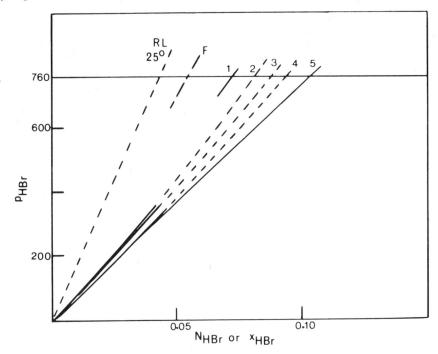

Fig. 94. Solubility of hydrogen bromide at 25°C. Registration of data on the R-line diagram. Line F indicates the position of what would be the "fugacity" line (see text). Line 1 is for the mole fraction N_{HBr} value for C_6H_6. Line 2 is the plot of N_{HBr} vs p_{HBr} for toluene, and line 3 is the corresponding mole ratio x_{HBr} plot. The full lines at the lower ends of lines 2 and 3 are from observed data; the broken lines are extrapolations to 1 atm. Line 4 is for x_{HBr} for o-$NO_2C_6H_4CH_3$, and line 5 is the linear extrapolation of the tangent to the lower end of the full line 4. The pressures p_{HBr} are in mm Hg.

the solubility of hydrogen chloride in a number of aromatic hydrocarbons at low temperatures (about −78°C). The aromatic hydrocarbon (1 mole) was dissolved in n-heptane (20 moles) to give a 5 mole % solution. They expressed the "Henry's law constant" as p_{HCl}/N_{HCl} for $N_{HCl} \to 0$ and used these constants to place the order of the strength of basic function as

$$C_6H_6 < C_6H_5CH_3 < C_6H_4(CH_3)_2 < C_6H_3(CH_3)_3 \text{ (mesitylene)}$$

The "constant" was therefore for the mole fraction form, and based on measurements for low p_{HCl} values. From the point of view of the solubility of hydrogen chloride, linear extrapolation to 1 atm would give a grossly misleading value of N_{HCl}. An extrapolation from $N_{HCl} = 0.021$ at p_{HCl} 6.27 mm Hg at −78.5°C, for example, would give the absurd value of $N_{HCl} = 2.55$ at 1 atm. Even at 20°C, it is clear from Fig. 92 that the N_{HCl} plot would not be linear.

As recorded, the data were not deemed to show any correlation with so-called polarity: despite the "greater polar properties" of chlorobenzene compared with benzene or mesitylene, the solubility of HCl is decidedly lower in the chloro compound than in the other two compounds.

Strohmeier and Echte (1957)[348] discussed the solubility of hydrogen chloride in organic compounds and the relative strength of the basic function of aromatic compounds and ethers. This is a difficult paper to analyze briefly because the authors measured p_{HCl} for solutions of 1 mmole of an ether (one of 10) in n-heptane (20 mmole), as in the procedure of Brown and Brady (1949, 1952).[35] The mole fraction N_{HCl} under these conditions was no more than about 0.055. Straight lines were obtained for the N_{HCl} vs p_{HCl} plots for diphenyl ether and anisole (both of relatively small intensity of basic function) up to p_{HCl} = 500 mm Hg, but dioxane and the simple aliphatic ethers gave curves with a decided tail-up at the higher pressures up to 500 mm Hg. The important point here is the use of the "Henry's law constant" concept. These authors took the mole fraction form of the law, $p_{HCl} = K \times N_{HCl}$. They recorded the "constant" K as that obtained by extrapolation to $N_{HCl} = 0$. The constant for carbon tetrachloride was given as 24,600, and that for tetrahydrofuran as 2100. Whatever the relationship there may be between these constants and the intensity of acid–base function, they should be used with great caution in indicating the N_{HCl} or x_{HCl} values at higher p_{HCl}; when used out of complete context, as is always a probability, they would be grossly misleading. In the examples of the di-n-butyl ether, for instance, the value of the so-called "Henry's constant" K for each of four of the observed p_{HCl} values is shown as follows:

$$K = 27/0.005 = 5400 \qquad K = 58.1/0.01 = 5810$$
$$K = 92.1/0.015 = 6140 \qquad K = 556/0.055 = 10{,}110$$

The lowest p_{HCl} recorded was 27 mm Hg, and the list value of $K = 5200$ is meant to emerge from the tangent to the curve as it closely approaches $p_{HCl} = 0$. If this value for K is reported out of *complete* context, it must of necessity mean that N_{HCl} at 760 mm Hg for $(n\text{-Bu})_2\text{O}$ is $760/5200 = 0.146$ and $x_{HCl} = 0.171$ at 0°C—widely removed from the observed value of $x_{HCl} = 1.048$ (see Table 13A). What was being implied in the paper is that no matter what the curvature really is over the observed range of p_{HCl} and N_{HCl} for each of the ethers named, essentially straight lines, $p_{HCl} = KN_{HCl}$, will form as $N_{HCl} \to 0$. There is no scientific content in this; it is simply a matter of scale and operational competency. To state that the so-called "Henry's law" "applies" to all systems as $N_A \to 0$ is stating no more than that $p_A = 0$ when $N_A = 0$.

Fontana and Herold[82] determined the solubility of hydrogen bromide in n-butane and n-hexane; for a pressure range of 0 to 35 lb/in.^{2}, the solubilities were "found to follow Henry's law quite well," and "Henry's law constants" were given in units of lb/in.^{2} mole fraction (HBr). Hence, Henry's law was taken to be the mole fraction form $p_{HBr} = KN_{HBr}$, and the mole ratio form $p_{HBr} = Kx_{HBr}$ was not mentioned. The plots of the log N_{HBr} vs $1/T \times 10^3$ for $p_{HBr} = 1$ atm were compared with that of the "ideal solubility calculated from Raoult's law."

Boedeker and Lynch[27] stated that the solubilities of hydrogen bromide in n-hexane, n-octane, and n-decane at 25, 35, and 45°C "obey Henry's law quite well." They expressed their "Henry's law constant" in units of mm HBr/mole HBr/1000 g solution. This form is, in principle, neither the N nor the x form, and although for n-hexane (mol. wt = 86 compared with 81 for HBr) the near linearity of the plot means a near linear plot for N_{HBr}, the N_{HBr} value for hexane being 0.04 at 25°C, there is a clear divergence for n-decane. Even for n-hexane, the p_{HBr} vs x_{HBr} plot is nonlinear.

Plots of moles/1000 g solution vs N_{HBr} and vs x_{HBr} for a liquid S having an assumed molecular weight of 100 show that the N_{HBr} line is not quite straight and the x_{HBr} line is decidedly curved convex upward. It was remarked that Fontana and Herold had "assumed Raoult's law to be valid," but they did *not*, in the sense implied; it was Henry's law which was assumed, and it was the "ideal solubility" which was calculated from Raoult's law.

Boedeker and Lynch gave two lines for the log (ideal N_{HBr}) vs $1/T \times 10^3$, one based on fugacity, $f/f°$, and one on pressure, $p/p°$. These workers[27] expected hydrogen bromide to exhibit its *true solubility behavior* in n-butane and n-hexane since polar effects and unsaturation are absent. This concept of "true solubility" has no scientific content. Low-pressure data for HBr and alkylbenzenes, similar to the HCl data,[35] were given by Brown and Wallace.[36]

The large solubility ($x_{HCl} = 2.35$ at 10°C) of HCl in tri-n-butyl phosphate is in accord with its description as "a strongly basic solvent" [Tuck and Diamond (1958)[359]]. Trialkyl phosphites of ordinary reactivity, e.g., $(n\text{-BuO})_3P$, quickly react at room temperature with HCl to give the dialkyl hydrogen phosphite and alkyl chloride. This reaction does not occur with triphenyl phosphite ($x_{HCl} = 0.69$ at 10°C) and has a low probability of occurrence for tris(trichloroethyl) phosphite ($x_{HCl} = 0.46$) at 10°C. The x_{HCl} value for phenol is 0.09 and that for Cl_3CCH_2OH is 0.087 at 10°C; it therefore appears that the lone pair of electrons on phosphorus in these phosphites have electron-donor function, and indeed the postulated mechanism for the quick reaction between the n-butyl phosphite and hydrogen chloride is based on this effect. The x_{HCl} value for dialkyl hydrogen phosphites of ordinary reactivity is about 2 at 10°C.

The simple tetraalkyloxysilanes have high x_{HCl} values at 10°C; for $(n\text{-BuO})_4Si$, it is 2.0; but replacement of the n-BuO group by a chlorine atom drastically lowers the value to 0.36, and on the replacement of two n-BuO groups by chlorine, the value falls still further to 0.12.

Borates provide a special feature because of the internal acid–base function, resulting in what has been called "back coordination" or "back-bonding."[94] It is understandable that the x_{HCl} value for $(CCl_3CH_2O)_3B$ is, by comparison, very low, 0.047 at 10°C.

There is a relatively small x_{HCl} value for compounds such as butyl-o-phenylene borate due to the aromatic ring, although x_{HCl} for the corresponding dimethylene borate is almost as low.[95] For n-butyl dichloroborinate, $n\text{-BuOBCl}_2$, x_{HCl} is only 0.346 even at −49°C.

9.2.5. Sulfur Compounds

Charalambous, Frazer, and Gerrard[48] determined the solubility of hydrogen chloride using, as examples, sulfones, sulfonates, and sulfonyl chlorides. The sulfones, e.g., $n\text{-}Pr_2SO$, show x_{HCl} values of about 1 at 0°C, whereas for sulfuryl chloride the x_{HCl} value corresponds almost to the R-line value, in accordance with the influence of chlorine already indicated. It will be noticed that sulfuric acid shows an x_{HCl} value of 0.02, which would appear to mean that HCl is unable to intervene in the system of hydrogen bonding in sulfuric acid itself. According to Peach and Waddington,[299] conductimetric titrations with boron trichloride in liquid hydrogen chloride show that dimethyl and diphenyl sulfones are "moderately strong bases" whereas sulfuryl chloride "has no basic properties."

	$n\text{-PrO}$ \\ S \\ $n\text{-PrO}$ / \\\\ O O	Cl \\ S \\ Cl / \\\\ O O	[HO \\ S \\ HO / \\\\ O O]$_n$
x_{HCl} at 0°C:	1.17	0.03	0.02

The x_{HCl} values for the sulfonates, $XSO_2OBu\text{-}n$, are in the order

$$X = Et > Me > p\text{-}MeC_6H_4 > Ph > p\text{-}ClC_6H_4 > Cl$$

The sulfonyl chlorides, XSO_2Cl, show the order

$$X = i\text{-}Pr > Et > Me > Ph > Cl$$

The question arises relating to the sites of basic function. Do the molecules of hydrogen chloride hydrogen-bond ($\geq$O$\cdots$HCl) exclusively on one of the atoms of oxygen, or is there a distribution, not only as between the two (S=O) oxygens, but also with regard to other potential basic-function sites? It does not follow that the more lone pairs or π sites there are, the greater the x_{HCl} value; internal competition by electronic effects, *delocalization*, could make it much more difficult for a molecule of hydrogen chloride to hold on long enough to any particular basic-function site. A sulfoxide (one oxygen atom) can have a strong basic function compared with the corresponding sulfone (two oxygen atoms).

Gerrard and Lines[106] found that the x_{HCl} values for n-butyl benzenesulfinate rise from 1.412 at 20°C to 2.630 at −17°C, the values being 1.663 at 20°C and 2.79 at −20°C for the toluene-p-sulfinate, and 1.054 at 20°C and 2.664 at −20°C for the p-chlorobenzenesulfinate. The high values for the sulf*in*ates as compared with the sulf*on*ates drew attention to a reversible fission at the R'S(O)—OR link giving R'S(O)—Cl+ROH,HCl, as happens with the dialkyl sulfites. Table 22 shows the comparison between the amount of HCl absorbed by the sulfinate, ROS(O)R', and the sum of the amounts absorbed by each of the units, ROH and ClS(O)R', and indicates the occurrence of the reversible reaction

$$ArS(O)OBu\text{-}n + 2HCl \rightleftharpoons ArS(O)Cl + n\text{-}BuOH_2Cl$$

Table 22. Comparison between Amount of HCl Absorbed by
the Sulfinate and the Sum of the Amounts Absorbed by Each of
the Units ROH and ClS(O)R'

Temperature, °C	x_{HCl} in n-BuOH	x_{HCl} in ArS(O)Cl	Found for ArS(O)OBu-n
20	0.866	0.127	1.412[b]
10	0.963	0.173	1.668[b]
0	1.065	0.225	1.987[b]
−8	1.158	0.269[a]	2.217[b]
−17	1.262	0.322[a]	2.630[b]
20	0.866	0.168	1.663[c]
10	0.963	0.226	1.927[c]
0	1.065	0.287	2.181[c]
−10	1.178	0.350[a]	2.49[c]
−20	1.303	0.423[a]	2.79[c]

[a] By extrapolation.
[b] Ar = C_6H_5.
[c] Ar = p-$CH_3C_6H_4$.

Although the x_{HCl} values for the mercaptans are much smaller than for the corresponding oxygen compounds, the basic function of sulfur is clearly indicated, and this is especially so for the dialkyl sulfides. The influence of the phenyl group is also clear. The x_{HBr} and x_{HI} values are remarkably large in the alkyl sulfides, e.g., in n-Bu_2S[89] (see Table 13C).

There has been some controversy about the electron-acceptor function of H_2S and simple mercaptans. Gordy and Stanford pointed out that Sidgwick's (1933) emphasis on the improbability of hydrogen bonding with an atom in the second period of the periodic table was not ratified by evidence subsequently observed. The usual spectroscopic approach is to observe the effects on the bonding of the group which "donates the proton."

It will be noticed that the compound providing the hydrogen for the hydrogen bonding is termed a "proton donor." The lesson to be learned here is that detection of acid–base function is a matter of delicacy of the probe used in the diagnostic procedure. Lassettre (1937)[227] stated that molecular-weight data show that SH groups do not form hydrogen bonds to "any appreciable extent," but "appreciation" will depend upon the probe.

As with alcohols, mercaptans must be considered from the aspect of the site of acid–base function; we must be concerned with the basic function at the sulfur site and the acid function at the hydrogen site (the thioethers do not have the corresponding acid-function site); but both mercaptans and thioethers can have other sites which must be considered from the aspect of acid–base function, e.g., the phenyl group. That "hydrogen sulfide and the mercaptans are not associated" is taken to mean that the linkage

$$\text{H—S}\cdots\text{H—S}$$

has not been observed. Gordy and Stanford reported a shift to lower frequency of the S—H band in certain systems illustrated by the data given here. The SH band of PhSH was shifted from 3.88 to 3.91μ by isopropyl ether, indicating the

$$Ph-S-H\cdots O=$$

hydrogen bonding. Pyridine caused a shift to 4.06μ. Apparently, n-butyl mercaptan experienced no observed shift in the presence of isopropyl ether; the shift with pyridine was 0.09μ.

Heafield, Hopkins, and Hunter (1942)[153] pointed out that there are various indications that the sulfur atom can form hydrogen bonds of the type

$$-S-H\cdots O= \quad \text{and} \quad -S-H\cdots N\equiv$$

which possess "considerable stability." It will be noticed that it is the hydrogen of the S—H group which is engaged in these examples, whereas when a hydrogen halide is involved, the hydrogen bonding may be deemed to engage the sulfur atom as follows:

$$\begin{array}{c} R \\ \diagdown \\ S\cdots H-X \\ \diagup \\ H \end{array} \qquad \begin{array}{c} R \\ \diagdown \\ S\cdots H-X \\ \diagup \\ R' \end{array}$$

Josien and her colleagues[194] reviewed the situation relating to the

$$S-H\cdots S-$$

hydrogen bonding and studied the shift of the 2592-cm^{-1} infrared band for gaseous thiophenol when the compound is in the liquid state or dissolved in CCl_4 and in a mixture of CCl_4 and other compounds (e.g., C_6H_6, C_4H_9Cl). The data were deemed to show that sulfur can form hydrogen bonds with itself, as well as with oxygen and halogen compounds, and with the benzene nucleus. A study of H_2S itself and in 19 different liquids led to the conclusion that H_2S forms "complexes moleculaires" with pyridine, ethers, and aromatic hydrocarbons.

In the purification of certain inorganic halides such as BCl_3, BBr_3, $SiCl_4$, $TiCl_4$ for technical purposes, the difficulty of removing the dissolved hydrogen halide has been referred to, but the amount of hydrogen halide believed to be involved was rarely indicated. In a patent specification,[306] it is stated that "the original silicon tetrachloride may contain between 5 and 16% by weight of hydrochloric acid." The figure "16%" corresponds to 0.88 mole of HCl per mole of $SiCl_4$. Even after treatment with tetrahydrofuran and a liquid amalgam, the HCl content was determined as 2.6% by weight, corresponding to $x_{HCl}=0.12$. The observed x_{HCl} value for $SiCl_4$ is about 0.04 at 0°C and only 0.06 at -20°C. That for BCl_3 is 0.02 at -10°C, but for $POCl_3$ the value is 0.320 at 0°C and 0.230 at 5°C, presumably due to the basic function sited on oxygen.

9.3. Aspect of Electrolytic Conductance

The review by Janz and Danyluk (1960)[189] on "Conductances of Hydrogen Halides in Anhydrous Polar Organic Solvents" is essential reading in the study of the complexity of phenomena and terminology related to the mechanisms of the process of dissolution. The general conclusion was that in the majority of organic compounds, with the exception of the lower aliphatic alcohols, the hydrogen halides are relatively poor electrolytes. The equivalent conductances for concentrations less than 0.01 N are lower than $1\,\Omega^{-1}$-cm^2. Such concentrations are less than $x_{HX} = 0.001$, and throughout the study of the review this point should be kept in mind. At 0°C and $p_{HX} = 1$ atm, the x_{HX} values are much larger than that small value.

Attempts at establishing a correlation between dielectric constant and electrolytic "strength" in these systems were described as unsuccessful. The term *phoreogram* was used to designate the plot of the equivalent conductance (Λ) vs the square root of the concentration ($C^{1/2}$). According to the following order of electrolytic "strengths," the intensity of the basic function of the alcohols is supposed to decrease as the number of carbon atoms increases:

Methanol: HX are strong electrolytes.
Ethanol: HX are moderately weak electrolytes.
n-Propanol: HX are still weaker electrolytes.
n-Butanol:⎫ HX are weak electrolytes; the phoreograms show the sharp
i-Butanol: ⎭ tail-up at low concentrations typical of weak electrolytes.

This correlation conflicts strongly with the interpretation of the x_{HX} values as shown in Tables 13A and 13D. For n-octanol, the equivalent conductance would probably be very small, and at 0°C and $p_{HX} = 1$ atm the proportion of the (about) 1 mole of HX per mole of n-octanol which gives rise to ions on a time-average basis must be very small.

According to Goldschmidt et al.,[132] Λ for a 3.91×10^{-4} N solution of hydrogen chloride in $C_nH_{2n+1}OH$ decreases as follows: 187.9 ($n = 1$); 76.9 ($n = 2$); 37.41 ($n = 3$); 20.80 ($n = 4$). n-Alcohols, CH_3OH to n-$C_8H_{17}OH$, have x_{HCl} values from about 0.9 to 1.0 at 10°C, 1 atm. It must be emphasized that dilutions of the order of 4×10^{-4} N are far removed from the much higher concentrations usually prevailing in a large field of experimental chemistry.

According to Guss and Kolthoff (1940),[143] the strong acid is completely ionized into ROH_2^+ and anions, and water is a much stronger base than the alcohols such as methanol, ethanol, and n-butanol. A scrutiny of the x_{HX} vs p_{HX} plots puts such statements in their proper focus in a remarkable way. At 0°C and $p_{HX} = \sim 5$ mm Hg, the x_{HX} value for water is larger than that for n-octanol; but at $p_{HX} = 1$ atm, the x_{XH} values for water are only about half of the values for the alcohol.

Janz and Danyluk referred to liquids such as nitrobenzene as "nonbasic," although that compound was described as polar. The "weakly acidic" properties of the nitro group were said to limit the utility of nitrobenzene and nitromethane as ionizing media for the hydrogen halides. It was also stated that hydrogen chloride and hydrogen bromide are very weak electrolytes in acetic acid.

Notice that the x_{HCl} values for 10°C, 1 atm are

$$CH_3OH, 0.857, \quad C_6H_5NO_2, 0.105, \quad CH_3COOH, 0.195$$

At 10°C, 1 mole of CH_3COOH absorbs about 0.19 mole of HCl at $p_{HCl} = 1$ atm, thus showing a definite basic function toward HCl. According to the Λ data, Λ for this strongest "solution" must be very small indeed, meaning that most of the HCl is hydrogen-bonded as

$$CH_3C(H)O_2\cdots H\text{---}Cl \quad \text{and/or} \quad CH_3C(H)O_2H^+\cdots Cl^-$$

Therefore, conductance data are not related in a simple way with acid–base function.

Hydrogen chloride is stated to be an extremely poor conductor in Et_2O (the dielectric constant is 4.33 at 20°C); Λ is less than 1 for 0.02–6.3 equivalents per liter at 18°C. In the phoreogram, Λ for 1 mole/liter appears to be (almost) zero, and Mounajed[267] believed that this was due to a combination of H^+ and Cl^- ions with undissociated hydrogen chloride.

Dorofeeva and Kudra[70] found the decomposition potential (2.16–2.18 V) from the polarographic curves, stated to be due to the discharge of "H" from un-ionized Et_2O,HCl. The potential dropped to 1 V in 12 days and was presumed to be due to formation of water and hydrated H^+ ions. The negative coefficient of conductance with temperature was attributed to a *larger dissociation of Et_2O,HCl into Et_2O + HCl at higher temperatures*. A significant point is that the change in conductance with concentration was attributed to a smaller formation of Et_2O,HCl at low concentrations of HCl and an increased ionization of that present. Based on solubility work, it may be concluded with confidence that almost every molecule of hydrogen chloride is at least hydrogen-bonded, and so the concentration of HCl, in the sense *apparently* meant by the statement given in italics, is a very small value, and what is changing is the hydrogen-bonded compound Et_2O,HCl which may be in equilibrium with some "ionic" form. The small absolute change of a small conductance ($\Lambda = 0.008$ at $C^{1/2} = 0.1$; $\Lambda =$ "0" at $C^{1/2} = 1.0$; $\Lambda = 0.025$ at $C^{1/2} = 2.0$) cannot rationally be explained by any straightforward way. Furthermore, a further lesson can be learned on the laxity which can give rise to irrational general statements in the form of overall stoichiometric equations. Mounajed attributed the temporal increase in Λ (0.0059 for 0.02 N solution changing to 0.085 Ω^{-1}-cm^2 during one month) to the reaction

$$Et_2O + 2HCl = 2EtCl + H_2O$$

Janz and Danyluk believed that the ether would simply be cleaved to chloroethane and ethanol in such "dilute" solutions:

$$Et_2O + HCl = EtCl + EtOH$$

Neither of these postulations come anywhere near the reaction patterns of the real systems. See Gerrard and Hudson (1965).[102a]

From a detailed study of infrared spectra, Seel and Sheppard (1968)[332] concluded that for the system $Me_2O + HX$ (X = Cl, Br, I) in the solid state, the adduct is formed by hydrogen bonding, $Me_2O\cdots HX$, as previously observed for the other phases, and not by proton transfer.

Maass and McIntosh[250] referred to the three compounds Et_2O,HCl (mp $-92°C$), $Et_2O,2HCl$ (mp $-88°C$), and $Et_2O,5HCl$ (mp $-89°C$). At $-89°C$ the pure hydrogen chloride was stated to have an extremely low conductance, which is rapidly increased on the addition of ether, reaching a maximum at the ratio $Et_2O:5HCl$. Maass and Danyluk (cited by Janz and Danyluk[189]) stated that a 1:1 mixture (Et_2O,HBr) at $-40°C$ had a specific conductance of $4.5 \times 10^{-3}\ \Omega^{-1}\text{-cm}^2$ in contrast to the corresponding value of $<10^{-6}\ \Omega^{-1}\text{-cm}^2$ for the pure hydrogen bromide at its mp. The mixture was therefore deemed to have "a fair degree of ionic character," attributed to the reaction

$$HX + D \rightleftharpoons DH^+ + X^- \text{ (solvated)}$$

Myers and Willett (1967)[272] declared that "hydrogen chloride molecules" are relatively undissociated in ethyl acetate "due to the low dielectric constant." (See comment about nitrobenzene, which has a "high" dielectric constant.) They refer to the greater basicity of the ester with respect to the hydrogen chloride molecule. When the "polar" ethanol is added to the hydrogen chloride–ester system, it becomes "protonated" to $C_2H_5OH_2{}^+Cl^-$, according to these workers. Note that x_{HCl} is 0.95 for C_2H_5OH and 0.66 for $CH_3CO_2C_2H_5$ at 10°C, 1 atm. From vapor pressure measurements in the vapor system at 30 to $-10°C$, Maass and Morrison[249] concluded that a 1:1 compound, $(CH_3)_2O,HCl$, existed in the vapor state, the dissociation of which increases with rising temperature. See also Russell and Maass.[320]

Maass and McIntosh[250] studied the methanol–hydrogen chloride system at $-89°C$. Only one compound, CH_3OH,HCl, mp $-62°C$, was isolated. (HCl is a liquid at $-89°C/1$ atm.)

Steele, McIntosh, and Archibald (1905)[341] examined the hydrogen halides as "conducting solvents." The pure "solvents" were found to be extremely poor conductors at low temperatures (about $-99°C$).

Waddington and Klanberg[366] used liquid hydrogen chloride as an ionizing solvent. The solubilities and conductivities of a number of substances, especially halides of Group III–V elements, were determined, and the compounds were classified as strong, moderate, and weak electrolytes according to the effect on the conductance of liquid HCl. Kilpatrick and Luborsky[203] used a conductometric method to study the "base strengths of aromatic hydrocarbons relative to hydrofluoric acid in anhydrous hydrofluoric acid as the solvent." The equivalent conductances of a series of methylated aromatic hydrocarbons were calculated from $\Lambda = 1000\ L/C$, where L, is the specific conductance at the equivalent concentration

C. Concentrations were in moles per 1000 g of liquid HF, i.e., molalities. The specific conductance L was derived from $L_{measured} - L_{solvent}$. According to these authors, this expression did not represent the actual relationship because of interaction between the "solvent" and "solute." A large part of the "solvent conductance" was attributed to traces of water and salts. Water was described as a strong electrolyte in liquid hydrogen fluoride, and also as a strong base; therefore, it will interfere with the acid–base equilibrium under examination.

Kilpatrick and Luborsky concluded that the methylbenzenes act as weak bases with reference to hydrogen fluoride as an acid, and the acid–base equilibrium was represented as

$$Ar + HF \rightleftharpoons ArH^+ + F^-$$

The "relative basicity" at $0.1\,m$ was recorded as: benzene, 0.09; toluene, 0.6; p-xylene, 1.0; m-xylene, 26; mesitylene, 13,000; and hexamethylbenzene, 97,000.

According to Andrews (1954)[3] solubilities can provide a qualitative measure of the capacity of the aromatic nucleus to react with electron acceptors. The solubility of benzene in anhydrous hydrogen fluoride was given as 2.25 wt- % at 0°C, although the alkanes and cycloalkanes are "virtually insoluble." This difference was attributed to the reaction

$$C_6H_6 + HF \rightleftharpoons C_6H_6H^+ + F^-$$

It was appreciated, however, that the matter was not so simple as this, for the order of solubility of benzene, toluene, and the xylenes in hydrogen fluoride was stated to be the inverse of that expected from other evidence of the order of electron-donor function. A weight percent of 2.25 means that the $x_{C_6H_6}$ value is only 0.006 (HF being taken as the molecule). Even if the weight percent of the other benzene hydrocarbons were 2.25, as for benzene, the x_A values would decrease as the molecular weight of the hydrocarbon increases. Now these x_A values are absolutely small, the predominant factor being the hydrogen-bonding structure of the hydrogen fluoride.

At room temperature, the solubility of hydrogen chloride in moles HCl per mole of aromatic hydrocarbon, benzene, toluene, etc. appears to fit in well with the "strength" of the electron-donor function of the hydrocarbon as determined by spectroscopic measurements. Furthermore, the formation of the ions $C_6H_6H^+ + Cl^-$ does not appear to occur. Yet is not HCl supposed to be a "strong acid" stronger than HF?

With reference to the equation

$$C_6H_6 + HF \rightarrow C_6H_6 \cdot H^+ + F^-$$

does this mean that every molecule of C_6H_6 is converted into $C_6H_6H^+$, giving separated ions? Or are we to assume an equilibrium ($\rightleftharpoons$), and if so, what are the species in the equilibrium system—free C_6H_6, free HF, and the ions? According to certain textbooks, hydrocarbons have no tendency to gain or lose protons. The complexity of terminology is further exemplified in the papers by Feakins and

Watson (1963, 1964)[79] in which was considered the determination of the free energies of transfer (ΔG_t°) of the pairs of ions H^+Cl^-, H^+Br^-, and H^+I^- from water to 10 wt. % and 43.12 wt. % methanol–water mixtures. It was concluded that the methanol molecule is more basic and less acidic than the water molecule.

9.4. Nitrile–Hydrogen Halide Systems

Gerrard and Maladkar[1] investigated certain hydrogen halide–nitrile systems. The passage of hydrogen chloride into acetonitrile at <6°C afforded white crystals of ($CH_3CN,2HCl$) as soon as about 1.1 mole of HCl per mole of nitrile had been added, and the formation of crystals continued without further passage of gas until the concentration of "HCl" fell to about 0.3 mole/mole. A liquid mixture containing 0.67 mole "HCl" per mole of nitrile gave no crystals when chilled to −20°C but did so immediately on being seeded at −5°C. At 0°C white crystals formed when the amount absorbed reached 2.3 mole/mole, and crystallization continued without further passage of gas. C_2H_5CN, n-C_3H_7CN, n-$C_6H_{13}CN$, and $C_6H_5CH_2CN$ did not form crystals even at −30°C, when as much as 3 moles HCl per mole nitrile had been absorbed. All six nitriles avidly absorbed hydrogen bromide and hydrogen iodide at room temperature, giving fine white crystals, RCN,2HX. The critical amounts of HX in mole/mole required to start the crystallization at 20°C were (R = Me, Et, etc.):

HBr: Me, 0.01; Et, 0.02; n-Pr, 0.04; n-hexyl, 0.03; $PhCH_2$, 0.01; Ph, 0.02
HI: Me, 0.01; Et, 0.009; n-Pr, 0.005; n-hexyl, 0.0075; $PhCH_2$, 0.008; Ph, 0.013

Gerrard and Maladkar (1969)[114] described a procedure for the quantitative absorption of effluent hydrogen halides based on the avidity of benzonitrile (low volatility) for the hydrogen halides. The urgency to form the crystal lattice is thus indicated. These solids were remarkably thermally stable, and in the stoppered flask they were stored for weeks in a desiccator without the development of the slightest excess pressure. The bromine compounds remained white, but the iodine compounds became yellow. The bromine compounds of the easily volatile nitriles (R = Me, Et, Pr) gradually lost HBr and nitrile at 20°C/18 mm Hg, but the constitution of the crystals remained at RCN,2HBr, there being no evidence of the formation of the monohydrohalide RCNHBr. The other bromine systems lost hydrogen bromide only, and the generated nitrile wetted the remaining crystals and tended to hinder loss of HBr, especially toward the end. Hantzsch (1931)[149] placed the dihydrobromide of p-tolyl cyanide in a desiccator containing KOH and P_2O_5 and plotted the loss in weight against the time to indicate the progress of the postulated reaction

$$MeC_6H_4CN,2HBr \rightarrow MeC_6H_4CN,HBr + HBr$$

A "scharfe Knick der Kurve" at the 1:1 ratio was taken to indicate the formation of the monohydrohalide. But experiments with benzonitrile throw doubt on the validity of the above conclusion. The shape of Hantzsch's curve is not clear enough to be convincing, and it could be attributed to a slowing down of the rate of loss of HBr from the solid because of wetting by the resulting free nitrile. The dihydroiodides showed the same general behavior at 20°C/18 mm Hg, but the rate of loss was much less than that for the hydrobromides. Again there was no evidence of the formation of the monohydrohalide.

Ethyl cyanoacetate showed x_{HCl} values which increased steeply between 0 and −5°C, reaching 2.0 at −20°C, 1 atm, but crystals were not formed. The x_{HCl} value for ethyl acetate is 0.76 at 0°C, 1 atm, and for ethyl chloroacetate it is 0.36, a result compatible with the reduction in intensity of the basic function at the oxygen site or sites in the chloroester due to the usual inductive influence of the chlorine atom. According to Jain and Rivest (1964),[186] the cyano group has a similar effect as chlorine. This conclusion would seem to indicate the probability that the bulk of the hydrogen chloride is associated with the nitrile group in the systems, showing about two or more moles of hydrogen chloride per mole of the cyano ester. The behavior of the hydrogen bromide system especially conforms to this view, for a fine white crystalline precipitate was immediately formed when the gas was passed into the cyano ester, and the empirical formula appeared to be $CNCH_2CO_2Et,2HBr$. At 0°C/18 mm Hg the compound lost hydrogen bromide, the ester remaining behind; but there was no indication of a monohydrohalide.

The hydrogen halide–nitrile systems have a long and confusing history. Gautier (1869)[92] believed that the passage of hydrogen chloride into acetonitrile gave a crystalline substance, described as the monohalide CH_3CN,HCl, but he was unable to obtain a stable specimen. He obtained the "bromohydrates" of CH_3CN and $EtCN$ as white crystalline substances by "saturating" the nitrile with dry hydrogen bromide at 0°C, and he described them as "sesquisels," $2RCN,3HBr$. There is an early mention of the compounds by Engler (1867),[77] and Blitz (1892)[25] described a number of 1:2 "addition compounds" formed from aqueous hydrogen iodide and aromatic nitriles and isolated by drying in a vacuum desiccator. To prepare the aliphatic compounds, dry hydrogen iodide had to be used. The compounds were assigned the formula

$$R-C\begin{matrix} \diagup I & \diagup H \\ -N & \\ \diagdown I & \diagdown H \end{matrix}$$

where R = Ph, o-, m-, and p-$CH_3C_6H_4$, o-, m- and p-$NO_2C_6H_4$, CH_3, Et, and $PhCH_2$. Naumann[284] listed the "qualitative" solubilities of a number of inorganic salts in acetonitrile and declared that the hydrogen halides (HCl, HBr, HI) were *leicht löslich*. Hantzsch (1931)[149] concluded that the "addition products" of hydrogen halides and nitriles are nitrilium salts $(RC\equiv NH)X,HX$. It was stated that, as a rule, the dihalides were precipitated in preference to the monohalides, which were said to be more soluble. The monohalides listed were CCl_3CN, HBr,

PhCH$_2$CN, HBr, PhCH=CHCN, HBr, although it was not clear how he prepared these.

Hinkel and Treharne (1945)[169] did not refer to Hantzsch's observation and stated that hydrogen chloride readily combines with acetonitrile at low temperatures (ice–salt) to form the crystalline "acetamide dichloride" MeCCl$_2$NH$_2$, mp 6°C, although it was stated that at 40°C/11 mm the solid dichloride lost its crystalline structure. At ordinary temperatures "methyl cyanide, acetamide chloride [CH$_3$C(Cl)=NH], and hydrogen chloride" were obtained, but there was no stated evidence for the existence of the monochloride.

By the aid of binary freezing-point diagrams, for temperatures about $-100°C$, Murray and Schneider (1955)[271] indicated evidence of the formation of the following molecular compounds of nitriles with chloroform and with hydrogen chloride (the mp in degrees centigrade is given in brackets): MeCN,HCl ($-63.2°$), 2MeCN,3HCl ($-88°$),* MeCN,5HCl ($-123.6°$), MeCN,7HCl ($-125.0°$), EtCN,HCl ($-97.2°$), 2EtCN,3HCl ($-117°$),* EtCN,5HCl ($-129°$), PrCN,HCl ($-80.6°$), 2PrCN,3HCl ($-109°$),* PrCN,5HCl ($-130°$), EtCN,CHCl$_3$ ($-90.5°$), PrCN,3CHCl$_3$ ($-91.5°$),* and PhCN,CHCl$_3$ ($-67.5°$).* The authors refer to the directional nature of the force field due to a well-directed lone pair orbital on the N atom, which may exhibit strong donor properties, and two π-orbitals which may exhibit weak donor properties. Chloroform was linked with hydrogen chloride as a "*good acceptor*" molecule. The authors were surprised in their failure to show the formation of a 1:1 complex in the CHCl$_3$—CH$_3$CN system, and this was attributed to the ability of acetonitrile to associate with itself more strongly than with chloroform. Berkeley and Hanna (1957, 1964)[22] recognized a "hydrogen-bond shift" when the "proton donor" CHCl$_3$ was mixed with the "proton acceptor" CH$_3$CN:

$$CHCl_3 + CH_3CN \rightleftharpoons CH_3CN\cdots H-CCl_3$$

Janz and Danyluk (1959)[188] described their conductance data for acetonitrile–hydrogen halide systems and declared that the hydrogen halides are weak electrolytes in acetonitrile, being only slightly ionized. Their data appear to place the acid function in the order HI > HBr > HCl, but the spacing is very different for the various sites of basic function. Assuming Λ_0 to be 176 for HBr, the degrees of dissociation of 0.0001, 0.001, and 0.01 M HBr solutions in acetonitrile were computed to be 0.13, 0.09, and 0.07, respectively.

It was mentioned that Pleskov[304] reported Λ for HCl as 178 for 0.001 m HCl, 200 times as large as Janz and Danyluk's value, a difference which could not be accounted for.

We have here clearly revealed the problem of terminology. From the reference to low degree of dissociation and the condition of being slightly ionized, it would appear that ionization and dissociation are taken as synonymous, and the formation of an ion pair by proton transfer is not ionization. Even for 1 molecule of added HCl

*Melts incongruently.

to 250,000 molecules of CH_3CN, the conductance data show a small "degree of dissociation," i.e., a small proportion of the "separate ions" compared with ion pairs and hydrogen-bonded (no proton transfer) forms. Whereas the solvation of the cationic part $CH_3C\overset{+}{B}H\cdots NCCH_3$ is conceivable in terms of electron donor–acceptor function ($CH_3C\overset{+}{N}H$ being a relatively "strong" electron acceptor as is apparent from the facts), we are up against the difficulty of solvating an ion Cl^-, an electron donor, by CH_3CN. Janz and Danyluk do not state how Cl^- (solvated) can come about, but they present us with a situation in which a molecule of hydrogen chloride prefers to hydrogen-bond (complex) not with Cl^- but with Cl^- (solvated), of which there are so few about, rather than become involved with CH_3CN (of which there are so very many molecules immediately available) in what may be termed the primary process:

$$CH_3CN + HCl \rightarrow CH_3CN\cdots H-Cl, \text{etc.}$$

The final question is: What is the significance of these conductivity data in the description of the interaction of hydrogen halides and nitriles when the two reactants are involved at approximately equimolecular proportions? Infrared spectra were discussed by Janz and Danyluk, and the work of Josien and Sourisseau (1955)[195] was cited.

There are many papers on the acid–base function of nitriles, showing an appalling conflict in terminology which defies summarization. The textbook opinion appears to be that "acetonitrile is useful as a solvent for promoting ionic reactions—it is aprotic, that is, although because of its dielectric properties (the dielectric constant is 37.5 at 20°C) it aids the separation of charges, it does not provide protons." Yet it is also stated that it "reacts mainly as if it contained active hydrogen:

$$CH_3C\equiv N \rightleftharpoons CH_2=C=NH"$$

From an examination of the refractivity of propionitrile–amine systems in water, Oda *et al.* (1960)[293] concluded that hydrogen bonding involved an α-hydrogen, and indeed isobutyronitrile (a "proton donor") was deemed to use an α-hydrogen in hydrogen-bonding with the electron donor site of benzonitrile (called a "proton acceptor"):

According to Saum (1960),[321] the viscosities of mononitriles indicate a degree of association which approaches but does not exceed the dimer level. The association was not deemed to involve α-hydrogen bonding (because t-BuCN is not anomalous) but to involve interaction between pairs of CN groups; it could be due to dipole–pair bonding (mainly), but could also involve random dipole interaction. The discussion led to the concept of a strong and rather specific interaction between pairs of CN groups in the same sense that one uses the term hydrogen bond.

Weimer and Prausnitz[369] refer to propionitrile as an "electron acceptor" with reference to p-xylene as "electron donor" in n-hexane and give the equilibrium constant as $K_c = 0.071$ liter/mole. The electron-acceptor (acid-function) site was not specified. Janz and Danyluk (1959) stated that acetonitrile behaves as an amphiprotic solvent; it has extremely weak acidic properties and slightly stronger basic properties.

The "acidic" site and the mechanism of its function were not specified, nor was an example given; but it is logical to infer that by the term *amphiprotic* hydrogen acidic function was meant. If the "basic properties" of acetonitrile are only *slightly* stronger than its "acidic properties," they must still be deemed very weak.

On the basis of lone pair dipoles, Schneider[328] predicted that cyano groups should form "stronger" hydrogen bonds than amines; but the NMR results of Korinek and Schneider[212] showed otherwise, and these authors referred to ionization potential "basicity."

Mitra[260] has made "infrared studies of nitriles as proton acceptors in hydrogen-bond formation."

According to Kolthoff and Coetzee (1957),[211] acetonitrile is a much weaker base and a much weaker acid than water. How, then, do we explain the ready formation of $CH_3CN,2HX$? Acetonitrile was deemed to be a poor solvating solvent. Acetic acid was classified as a strong hydrogen bonder. It should be kept in mind, however, that we should be concerned with functions—acid–base functions, hydrogen-bonding functions. Acetic acid can form hydrogen bonds via its carboxylic hydrogen, but it can do so via its carboxylic oxygen atoms, the hydrogen part being provided by a compound such as hydrogen chloride.

9.5. Solubility in Water

Of particular relevance here are two papers published at the time when the puzzles relating to strong electrolytes were being considered. Dobson and Masson (1923)[67] felt compelled to believe that in a solution of hydrogen chloride in water, not even the "uncombined" water was in the state in which it exists in pure water. Dunn and Rideal (1923)[71] considered the vapor pressure of "hydrochloric acid" and gave a description of the mechanisms of dissolution. They visualized the solvation of the molecule of hydrogen chloride, making it "more polar," followed by dissociation through solvation and "thermal agitation" of the solvated molecules to give solvated ions. As hydrogen chloride is added, a stage is "rapidly attained" where no "free water" is left in the solution. The extent of solvation of the molecules begins to fall, and this process is followed by the reduction in solvation of the ions. For all concentrations, there was visualized a system in dynamic equilibrium, entailing unsolvated molecules and solvated molecules and ions in various states of solvation. They drew attention to the difference between the activity of "hydrochloric acid" as measured by vapor pressure and that calculated from the electromotive force of concentration cells.

It is an informative exercise to consider these papers in the light of the plots of x_{HX} (or N_{HX}) vs p_{HX} for water and other liquids S.

Rodebush and Ewart (1932)[314] gave data for hydrogen chloride and acetic acid for 24.8°C but for p_{HCl} not higher than 93 mm Hg. They reported that the p_{HCl} line over the range of pressures observed showed a straight-line function of the mole fraction. Since they found that, for $N_{HCl} = 0.011$, p_{HCl} is 50.9 mm Hg for acetic acid and only 0.0001 mm Hg for water, they concluded that the relative activity in acetic acid is more than 100,000 times greater than in water. However, one must consider the following points. A linear extrapolation of their low-pressure data gives $N_{HCl} = 0.17$ and $x_{HCl} = 0.205$ for $p_{HCl} = 1$ atm for acetic acid. The observed value in the work of Gerrard et al. is $x_{HCl} = 0.12$ for 25°C. The N_{HCl} vs p_{HCl} plot for n-$C_5H_{11}CO_2H$ is concave upward, and the situation is similar to that for $(ClCH_2CH_2)_2O$ (Fig. 92). At 20°C and $p_{HX} = 1$ atm, N_{HCl} for acetic acid is 0.130, and for water it is 0.263; the corresponding N_{HBr} values are 0.324 and 0.307. What about activities now?

Rodebush and Ewart declared that whereas the solubility of hydrogen chloride in water is large, it is no more than that predicted for an "ideal solution" in benzene, a liquid which "cannot be coordinated by the hydrogen ion." However, one must contrast this with the data for hydrogen iodide. At 20°C and $p_{HI} = 1$ atm, the N_{HI} value for benzene is a little larger than that for water.

Saylor (1937)[323] determined the "moles HCl/1000 g C_6H_6" for p_{HCl} up to 570 mm Hg at 30°C. I have shown the N_{HCl} vs p_{HCl} plot in Fig. 92. See also Fig. 95.

A study of the N_{HX} vs p_{HX} plots for water at 0°C shows an involvement of HX with the water structure as HX is added, p_{HX} remaining low until $N_{HX} = \sim 0.2$ has been added. This effect may be interpreted in a simple way as follows (the regular arrangement given is merely for convenience in delineation):

$$\begin{array}{c} H-O\cdots H-O\cdots H-O\cdots \\ \backslash \backslash \backslash \\ H H H \\ Cl Cl^- \\ \backslash \backslash \\ H H \\ O O-H\cdots \\ HHH \end{array}$$

About four molecules (H_2O units) of water are required to produce what may be deemed to be the maximum concentration of "ions" per mole of HCl. Further addition of HX causes a more rapid rise in p_{HX}, and finally for $p_{HX} = 1$ atm, at least two (H_2O) units are required to hold one HX unit, the second half of the HX being

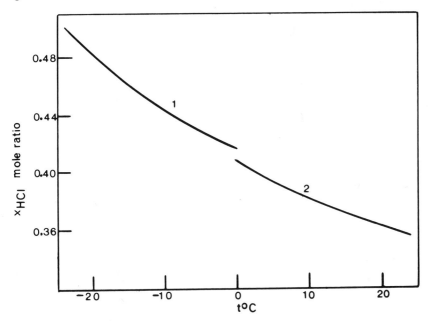

Fig. 95. Mole ratio x_{HCl} solubility of hydrogen chloride in water at 1 atm and different temperatures $t°C$. Line 1 is from data by Roozeboom, and line 2 from data by Roscoe and Dittmar.

mainly involved in hydrogen bonding without "proton transfer." I visualize the whole process from beginning to end as an involvement of HCl in the entire polymeric structure of water in a complex pattern of dynamic equilibria.

10

Ammonia and the Three Methylamines

10.1. Ammonia

10.1.1. Solubility in Water

The conventional statement is: "For a weak base such as ammonia NH_4OH, $BOH = B^+ + OH^-$,

$$\frac{[B^+][OH^-]}{[BOH]} = \frac{a^2}{(1-a)V} = K_b$$

where K_b is the dissociation constant of the base." The so-called "0.88 ammonia" is about 18.0 M, 31.3 m, and contains about 0.564 mole per mole (H_2O) of water at 12°C. At 0°C, x_{NH_3} is 0.953 for $p_{NH_3} = 1$ atm, and this corresponds to 52.9 m and should, in accordance with the conventional statement, be looked upon as almost "pure NH_4OH."

The R-line x_{NH_3} value at 1 atm and 0°C is 0.31 for ammonia, bp −33°C/1 atm. In comparison, the corresponding x_{HI} value for hydrogen iodide, bp −35°C/1 atm, is 0.37. The x_{NH_3} value for water at 0°C, $p_{NH_3} = 1$ atm is more than *twice* the corresponding x_{HI} value. In an aqueous solution of ammonia the degree of dissociation at dilution V liters is said to be given by $\Lambda/\Lambda\infty$, where $\Lambda\infty$ is the molar conductance at infinite dilution, i.e., 252 at 25°C. Now this argument is built up from data for very dilute solutions, and the degree of dilution is seen from the following. The range of concentrations for which the conductance data are given is for V from 8 liters to something like 200 liters. In Fig. 96 is given the plot of Λ and V, values for α being shown for certain values of V. For a solution containing 1 mole of "NH_3" in 8 liters, α is 0.0135, and the x_{NH_3} value is only 0.002. What meaning can α have for $x_{NH_3} = 0.95$? According to the Pauli principle, the compound designated "NH_4OH" cannot exist, and therefore *any* function attributed to that imaginary object must be fictitious.

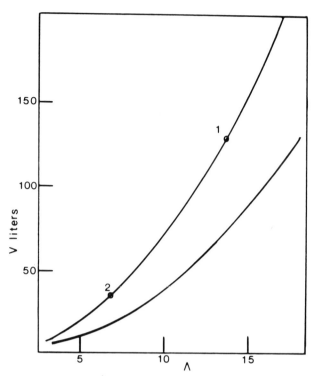

Fig. 96. The ammonia–water system. Line 1–2 is a plot of the equivalent conductance Λ vs concentration, expressed as one mole of ammonia in V liters of solution. At point 1 the "degree of dissociation α" is 0.0536; at point 2 it is 0.0266; the temperature is 25°C. When $V = 8$, the mole ratio of ammonia is $x_{NH_3} = 0.002$. The other line is for acetic acid, and is drawn for comparison.

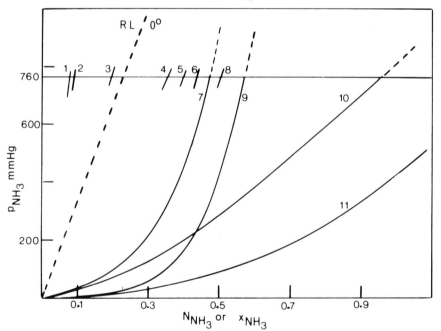

Fig. 97. Solubility of ammonia at 0°C. Registration of mole fraction N_{NH_3} values on the R-line diagram. The liquids are: (1) $C_6H_5CH_3$; (2) $(n\text{-}C_8H_{17})_2O$; (3) $n\text{-}C_8H_{17}Cl$; (4) $n\text{-}C_8H_{17}OH$; (5) C_2H_5OH; (6) CH_3OH; (7) water; (8) $ClCH_2CH_2OH$; (9) Cl_3CCH_2OH. Lines 1–9 are mole fraction N_{NH_3} vs p_{NH_3} plots. Line 10 is the corresponding mole ratio x_{NH_3} plot for water and line 11 is the corresponding mole ratio x_{NH_3} plot for Cl_3CCH_2OH.

Look at Fig. 97, which shows the plot of x_{NH_3}/p_{NH_3} for water at 0°C. As ammonia is added to water, there is a clearly discernible rise in p_{NH_3}, and the plot is concave upward. It is quite different from the x_{HX}/p_{HX} plots. Even for values of x_{NH_3} about 0.0001, the main effect seems to be hydrogen bonding, proton transfer, e.g.,

$$[H_2O]_n \cdots HOH \cdots NH_3 \rightleftharpoons [H_2O]_n \cdots HO^- \cdots HNH_3^+$$

being a relatively rare occurrence, judging from the conductance data. About $x_{NH_3} = 0.4$ (there being one molecule of "NH_3" for every 2.5 molecules of water (as "H_2O"), the slope begins to get more concave upward, pointing to a decrease of ease of accommodation. From about $x_{NH_3} = 0.5$ to the value of 0.95, the plot is almost linear. The corresponding N_{NH_3} plot is shown in contrast. The structure of the system at that stage seems to be equivalent to the simple one depicted as follows:

$$\begin{array}{cccc} H-O\cdots H-O\cdots H-O\cdots H-O\cdots \\ \backslash & \backslash & \backslash & \backslash \\ H & H & H & H \\ \vdots & \vdots & \vdots & \vdots \\ NH_3 & NH_3 & NH_3 & NH_3 \end{array}$$

Just below 0°C the x_{NH_3} value is actually 1.0. Certain chemists will deem this pattern much too simple, but it gives the mole ratio we have to accommodate.

The current textbook treatment of the ammonia–water system is in urgent need of revision for it retains clearly demonstrable inconsistences in terminology and disregard of facts. In a book of 1971 vintage, ammonia is classified as a base because its addition to water increases the hydroxyl-ion concentration:

$$NH_3 + H_2O \rightarrow NH_4^+ + OH^-$$

"Ammonia is very soluble in water," the authors declared, "which is easily explained by the fact that both NH_3 and H_2O are polar molecules." "Not so easy to explain is the basic character of the aqueous solutions formed."

The $x_{C_4H_{10}}$ value for n-decane at 5°C and $p_{C_4H_{10}} = 1$ atm is *four* times the x_{NH_3} value for water at 5°C and $p_{NH_3} = 1$ atm. Under what I take to be the usual connotation of the term *polar*, can any compound be much less "polar" than these hydrocarbons? Furthermore, sulfur dioxide appears to be about as "polar" as ammonia, and yet the x_{SO_2} value for water (0.06 at 0°C, $p_{SO_2} = 1$ atm) is less than that for any other examined liquid. From the aspect of the term "polar," I see little difference between ammonia and hydrogen chloride, and yet there is a large quantitative difference in the number of ions produced for $x_{gas} = 0.2$, for example, by these two compounds.

10.1.2. Nonaqueous Liquids S

Gerrard and Maladkar[115] reported a number of x_{NH_3} values, and the N_{NH_3} vs p_{NH_3} plots of those data and of new data by Gerrard are exemplified in Figs. 98 and

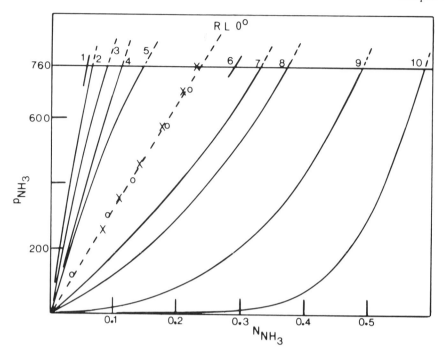

Fig. 98. Solubility of ammonia in different liquids at 0°C and different pressures p_{NH_3} in mm Hg. Mole fraction N_{NH_3} vs p_{NH_3} plots on the R-line diagram. The liquids are: (1) n-$C_{10}H_{22}$; (2) 1,3,5-$C_6H_3(CH_3)_3$; (3) $C_6H_5N(C_2H_5)_2$; (4) (n-$C_{18}H_{17})_2O$; (5) n-$C_8H_{17}NH_2$; (6) $CHCl_3$; (7) $C_6H_5NH_2$; (8) n-C_6H_7OH; (9) $C_6H_5CH_2OH$; (10) Cl_3CCH_2OH. The crosses near the R-line are for $C_6H_5NHC_2H_5$, and the circles near the R-line are for $HCON(CH_3)_2$.

99. The alcohols n-$C_8H_{17}OH$, $PhCH_2OH$, and Cl_3CCH_2OH give lines which are strongly concave upward on the right of the R-line for 0°C, indicating the operation of the basic function of ammonia and the acid function sited on hydrogen of the alcohol. It is seen that the N_{NH_3} value for Cl_3CCH_2OH is decidedly larger than the corresponding value for water. Even for $ClCH_2CH_2OH$, the N_{NH_3} value is a little larger than that for water. The influence of chlorine in reducing the intensity of the basic function of the oxygen and increasing that of the acid function sited on hydrogen is seen in Table 23.

The line for aniline is well on the right, probably due to the hydrogen-bonding function $Ph(H)N–H\cdots NH_3$, since the line for diethylaniline is well over on the left. Mesitylene and n-decane are still further on the left. Lines for monoethylaniline and dimethylforamamide are almost coincident with the R-line. Are we to regard these mixtures as almost *ideal*? I should say not. The acid function sited on the hydrogen of chloroform appears to be clearly involved to give the position at 760 mm Hg.

Table 24 shows the N_{NH_3} values obtained by Bell[20] for 20°C, 1 atm. See Fig. 99 for the position of these values in the R-line diagram. There does not appear to be any systematic correlation between N_{NH_3} and the dielectric constant of S.

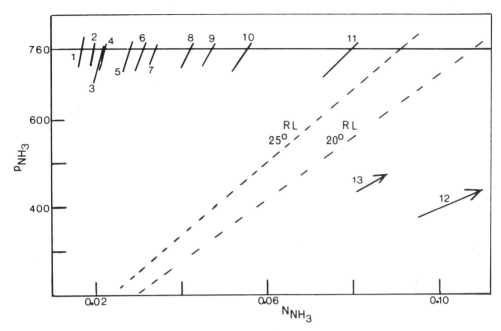

Fig. 99. Solubility of ammonia in various liquids at 760 mm Hg and 20°C. Registration of Bell's mole fraction N_{NH_3} values on the R-line diagram. The liquids are: (1) octane; (2) dodecane; (3) cetane; (4) hexane; (5) CCl_4; (6) $C_6H_5CH_3$; (7) C_6H_5Br; (8) C_6H_5Cl; (9) C_6H_6; (10) $C_6H_5CH_2Cl$; (11) $(CH_2Cl)_2$. The N_{NH_3} value for $CHCl_3$ was given as 0.193, and the approximate position on this scale of diagram is shown by line 12 at about 400 mm Hg; line 13 shows the corresponding position of Seward's (1932) value for 25°C.

Table 23. x_{NH_3} Values for 0°C, 1 atm

S	X_{NH_3}	X_{HCl}	X_{HBr}	X_{HI}
Water	0.953	0.41	0.49	0.39
MeOH	0.780	0.95		
EtOH	0.664	0.98		
n-OctOH	0.56	1.03	1.28	1.65
$(n\text{-Oct})_2O$	0.10	1.06	2.32	2.97
$MeCO_2C_8H_{17}\text{-}n$	0.285	1.02	1.90	
$n\text{-}C_8H_{17}Cl$	0.246	0.12	0.24	0.65
$CHCl_3$	0.24*	0.034	0.055	0.27
MeC_6H_5	0.085	0.09	0.17	0.56
$ClCH_2CH_2OH$	1.03	0.55		
Cl_2CHCH_2OH	1.31	0.18	0.53	
Cl_3CCH_2OH	1.33	0.09	0.17	

Table 24. x_{NH_3} Values for 20°C and $p_{NH_3} = 1$ atm from the N_{NH_3} Values of Bell[20]

S	N_{NH_3}	x_{NH_3}
C_6H_{14}	0.0223	0.0228
CCl_4	0.0281	0.0289
C_6H_6	0.0474	0.0498
$C_6H_5CH_3$	0.0313	0.0323
$CHCl_3$	0.193	0.239
C_6H_5Br	0.0340	0.0352
C_6H_5Cl	0.0423	0.0442
$C_6H_5CH_2Cl$	0.0556	0.0589
$ClCH_2CH_2Cl$	0.0797	0.0866

10.2. Methylamines

10.2.1. Significance of Solubility Data

Except for seemingly inaccurate data on trimethylamine given in the International Critical Tables (1928),[182] data are remarkably sparse; yet thousands of papers on the properties of these amines have appeared. The data presented herein show the need to consider the *inter*molecular effects which may influence the course of reactions in organic chemistry. Catalytic effects are especially to be looked for.

10.2.2. Comments on Electronic Effects

Hydrogen bonding between the N—H hydrogen with N or O in another molecule of the same or different compound, as well as other aspects of acid–base function entailing amines, is of supreme significance in biotechnology. Various probes show the complexity of the phenomena, and conclusions are far from unequivocal. The change in dipole moment (NH_3, 1.47; $MeNH_2$, 1.28; Me_2NH, 1.03; Me_3N, 0.61, and Et_3N, 0.61 D) and the muddled order of "base strength" (reference to water) have been accounted for by acknowledging a high lone-pair moment for ammonia, a rapid decrease in lone-pair moment as the polarizability of the "central atom" is increased, and the idea that "base strength" is related to both the lone-pair moment and the electron-cloud polarizability (Weaver and Parry, 1966).[368] See also Yoneda.[387] "Base strength"[368] is deemed to be a measure of the ability of electrons to move under the influence of a polarizing field into a position for bonding. Stevenson (1962)[345] used ultraviolet spectra to show that the tendency of the amines NH_3, $EtNH_2$, Et_3N to "complex with water" is "completely antiparallel" to the "basicity" of these amines as measured by K_b. The spectral shifts were not attributed to a "general medium effect" or to the presence of a large dipole in the solvent since the absorption band of Et_3N in acetonitrile ($\mu = 3.9$ D, $\varepsilon = 36.7$;

μ is the dipole moment and ε is the dielectric constant) was almost coincident with that in diethyl ether (μ 1.3 D, ε 4.3). Stevenson and Coppinger (1962)[346] concluded that Et_3N and chloroform in isooctane ($\mu = 0$, $\varepsilon = 2.0$) react to form a 1 : 1 "charge transfer complex" or to give rise to "contact charge-transfer spectra." Conditions relating to the formation of the N—H···N bond in secondary aliphatic amines themselves were studied by Murphy and Davis (1968)[270] by means of proton magnetic resonance. Of the three types of amines, Pearson and Vogelsong (1958)[301] concluded that the tertiary aliphatic amine (Et_3N) is "the strongest catalytic base" in chloroform and chlorobenzene; but in benzene the secondary amine (n-Bu_2NH) is the strongest, whereas in dioxane and ethyl acetate the primary one (n-hexyl NH_2) is the strongest.

From a rudimentary aspect, it may be stated that, whereas the basic function sited on nitrogen is concentrated there, the corresponding acidic function is dissipated over the three hydrogens attached to nitrogen in NH_3, over the two in CH_3NH_2, and confined to the one in $(CH_3)_2NH$. To what extent the hydrocarbon hydrogens can be involved is left as an open question. Zellhoefer et al. (1938)[392] (see also refs. 58 and 59) attributed the "large negative deviations from Raoult's law" shown by volatile haloforms such as $CHCl_3$ in "donor solvents" free from hydroxyl or amide hydrogens to the formation of hydrogen bonds:

$$X_3C-H \cdots N\equiv \quad \text{or} \quad X_3C-H \cdots O=$$

With reference to water and alcohols, the hydrogen-bonding forms

$$=O \cdots H-N= \quad \text{and} \quad -O-H \cdots N\equiv$$

must both be considered.

10.2.3. Experimental Details for the Methylamines

The mass of gas A absorbed by a known mass of liquid S at a measured total pressure, assumed to be very nearly p_A, was measured at a chosen temperature $t°C$ by the bubbler tube-manometer procedure previously described. The liquids S were purified and attested by conventional procedures. After the final removal of gas on the manometer assembly, the mass of each residual liquid was essentially the same and gave the same refractive index and infrared spectrum as the initial liquid weighed. The methylamines were obtained in cylinders from British Drug Houses and Cambrian Gases. Each gas was passed through a long column of pellets of potassium hydroxide, which showed no visible change. The change in vapor pressure p_A of each liquefied gas from $-12°C$ to the bp was determined by the bubbler tube-manometer assembly. The results conformed with those recorded in the literature. These and the literature data, $MeNH_2$ (bp $-6.3°C$), Me_2NH (bp $6.88°C$), and Me_3N (bp $2.87°C$),[357] enabled the R-lines to be drawn for the observational temperatures. Many of the observations for a gas A from one of the two sources were repeated on the specimen of gas from the other source; the results were concordant.

168 Chapter 10

10.2.4. Data for Methylamine, bp −6.3°C/1 atm

Table 25 gives N_A and x_A values for methylamine at different pressures and temperatures. Representative mole fraction–pressure plots are shown in Fig. 100 for 0°C, Fig. 101 for 10°C, and Fig. 102 for 20°C. Figure 103 illustrates the marked difference between the mole fraction, N_A, and the mole ratio, x_A, plots. Representative plots of x_A and temperature are shown in Fig. 104. The results for the alcohols appear to emerge substantially from the hydrogen-bonding function

$$\equiv N \cdots H - O - R$$

The N_A/p_A lines for 2,2,2-trichloroethanol, glycerol, ethylene glycol, and benzyl alcohol are strongly concave upward; the lines for n-octanol and water are less so. It appears that the hydrogen bonding

$$= O \cdots H - N =$$

is a minor function since the ethers give lines convex upward on the left of the R-line. The functioning of the hydrogen bonding

$$C_6H_5(R)N - H \cdots N \equiv$$

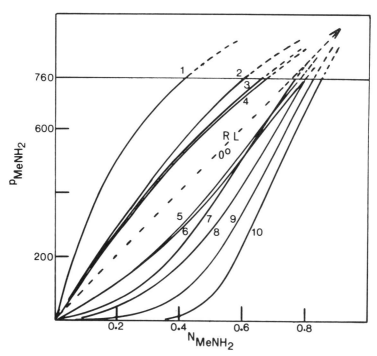

Fig. 100. Solubility of methylamine MeNH$_2$ in various liquids at 0°C and different pressures p_{MeNH_2}. Registration of the mole fraction N_{MeNH_2} vs p_{MeNH_2} plots on the R-line diagram. The liquids are: (1) n-C$_{10}$H$_{22}$; (2) (n-C$_8$H$_{17}$)$_2$O; (3) C$_6$H$_5$N(C$_2$H$_5$)$_2$; (4) C$_6$H$_5$N(CH$_3$)$_2$; (5) C$_6$H$_5$NHC$_2$H$_5$; (6) C$_6$H$_5$NHCH$_3$; (7) water; (8) n-C$_8$H$_{17}$OH; (9) C$_6$H$_5$CH$_2$OH; (10) Cl$_3$CCH$_2$OH. The pressures p_{MeNH_2} are in mm Hg.

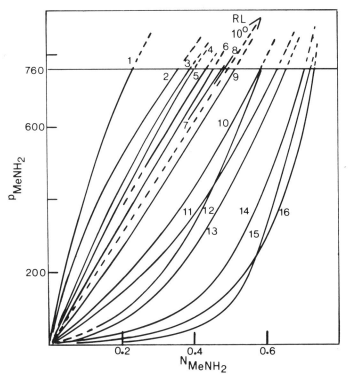

Fig. 101. Solubility of methylamine MeNH$_2$ in various liquids at 10°C and different pressures p_{MeNH_2}. Registration of the mole fraction N_{MeNH_2} vs p_{MeNH_2} plots on the R-line diagram. The liquids are: (1) n-C$_{10}$H$_{22}$; (2) m-C$_6$H$_4$(CH$_3$)$_2$ (curve for MeNH$_2$ in mesitylene is similar to 2); (3) m-BrC$_6$H$_4$CH$_3$; (4) C$_6$H$_5$N(CH$_3$)$_2$ (curves for MeNH$_2$ in o-NO$_2$C$_6$H$_4$CH$_3$ and 1-BrC$_{10}$H$_7$ are similar to 4); (5) C$_6$H$_5$OC$_2$H$_5$; (6) C$_9$H$_7$N; (7) 1,4-dioxane; (8) C$_5$H$_5$N; (9) C$_6$H$_5$CH$_2$NH$_2$; (10) C$_6$H$_5$NHC$_2$H$_5$; (11) C$_6$H$_5$NH$_2$; (12) water; (13) n-C$_8$H$_{17}$OH; (14) (CH$_2$OH)$_2$; (15) Cl$_3$CCH$_2$OH; (16) (CH$_2$OH)$_2$CHOH. The pressures p_{MeNH_3} are in mm Hg.

appears evident from the clustered position of the lines for C$_6$H$_5$NHCH$_3$ and C$_6$H$_5$NHC$_2$H$_5$ (concave upward on the right of the R-line) and the clustered position of those for C$_6$H$_5$N(CH$_3$)$_2$ and C$_6$H$_5$N(C$_2$H$_5$)$_2$ (convex upward, decidedly on the left of the R-line). The line for n-octylamine is just on the left, and that for benzylamine is on the right of the R-line. The line for pyridine is just on the left, and that for quinoline still more on the left of the n-octylamine line. The lines for ethyl benzoate and benzonitrile are near, and that for dimethylformamide is almost coincident with the R-line.

The benzene hydrocarbons give a cluster of lines well over on the left of the R-line, and lines for nitrobenzene and certain other benzene derivatives are in the mid-position on the left, i.e., are convex upward. The line for n-decane is well over on the left. The N_{MeNH_2} value for chloroform shows the functioning of the hydrogen bonding

$$Cl_3C-H\cdots N\equiv$$

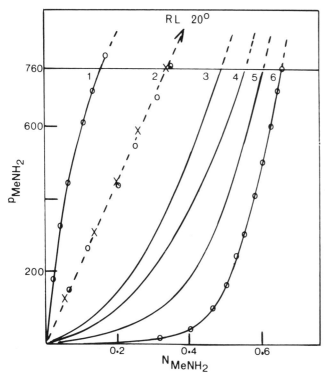

Fig. 102. Solubility of methylamine $MeNH_2$ in different liquids at 20°C and at different pressures p_{MeNH_2}. Registration of the plots of mole fraction N_{MeNH_2} vs p_{MeNH_2} on the R-line diagram. The liquids are: (1) $n\text{-}C_{10}H_{22}$; (2) the R-line (the crosses are for $C_6H_5COCH_3$; the circles are for $C_6H_5CO_2C_2H_5$); (3) water; (4) $n\text{-}C_8H_{17}OH$; (5) $C_6H_5CH_2OH$; (6) Cl_3CCH_2OH. The circles on lines 1 and 6 are the experimental points and are inserted to show the smoothness of the experimental line obtained by the bubbler-tube procedure. The pressures p_{MeNH_2} are in mm Hg.

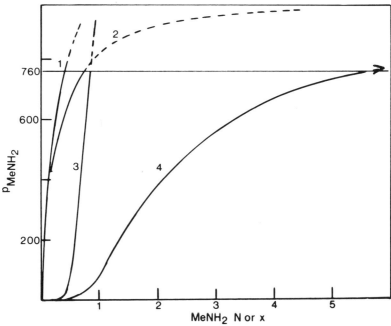

Fig. 103. Comparison of the mole fraction N_{MeNH_3} vs pressure p_{MeNH_2} and mole ratio x_{MeNH_2} vs p_{MeNH_2} plots for methylamine $MeNH_2$ at 0°C. Line 1 is the N_{MeNH_2}, and line 2 the x_{MeNH_2}, plot for $n\text{-}C_{10}H_{22}$; line 3 is the N_{MeNH_2}, and line 4 the x_{MeNH_2} plot for Cl_3CCH_2OH. The pressures p_{MeNH_2} are in mm Hg.

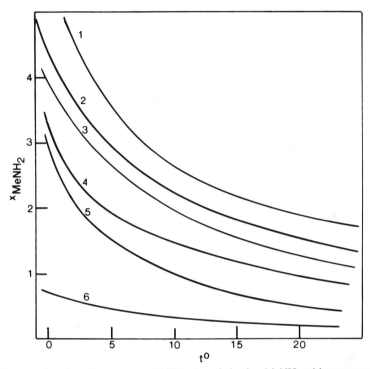

Fig. 104. Change of mole ratio x_{MeNH_2} solubility of methylamine $MeNH_2$ with temperature $t°C$ at $p_{MeNH_2} = 1$ atm. The liquids are: (1) Cl_3CCH_2OH; (2) $C_6H_5CH_2OH$; (3) $n\text{-}C_8H_{17}OH$; (4) water; (5) $HCON(CH_3)_2$; (6) $n\text{-}C_{10}H_{22}$.

Table 25. N_{MeNH_2} Values for Methylamine at Different Pressures p_{MeNH_2} and Temperatures t, °C

Liquid S	t, °C	\multicolumn{8}{c}{p_{MeNH_2} in mm Hg}	x_{MeNH_2} at 760 mm Hg
		100	200	300	400	500	600	700	760	
Cl$_3$CCH$_2$OH	0	0.524	0.581	0.628	0.676	0.722	0.766	0.815	0.848	5.580
	5	—	—	—	—	—	—	0.760	0.781	3.559
	10	0.494	0.552	0.588	0.624	0.655	0.683	0.710	0.723	2.616
	20	0.464	0.515	0.548	0.576	0.600	0.622	0.642	0.653	1.882
C$_6$H$_5$CH$_2$OH	0	0.420	0.506	0.571	0.630	0.680	0.731	0.785	0.816	4.430
	5	—	—	—	—	—	—	—	0.747	2.961
	10	—	—	—	—	—	—	—	0.690	2.222
	20	0.312	0.410	0.460	0.500	0.533	0.560	0.588	0.605	1.532
n-C$_8$H$_{17}$OH	0	0.300	0.425	0.511	0.582	0.644	0.705	0.763	0.796	3.900
	5	—	—	—	—	—	0.636	0.692	0.728	2.679
	10	0.255	0.355	0.424	0.485	0.537	0.586	0.635	0.663	1.970
	20	0.200	0.292	0.357	0.412	0.458	0.498	0.534	0.554	1.240
n-C$_4$H$_9$OH	10	—	—	—	—	—	0.580	0.630	0.661	1.945
	20	—	—	—	—	—	—	—	0.554	1.242
(CH$_2$OH)$_2$	10	0.375	0.476	0.537	0.583	0.623	0.660	0.691	0.709	2.436
(CH$_2$OH)$_2$CHOH	10	0.448	0.546	0.603	0.645	0.680	0.708	0.727	0.736	2.788
C$_6$H$_5$COCH$_3$	20	0.045	0.090	0.132	0.180	0.244	0.270	0.315	0.342	0.520
C$_6$H$_5$CO$_2$C$_2$H$_5$	0	0.093	0.187	0.281	0.376	0.466	0.560	0.656	0.715	2.510
	10	0.062	0.126	0.188	0.256	0.318	0.373	0.444	0.490	0.961
	20	0.044	0.088	0.135	0.180	0.224	0.272	0.320	0.348	0.534
C$_6$H$_5$OC$_2$H$_5$	10	—	—	—	—	0.285	0.342	0.402	0.436	0.773
C$_6$H$_5$CN	10	0.075	0.140	0.204	0.273	0.341	0.411	0.497	0.516	1.066
(n-C$_8$H$_{17}$)$_2$O	0	0.076	0.144	0.215	0.284	0.356	0.440	0.530	0.604	1.525
	10	0.040	0.084	0.135	0.185	0.237	0.291	0.350	0.388	0.633
(n-C$_5$H$_{11}$)$_2$O	0	0.075	0.143	0.212	0.290	0.373	0.474	0.586	0.680	2.124
1,4-Dioxane	0	—	—	—	—	—	—	0.429	0.481	0.927
	10	—	—	—	—	—	—	0.391	0.436	0.772
C$_6$H$_5$NO$_2$	10	0.053	0.105	0.158	0.209	0.264	0.324	0.367	0.408	0.689
o-NO$_2$C$_6$H$_4$CH$_3$	10	0.050	0.103	0.153	0.205	0.257	0.311	0.604	0.634	1.736
C$_6$H$_5$NH$_2$	10	0.153	0.269	0.363	0.436	0.496	0.551	0.742	0.788	3.717
C$_6$H$_5$NHCH$_3$	0	0.175	0.309	0.421	0.509	0.588	0.666			

$C_6H_5NHC_2H_5$	10	0.126	0.226	0.312	0.384	0.447	0.507	0.564	0.596	1.477
	0	0.168	0.300	0.410	0.495	0.572	0.650	0.730	0.780	3.460
$C_6H_5N(CH_3)_2$	10	0.117	0.212	0.300	0.382	0.438	0.500	0.557	0.592	1.450
	0	0.076	0.149	0.223	0.302	0.389	0.488	0.600	0.672	2.048
$C_6H_5N(C_2H_5)_2$	10	0.050	0.102	0.153	0.205	0.256	0.309	0.366	0.404	0.678
	0	0.073	0.144	0.215	0.290	0.376	0.476	0.584	0.655	1.900
C_9H_7N	10	0.044	0.088	0.130	0.174	0.217	0.263	0.316	0.360	0.562
C_5H_5N	10	0.060	0.120	0.181	0.243	0.302	0.362	0.421	0.456	0.838
$C_6H_5CH_2NH_2$	10	—	—	—	—	0.308	0.372	0.444	0.488	0.953
	10	0.084	0.156	0.224	0.289	0.355	0.419	0.482	0.520	1.083
n-$C_8H_{17}NH_2$	10	0.066	0.130	0.196	0.259	0.325	0.390	0.455	0.493	0.972
$HCON(CH_3)_2$	0	0.098	0.197	0.296	0.395	0.494	0.593	0.693	0.752	3.026
	5	0.080	0.160	0.240	0.335	0.398	0.478	0.560	0.601	1.502
	10	0.065	0.130	0.196	0.262	0.326	0.394	0.454	0.500	1.000
	20	0.040	0.084	0.127	0.172	0.216	0.264	0.311	0.340	0.515
m-$BrC_6H_4CH_3$	10	0.044	0.089	0.136	0.185	0.235	0.288	0.350	0.395	0.653
1-$BrC_{10}H_7$	10	0.048	0.095	0.144	0.191	0.242	0.295	0.356	0.397	0.659
C_6H_5Br	10	0.055	0.105	0.158	0.212	0.270	0.333	0.404	0.454	0.830
n-$C_{10}H_{22}$	10	0.028	0.061	0.097	0.141	0.193	0.262	0.350	0.413	0.704
	0	0.022	0.047	0.075	0.109	0.144	0.190	0.256	0.318	0.467
	5	0.020	0.044	0.070	0.100	0.132	0.168	0.206	0.253	0.307
	10	0.011	0.023	0.037	0.055	0.078	0.103	0.134	0.156	0.185
C_6H_6	20	—	—	—	—	—	—	—	0.408	0.690
$C_6H_5CH_3$	10	—	—	—	—	—	—	—	0.393	0.649
m-$C_6H_4(CH_3)_2$	10	0.027	0.064	0.100	0.145	0.196	0.275	0.320	0.358	0.556
$1,3,5$-$C_6H_3(CH_3)_2$	10	0.032	0.068	0.109	0.155	0.200	0.253	0.310	0.326	0.484
H_2O	0	0.255	0.380	0.464	0.530	0.595	0.668	0.731	0.770	3.348
	10	0.220	0.316	0.384	0.436	0.484	0.527	0.572	0.600	1.502
	20	0.154	0.245	0.307	0.355	0.392	0.428	0.464	0.485	0.943
$CHCl_3$	0	—	—	—	—	—	—	—	0.778	3.510
	10	—	—	—	—	—	—	0.614	0.644	1.809
CCl_4	10	—	—	—	—	—	—	—	0.400	0.666

Table 26. N_{Me_2NH} Values for Dimethylamine at Different Pressures p_{Me_2NH} and Temperatures t, °C

Liquid S	t, °C	p_{Me_2NH} in mm Hg								x_{Me_2NH} at 760 mm Hg
		100	200	300	400	500	600	700	760	
Cl$_3$CCH$_2$OH	10	0.518	0.588	0.644	0.690	0.740	0.800	0.861	0.900	9.000
	15	—	—	—	—	—	—	—	0.826	4.742
	20	0.474	0.537	0.582	0.621	0.658	0.695	0.733	0.756	3.103
	25	—	—	—	—	—	—	—	0.706	2.400
C$_6$H$_5$CH$_2$OH	10	0.388	0.486	0.558	0.625	0.696	0.760	0.830	0.875	7.000
	15	—	—	—	—	—	—	—	0.783	3.610
	20	0.340	0.422	0.480	0.536	0.587	0.631	0.675	0.705	2.380
n-C$_8$H$_{17}$OH	10	0.332	0.442	0.530	0.616	0.696	0.775	0.848	0.894	8.433
	20	0.261	0.371	0.455	0.520	0.575	0.634	0.686	0.719	2.559
n-C$_4$H$_9$OH	20	—	—	—	—	—	—	—	0.709	2.441
(CH$_2$OH)$_2$	20	0.267	0.382	0.455	0.524	0.584	0.637	0.685	0.719	2.555
(CH$_2$OH)$_2$CHOH	20	0.300	0.430	0.520	0.584	0.636	0.690	0.736	0.759	3.149
C$_6$H$_5$COCH$_3$	20	0.068	0.135	0.204	0.276	0.344	0.418	0.500	0.559	1.267
C$_6$H$_5$CO$_2$C$_2$H$_5$	20	0.072	0.146	0.218	0.291	0.367	0.442	0.523	0.572	1.337
C$_6$H$_5$OC$_2$H$_5$	20	0.067	0.136	0.206	0.278	0.350	0.423	0.502	0.554	1.243
C$_6$H$_5$CN	10	0.100	0.200	0.310	0.424	0.540	0.660	0.780	0.848	5.580
	20	0.075	0.150	0.224	0.296	0.368	0.443	0.524	0.576	1.358
(n-C$_8$H$_{17}$)$_2$O	20	0.090	0.173	0.253	0.330	0.404	0.477	0.559	0.605	1.515
(n-C$_5$H$_{11}$)$_2$O	20	0.080	0.160	0.240	0.318	0.397	0.477	0.554	0.596	1.475
1,4-Dioxane	20	—	—	—	—	—	—	—	0.549	1.217
C$_6$H$_5$NO$_2$	20	0.053	0.110	0.169	0.233	0.300	0.375	0.456	0.506	1.026
o-NO$_2$C$_6$H$_4$CH$_3$	20	0.060	0.117	0.178	0.240	0.303	0.370	0.450	0.505	1.010
C$_6$H$_5$NH$_2$	20	0.137	0.253	0.350	0.437	0.515	0.584	0.648	0.687	2.194
C$_6$H$_5$NHCH$_3$	20	—	0.240	0.334	0.414	0.482	0.547	0.611	0.650	1.857
C$_6$H$_5$NHC$_2$H$_5$	20	—	0.238	0.332	0.411	0.480	0.547	0.611	0.650	1.857
	10	0.086	0.180	0.280	0.386	0.508	0.638	0.780	0.848	5.580
C$_6$H$_5$N(CH$_3$)$_2$	20	0.061	0.127	0.192	0.257	0.330	0.404	0.485	0.538	1.165

Compound	t									
$C_6H_5N(C_2H_5)_2$	20	0.072	0.141	0.211	0.281	0.354	0.429	0.510	0.557	1.255
C_9H_7N	20	0.060	0.126	0.191	0.255	0.322	0.394	0.472	0.522	1.092
C_5H_5N	20	—	—	—	—	—	—	0.507	0.564	1.293
$C_6H_5CH_2NH_2$	10	—	—	—	—	—	—	—	0.815	4.417
	20	0.075	0.145	0.220	0.296	0.375	0.452	0.532	0.580	1.381
$n\text{-}C_8H_{17}NH_2$	10	0.120	0.242	0.363	0.486	0.603	0.713	0.825	0.888	7.911
	20	0.093	0.183	0.266	0.347	0.428	0.506	0.583	0.626	1.674
$HCON(CH_3)_2$	10	0.066	0.155	0.255	0.364	0.490	0.625	0.766	0.848	5.590
	15	—	—	—	—	—	—	—	0.686	2.226
	20	—	0.102	0.160	0.221	0.294	0.375	0.461	0.511	1.044
	25	—	—	—	—	—	—	—	0.412	0.700
$m\text{-}BrC_6H_4CH_3$	20	0.065	0.132	0.203	0.275	0.349	0.428	0.510	0.563	1.287
$1\text{-}BrC_{10}H_7$	20	0.060	0.117	0.175	0.237	0.300	0.368	0.446	0.500	1.000
C_6H_5Cl	20	—	—	—	—	—	0.429	0.520	0.575	1.354
C_6H_5Br	20	—	—	—	—	—	—	0.532	0.580	1.381
C_6H_5I	20	0.097	0.132	0.203	0.341	0.417	0.495	0.570	0.612	1.578
$n\text{-}C_{10}H_{22}$	10	0.068	0.140	0.235	0.338	0.456	0.598	0.755	0.840	5.330
	15	—	—	—	—	—	—	—	0.665	1.980
	20	0.065	0.110	0.168	0.225	0.288	0.352	0.436	0.501	1.008
	25	—	—	—	—	—	—	—	0.390	0.640
C_6H_6	20	—	—	—	—	—	—	—	0.541	1.180
$C_6H_5CH_3$	20	—	—	—	—	—	—	—	0.580	1.380
$m\text{-}C_6H_4(CH_3)_2$	10	0.070	0.157	0.262	0.382	0.517	0.651	0.790	0.862	6.246
	20	—	—	—	0.260	0.336	0.415	0.504	0.564	1.290
$1,3,5\text{-}C_6H_3(CH_3)_3$	20	0.060	0.124	0.188	0.255	0.327	0.404	0.492	0.549	1.217
$CHCl_3$	20	—	—	—	—	—	0.610	0.672	0.708	2.424
CCl_4	20	—	—	—	—	—	—	0.550	0.596	1.475
H_2O	20	0.117	0.215	0.300	0.375	0.440	0.502	0.566	0.606	1.538

10.2.5. Data for Dimethylamine, bp 6.9°C/1 atm

Table 26 gives N_A and x_A values for dimethylamine at different pressures and temperatures. Representative plots of N_A/p_A for 10 and 20°C are shown in Figs. 105 and 106. Representative plots of $x_A/t°C$ are shown in Fig. 107.

As for monomethylamine, the lines for alcohols are strongly concave upward well over on the right side of the R-line, but the lines for the ethers $(n\text{-}C_8H_{17})_2O$, $(n\text{-}C_5H_{11})_2O$, and $C_6H_5OC_2H_5$ are just on the right of the R-line; the line for dioxane, however, is on the left. Two notable trends are exemplified by the line for water, now much nearer the R-line but still on the right of it, and that for dimethylformamide, now decidedly convex upward on the left of the R-line. These trends are continued in the trimethylamine systems. The lines for aniline and the alkylanilines are similar to those for the monomethylamine systems; benzene and its homologs give a close cluster of lines over on the left of the R-line, and the line for n-decane is on the left of these. Chloroform again reveals its decided acidic function.

10.2.6. Data for Trimethylamine, bp 2.9°C/1 atm

Table 27 gives N_A and x_A values for trimethylamine at different pressures and temperatures. Representative plots of N_A/p_A for 5, 10, and 25°C are shown in Figs.

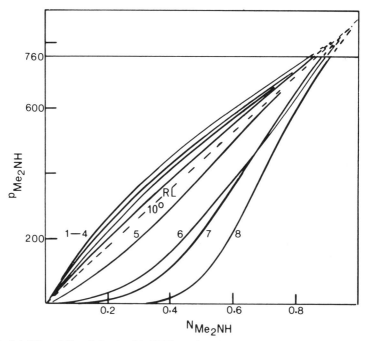

Fig. 105. Solubility of dimethylamine Me_2NH in various liquids at 10°C and different pressures p_{Me_2NH} in mm Hg. Registration of the mole fraction N_{Me_2NH} vs p_{Me_2NH} plots on the R-line diagram. The lliquids are: (1) $n\text{-}C_{10}H_{22}$; (2) $HCON(CH_3)_2$; (3) $C_6H_5N(CH_3)_2$; (4) C_6H_5CN; (5) water; (6) $n\text{-}C_8H_{17}OH$; (7) $C_6H_5CH_2OH$; (8) Cl_3CCH_2OH. The pressures p_{Me_2NH} are in mm Hg.

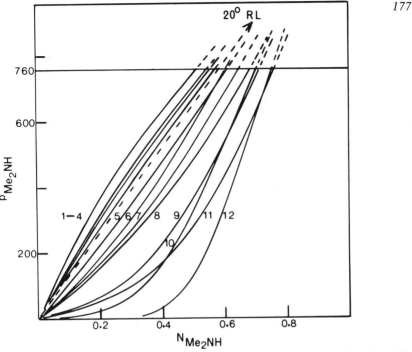

Fig. 106. Solubility of dimethylamine Me$_2$NH at 20°C at different pressures p_{Me_2NH}. Registration of the mole fraction N_{Me_2NH} vs p_{Me_2NH} plots on the R-line diagram. The liquids are: (1) HCON(CH$_3$)$_2$; (2) 1,3,5-C$_6$H$_3$(CH$_3$)$_3$; (3) m-BrC$_6$H$_4$CH$_3$; (4) C$_6$H$_5$CO$_2$C$_2$H$_5$; (5) C$_6$H$_5$I; (6) water (7) C$_6$H$_5$NHC$_2$H$_5$ (and C$_6$H$_5$NHCH$_3$); (8) C$_6$H$_5$NH$_2$; (9) n-C$_8$H$_{17}$OH; (10) C$_6$H$_5$CH$_2$OH; (11) (CH$_2$OH)$_2$CHOH; (12) Cl$_3$CCH$_2$OH. The pressures p_{Me_2NH} are in mm Hg.

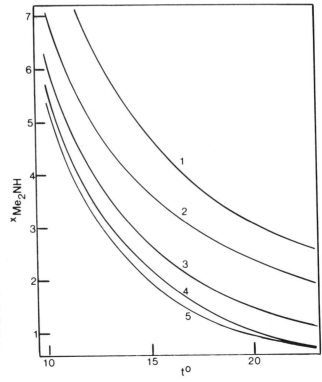

Fig. 107. Mole ratio x_{Me_2NH} solubility of dimethylamine Me$_2$NH at 1 atm and different temperatures t°C. The liquids are: (1) Cl$_3$CCH$_2$OH; (2) C$_6$H$_5$CH$_2$OH; (3) water; (4) HCON(CH$_3$)$_2$; (5) n-C$_{10}$H$_{22}$.

Table 27. N_{Me_3N} Values for Trimethylamine at Different Pressures p_{Me_3N} and Temperatures t, °C

S	t, °C	N_{Me_3N} at different pressures								x_{Me_3N} at 760 mm Hg
		100	200	300	400	500	600	700	760	
Cl_3CCH_2OH	5	0.478	0.545	0.600	0.660	0.722	0.800	0.875	0.930	13.280
	10	0.454	0.515	0.567	0.613	0.660	0.713	0.780	0.820	4.550
	20	—	—	—	—	—	—	—	0.689	2.216
	25	—	—	—	—	—	—	—	0.654	1.892
$C_6H_5CH_2OH$	5	0.315	0.407	0.484	0.557	0.644	0.742	0.866	0.925	12.330
	10	0.293	0.372	0.448	0.508	0.574	0.644	0.725	0.776	3.460
	20	—	—	—	—	—	—	—	0.585	1.409
	25	0.224	0.301	0.353	0.395	0.436	0.475	0.516	0.541	1.179
n-$C_8H_{17}OH$	5	0.263	0.401	0.510	0.607	0.691	0.790	0.882	0.936	14.620
	10	0.240	0.392	0.472	0.538	0.602	0.675	0.752	0.800	4.000
	20	—	—	—	—	—	—	—	0.636	1.750
	25	0.157	0.251	0.320	0.382	0.436	0.486	0.538	0.568	1.314
n-C_4H_9OH	20	—	—	—	—	—	—	—	0.571	1.340
	25	—	—	—	—	—	—	0.488	0.510	1.040
$(CH_2OH)_2$	10	0.116	0.216	0.304	0.384	0.455	0.531	0.616	0.672	2.048
	25	0.058	0.120	0.180	0.232	0.285	0.331	0.378	0.396	0.656
$(CH_2OH)_2CHOH$	10	0.134	0.238	0.316	0.395	0.464	0.530	0.604	0.657	1.918
	25	0.075	0.129	0.177	0.224	0.272	0.322	0.376	0.408	0.689
$C_6H_5COCH_3$	25	0.039	0.080	0.117	0.157	0.202	0.252	0.310	0.353	0.547
$C_6H_5CO_2C_2H_5$	20	—	—	—	—	—	—	—	0.506	1.025
$C_6H_5OC_2H_5$	25	0.051	0.103	0.152	0.205	0.256	0.292	0.376	0.423	0.734
C_6H_5CN	10	0.076	0.160	0.255	0.354	0.460	0.566	0.680	0.750	3.000
$(n$-$C_8H_{17})_2O$	10	0.067	0.140	0.217	0.302	0.400	0.507	0.640	0.720	2.571
	20	—	—	—	0.376	0.452	0.524	0.595	0.630	1.703
1,4-Dioxane	25	0.148	0.232	0.302	0.366	0.426	0.482	0.536	0.567	1.308
$C_6H_5NO_2$	10	—	—	—	0.260	0.356	0.464	0.595	0.680	2.140
	10	0.060	0.119	0.184	—	—	—	—	0.694	2.268
	25	0.032	0.070	0.110	0.151	0.196	0.243	0.293	0.326	0.483
o-$NO_2C_6H_4CH_3$	5	0.071	0.146	0.238	0.351	0.482	0.647	0.828	0.920	11.500

Solvent	t (°C)										
$C_6H_5NH_2$	10	0.064	—	0.132	0.202	0.284	0.380	0.486	0.605	0.676	2.086
	20	—	—	—	—	—	—	—	—	0.406	0.684
	25	0.040	0.080	0.217	0.118	0.160	0.204	0.250	0.301	0.331	0.497
$C_6H_5NHCH_3$	10	0.111	0.217	0.327	0.432	0.531	0.626	0.720	0.775	—	3.444
$C_6H_5NHC_2H_5$	10	0.124	0.234	0.336	0.433	0.526	0.620	0.716	0.775	—	3.444
$C_6H_5N(CH_3)_2$	10	0.124	0.233	0.335	0.432	0.526	0.612	0.712	0.771	—	3.367
$C_6H_5N(C_2H_5)_2$	10	0.071	0.149	0.235	0.328	0.422	0.535	0.660	0.744	—	2.907
C_7H_7N	10	0.083	0.165	0.255	0.352	0.454	0.560	0.673	0.750	—	3.000
C_5H_5N	10	0.051	0.112	0.179	0.248	0.321	0.413	0.545	0.648	—	1.844
$C_6H_5CH_2NH_2$	10	—	—	—	—	—	—	0.65	0.70	—	2.33
	5	0.060	0.128	0.203	0.284	0.380	0.492	0.640	0.732	—	2.731
$n\text{-}C_8H_{17}NH_2$	10	0.141	0.265	0.387	0.507	0.628	0.750	0.866	0.930	—	13.280
	20	0.106	0.213	0.317	0.418	0.516	0.617	0.717	0.776	—	3.450
	25	—	—	—	—	—	—	—	0.585	—	1.407
$HCON(CH_3)_2$	5	0.075	0.141	0.210	0.277	0.340	0.400	0.463	0.500	—	1.000
	10	0.028	0.072	0.128	0.202	0.305	0.442	0.700	0.880	—	7.679
	25	0.024	0.060	0.098	0.145	0.205	0.288	0.410	0.511	—	0.1046
$m\text{-}BrC_6H_4CH_3$	10	0.017	0.037	0.060	0.084	0.106	0.132	0.165	0.191	—	0.236
$1\text{-}BrC_{10}H_7$	25	0.089	0.171	0.261	0.362	0.476	0.590	0.704	0.772	—	3.386
$1\text{-}ClC_{10}H_7$	25	0.050	0.096	0.141	0.192	0.240	0.293	0.345	0.380	—	0.613
C_6H_5Br	10	0.043	0.090	0.132	0.178	0.228	0.278	0.332	0.365	—	0.575
C_6H_5I	10	0.088	0.182	0.290	0.386	0.496	0.600	0.710	0.770	—	3.348
$n\text{-}C_{10}H_{22}$	10	0.100	0.202	0.303	0.398	0.502	0.606	0.715	0.780	—	3.546
C_6H_6	10	0.108	0.211	0.313	0.416	0.517	0.624	0.728	0.786	—	3.673
$C_6H_5CH_3$	10	—	—	—	—	—	—	0.69	0.74	—	2.85
$m\text{-}C_6H_4(CH_3)_2$	10	—	—	—	—	—	—	0.718	0.768	—	3.311
$1,3,5\text{-}C_6H_3(CH_3)_3$	10	—	—	—	—	—	—	0.716	0.768	—	3.311
	10	0.096	0.192	0.284	0.382	0.488	0.600	0.700	0.756	—	3.098
H_2O	5	0.076	0.150	0.230	0.314	0.405	0.532	0.606	—	0.544	—
	10	0.056	0.116	0.178	0.246	0.317	0.397	0.485	0.554	—	1.183
$CHCl_3$	25	—	—	—	—	—	—	—	—	—	1.241

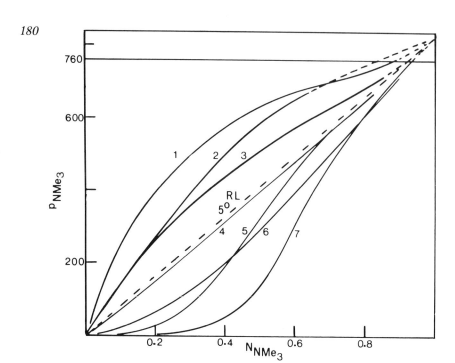

Fig. 108. Solubility of trimethylamine NMe$_3$ at 5°C and different pressures p_{NMe_3} in mm Hg. Registration of mole fraction N_{NMe_3} vs p_{NMe_3} on the R-line diagram. The liquids are: (1) HCON(CH$_3$)$_2$; (2) water (3) o-NO$_2$C$_6$H$_4$CH$_3$; (4) n-C$_8$H$_{17}$NH$_2$; (5) C$_6$H$_5$CH$_2$OH; (6) n-C$_8$H$_{17}$OH; (7) Cl$_3$CCH$_2$OH.

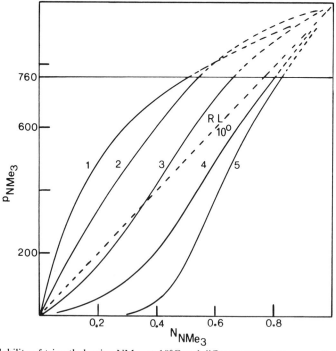

Fig. 109. Solubility of trimethylamine NMe$_3$ at 10°C and different pressures p_{NMe_3}. Registration of the mole fraction N_{NMe_3} vs p_{NMe_3} plots of the R-line diagram. The liquids are: (1) HCON(CH$_3$)$_2$; (2) water; (3) (CH$_2$OH)$_2$CHOH [curve for NMe$_3$ in (CH$_2$OH)$_2$ is similar to 3]; (4) n-C$_8$H$_{17}$OH; (5) Cl$_3$CCH$_2$OH. The pressures p_{NMe_3} are in mm Hg.

Ammonia and the Three Methylamines 181

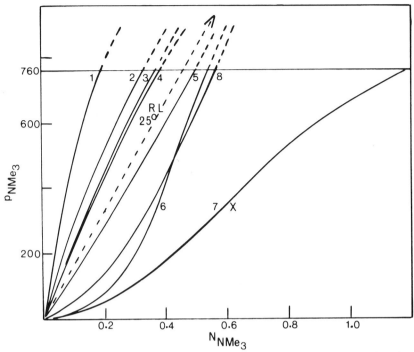

Fig. 110. Solubility of trimethylamine NMe$_3$ at 25°C and different pressures p_{NMe_3}. Registration of the mole fraction N_{NMe_3} vs p_{NMe_3} on the R-line diagram. The liquids are: (1) HCON(CH$_3$)$_2$; (2) C$_6$H$_5$NO$_2$ (curve for NMe$_3$ in o-NO$_2$C$_6$H$_4$CH$_3$ is immediately on the right of 2); (3) 1-ClC$_{10}$H$_7$; (4) 1-BrC$_{10}$H$_7$; (5) n-C$_8$H$_{17}$NH$_2$; (6) C$_6$H$_5$CH$_2$OH; (7, X) the mole ratio x_{NMe_3} plot corresponding to 6; (8) n-C$_8$H$_{17}$OH. The pressures p_{NMe_3} are in mm Hg.

108–110. Representative plots of x_A/t°C are shown in Fig. 111. The marked difference in the N_A/p_A and x_A/p_A plots is illustrated in Fig. 112 for the example of benzyl alcohol at 5°C. Lines for chloro-, bromo-, and iodobenzene and for the benzene homologs are clustered near the R-line. Again the lines for dimethyl- and diethylaniline form a pair convex upward on the left, and those for the corresponding monoalkylanilines are concave upward on the right of the R-line; aniline is again on the right. Whereas the line for benzylamine is just on the left, that for n-octylamine is just on the right of the R-line. The monohydric alcohols maintain their positions well over on the right; the deep sagging of the lines for benzyl alcohol, still further accentuated in the line for 2,2,2-trichloroethanol, is strongly indicative of the acidic function presumably entailing the interposition of the hydrogen bonding

$$\equiv N \cdots H - OR$$

in the hydrogen-bonding structure of the alcohol. Water now gives a line pronouncedly convex upward well on the left of the R-line, and the line for dimethylformamide is still further on the left; the three methyl groups in trimethylamine evidently hinder an intrusion, a fitting into the hydrogen-bonding structure of these two liquids. A similar reason may be given for the lines for ethylene glycol and glycerol, which now wander near the R-line in a peculiar way.

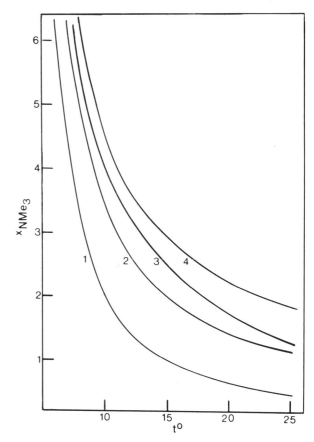

Fig. 111. Solubility of trimethylamine NMe_3 at 1 atm and different temperatures $t°C$. The mole ratio x_{NMe_3} vs $t°C$ plots are for the liquids: (1) o-$NO_2C_6H_4CH_3$; (2) $C_6H_5CH_2OH$; (3) n-$C_8H_{17}OH$; (4) Cl_3CCH_2OH.

10.2.7. Apparent Errors in Published Data for Trimethylamine

An example of the use of the R-line diagram in checking published data is afforded by the trimethylamine data of Halban (1913).[146]

The solubilities of this gas in 18 organic liquids S at 25°C and p_{Me_3N} = 760 mm Hg are recorded in the International Critical Tables (1928)[182] as the Bunsen absorption coefficients (α) derived from the respective Oswald coefficients L, taken as the volume of gas absorbed by one volume of S but presented in the original paper[146] as "concentration of gas in a liter of solution" divided by "concentration of gas per liter in the gas phase." In conventional treatment these two forms of L are taken to be equivalent or essentially so, but there are many examples in which this is far from being true. The point is well illustrated by the example of benzyl alcohol S, for which Halban found that "0.095 moles per liter of solution" was the "concentration" at p_{Me_3N} = 1.35 mm Hg. Assuming "compliance with Henry's law," he calculated the "concentration in moles per liter of solution" at 760 mm Hg as $0.095 \times (760/1.35)$ and thence the L (concentration form) = 1308, taken as L (vol. A/vol. S),[182] to give $\alpha = 1308 \times (273/298) = 1198$. This number is equivalent to

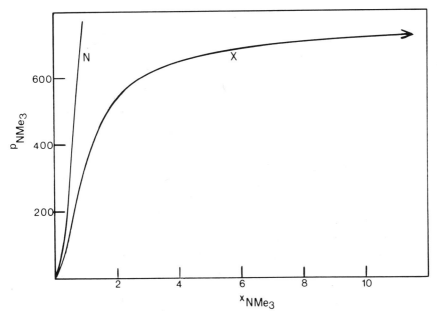

Fig. 112. A comparison of the mole fraction N_{NMe_3} vs pressure p_{NMe_3} and mole ratio x_{NMe_3} vs p_{NMe_3} plots for benzyl alcohol at 5°C. The x_{NMe_3} lines reaches the horizontal at 1 atm at $x_{NMe_3} = 24.00$ ($N_{NMe_3} = 0.96$). The pressures p_{NM_3} are in mm Hg.

5.78 for x_{Me_3N} ($N_{Me_3N} = 0.852$) at 25°C and 760 mm Hg, for which the volume of *solution* would be at least three times that of the benzyl alcohol taken. The value found in the present study is $x_{Me_3N} = 1.179$ ($N_{Me_3N} = 0.541$), a discrepancy explained as follows (see Fig. 110). At $p_{Me_3N} = 1.35$ mm Hg, the point on the *x* vs *p* plot is near the origin, and 0.095 moles of Me_3N "per liter of solution" may be read as approximately moles per liter of original liquid S, equivalent to the mole ratio form. The extrapolation $0.095 \times (760/1.35)$ is therefore in the mole *ratio* form of "Henry's law," and it gives $x_{Me_3N} = 5.78$. In the present work, the N_{Me_3N} plot is strongly concave upward and is on the right side of the R-line (see Fig. 110). The x_{Me_3N} plot is of the reclining S form, concave upward at the lower pressures, convex upward as $p_{Me_3N} = 760$ mm Hg is approached, and reaching that horizontal at $x_{Me_3N} = 1.179$. Halban gave "concentration" values for six pressures up to 25 mm Hg; the calculated x_{Me_3N} values fall on a line definitely concave upward. On the other hand, Halban's data for nitrobenzene, converted into $\alpha = 48.6$,[182] corresponds to $x_{Me_3N} = 0.227$, in contrast with 0.483 found in the present work. This is because the N_{Me_3N} plot is convex upward on the left of the R-line, and the x_{Me_3N} plot is likewise convex upward. Therefore, for this liquid the x_{Me_3N} value increases more rapidly with p_{Me_3N} than the linear function would allow. For the same reason, Halban's data give $x_{Me_3N} = 0.264$ for 1-bromonaphthalene, whereas in the present work it is 0.613. The α values of the remaining 15 liquids are open to the same objection, as also appear to be the values $\alpha_{25} = 366$, now calculated as $x_{Et_3N} = 0.881$ ($N_{Et_3N} = 0.468$) for nitromethane, and $\alpha_{25} = 1979$, $x_{Et_3N} = 11.5$ ($N_{Et_3N} = 0.920$) for hexane, based also on Halban's data. Halban determined his solubility data in connection with his main study which was on the effect of solvents on the kinetics of the *p*-nitrobenzyl chloride–trimethylamine (and triethylamine) systems; therefore, his conclusions require reconsideration.

10.3. Appreciation of the Essential Pattern of Data

The predictability factor is always in the mind of the operational chemist. The Hildebrand solubility parameter equation was never thought of as covering the systems described in this chapter. I know of no mathematical approach to the prediction of solubility data for such systems. There is a need for more factual data, and this is simply a matter of man-hours; but a study of the R-line diagrams given herein, a consideration of the general concept of acid–base function, as suggested herein, and the exercise of chemical intuition should now enable one to appreciate the essential pattern of data and entertain useful speculation, on a rational basis, about the probable N_{gas} value for a certain gas A and liquid S at a pressure p_{gas} and temperature $t°C$.

The first factor to consider is the bp of the gas A; this indicates the position of the R-line in the whole panorama of the spectrum and refers to the tendency of A to condense at the operational temperature $t°C$ at pressure p_A. On this factor alone, we may expect the N_{NH_3} values to be decidedly less than the corresponding values for the three methylamines. The intermolecular structure of the liquid S will be a second, but constant, factor for the same S at the same $t°C$, but the mode of interaction of ammonia or the individual amines will not be strictly parallel. A scrutiny of the diagrams gives a useful appreciation of the degree to which these differences may be discerned.

It is left to the reader to predict the solubility patterns for the three ethylamines in the liquids S named in this chapter.

10.4. Recent Studies on the Interaction of Aliphatic Amines with Water Structure

Kaulgud and Patil (1974)[400] have described their work on partial molal volumes and apparent compressibility of the amines $MeNH_2$, Me_2NH, $EtNH_2$, Et_2NH, *n*-$PrNH_2$, and *n*-$BuNH_2$ in water at mole fractions N from 0 to 0.25 at 20°C. They wrote of the "negative deviations from Raoult's law," shown by $MeNH_2$, Me_2NH, and $EtNH_2$, which were stated to have activity coefficients less than unity and deemed to indicate "affinity for water molecules" and "substitutional dissolution." The other amines, however, showed "activity coefficients greater than unity," due to "interstitial dissolution" and the occupation of "cavities existing in the open

Ammonia and the Three Methylamines

water structure" or of "cavities which are *created on demand*," thus "forcing the water into an ordered arrangement (hydrophobic hydration)." With these aspects in mind, it is relevant to look again at the data for the three methylamines.

In Fig. 102 I show the N_{MeNH_2}/p_{MeNH_2} line for water at 20°C. It is concave upward on the right of the R-line. For $N_{MeNH_2} = 0.245$, near the upper limit of concentration examined by Kaulgud and Patil, the p_{MeNH_2} is only 200 mm Hg. At $p_{MeNH_2} = 760$ mm Hg, N_{MeNH_2} is 0.485, i.e., there is 0.943 mole of $MeNH_2$ per mole of water. At 0°C there are 3.348 molecules of $MeNH_2$ for each molecule of water (H_2O) for $p_{MeNH_2} = 760$ mm Hg. In Fig. 106 I show the N_{Me_2NH}/p_{Me_2NH} line for water at 20°C. The line is still concave upward but is now nearer to the R-line. For $N_{Me_2NH} = 0.215$, p_{Me_2NH} is only 200 mm Hg. At $p_{Me_2NH} = 760$ mm Hg, $N_{Me_2NH} = 0.606$, equivalent to 1.538 molecules of amine for each molecule of water (H_2O). In Fig. 109 I show the N_{Me_3N}/p_{Me_3N} line for water at 10°C, the line for 20°C not having been observed. It is *convex* upward, on the left of the R-line. For $N_{Me_3N} = 0.246$, p_{Me_3N} is 400 mm Hg. For $p_{Me_3N} = 760$ mm Hg, N_{Me_3N} is 0.544, i.e., there are 1.183 molecules of amine for each molecule of water (H_2O). How does the "interstitial dissolution" model fit in with this?

11

Dimethyl Ether

Apart from data by Horiuti[177] for CCl_4, C_6H_6, C_6H_5Cl, CH_3COCH_3, and $CH_3CO_2CH_3$ for 25°C, most of the available data are by Gerrard.[99] I have previously[99] given the R-line diagrams based on Horiuti's data, and examples of x_{Me_2O} vs p_{Me_2O} plots. I now give examples of N_{Me_2O} vs p_{Me_2O} plots, (Figs. 113, 114), as well as x_{Me_2O} vs $t°C$ plots (Fig. 115).

As for the diethyl ether mentioned in Chap. 3, the acid–base function designated as $n\text{-}C_5H_{11}CO_2H\cdots O(CH_3)_2$ and especially $Cl_3CCH_2OH\cdots O(CH_3)_2$ is clearly discernible; but n-octanol and dimethylformamide are not especially accommodating.

Dimethyl ether is a member of the trio CH_3OCH_3, $CH_3OC_2H_5$, $C_2H_5OC_2H_5$, for which the main difference in N_A values for corresponding liquids S will be due to the increase in condensing tendency in passing from the dimethyl to the diethyl ether. Fox and Lambert[84] stated that chloroform and ether may be expected to form a loose intermediate compound by hydrogen bonding, $Et_2O\cdots HCCl_3$. I have already mentioned the ether–sulfuric acid systems (see Fig. 38). The addition of $(CH_3)_2O$ to sulfuric acid at 20°C appears to entail a gradual interference with the hydrogen-bonding structure of sulfuric acid; p_{Me_2O} remains low until N_{Me_2O} (added) reaches about 0.4, and then the N_{Me_2O} vs p_{Me_2O} plot bends round and rises steeply at about $N_{Me_2O} = 0.5$ (i.e., $x_{Me_2O} \approx 1$). Absorption continues as the p_{Me_2O} is increased, and eventually N_{Me_2O} becomes nearer to unity as $p_{Me_2O} \to p°_{Me_2O}$. In textbooks this system is represented as

$$R_2O + H_2SO_4 \to R_2OH^+ + HSO_4^-$$

It is stated that this reaction goes virtually to completion. Must I conclude from this that, when p_{Et_2O} is about 40 mm Hg, the system is virtually $(Et_2OH^+HSO_4^-)$ and nothing else? However, as p_{Et_2O} is increased, N_{Et_2O} (added) continues to increase, as shown in Fig. 38. By the stage where p_{Et_2O} is about 280 mm, a total of 2 moles of Et_2O have been absorbed for 1 mole of the original acid. Have I now $(2Et_2OH^+SO_4^{2-})$? The low conductance in the HCl–ether systems referred to in Chap. 9 appears to contradict this. However, ether is still further absorbed as p_{Et_2O} is increased. It appears to me more likely that the ether gradually becomes involved in

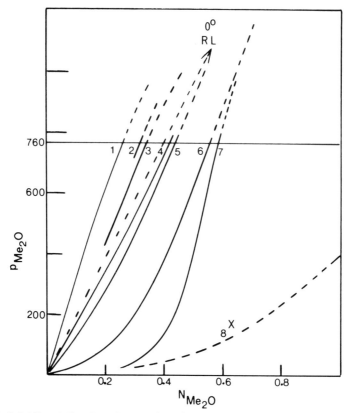

Fig. 113. Solubility of dimethyl ether Me$_2$O at 0°C and different pressures p_{Me_2O}. Registration of mole fraction N_{Me_2O} vs p_{Me_2O} plots on the R-line diagram. The liquids are: (1) HCON(CH$_3$)$_2$; (2) C$_6$H$_5$NO$_2$; (3) o-HOC$_6$H$_4$CO$_2$CH$_3$; (4) (n-C$_8$H$_{17}$)$_2$O; (5) n-C$_5$H$_{11}$CO$_2$H; (6) Cl$_3$CCH$_2$OH; (7) Cl$_2$CHCO$_2$H at 15°C. Line 8 (X) is the mole ratio x_{Me_2O} plot for 7. The pressures p_{Me_2O} are in mm Hg.

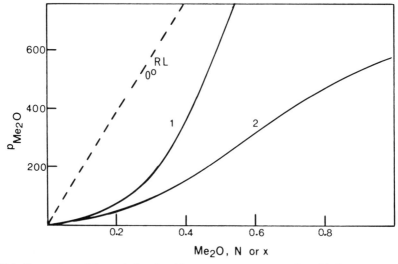

Fig. 114. Comparison of the mole fraction N_{Me_2O} vs pressure p_{Me_2O} plot with the corresponding mole ratio x_{Me_2O} plot for trichloroethanol (Cl$_3$CCH$_2$OH) at 0°C. The pressures p_{Me_2O} are in mm Hg.

Dimethyl Ether

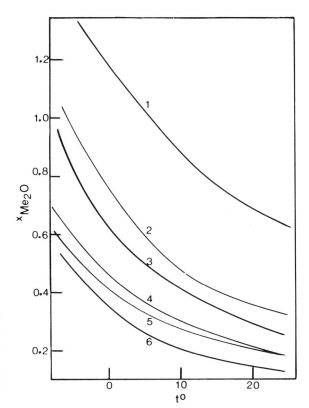

Fig. 115. Solubility of dimethyl ether Me$_2$O at 1 atm. Plots of the mole ratio x_{Me_2O} vs temperature $t°C$. The liquids are: (1) Cl$_3$CCH$_2$OH; (2) n-C$_5$H$_{11}$CO$_2$H; (3) (n-C$_8$H$_{17}$)$_2$O; (4) o-HOC$_6$H$_4$CO$_2$CH$_3$; (5) n-C$_8$H$_{17}$OH; (6) HCON(CH$_3$)$_2$.

the hydrogen-bonding structure of sulfuric acid, and as a consequence some measure of proton transfer may occur (see Fig. 116). On the other hand, the x_{Me_2O} value for water is only 0.03 at 0°C; an explanation of the large difference in the x_{Me_2O} value for sulfuric acid as compared with that for water is to be found partly in the unsymmetrical pattern of the hydrogen bonding in the sulfuric acid. Dimethyl and diethyl ethers can break up the pattern in the way shown, relying on the acidic function sited on hydrogen. Hydrogen chloride has a x_{HCl} value of only 0.02 for sulfuric acid, in which the basic function is dispersed over five sites. In water, hydrogen chloride has a better chance, and so the x_{HCl} value is 0.41 at 0°C, p_{HCl} = 1 atm. Notice that the x_{NH_3} value for water at 0°C is 0.95 at p_{NH_3} = 1 atm; this would appear to be due to the intensive-enough basic function of ammonia, the formation of ions being a very minor function.

As with sufur dioxide, methyl chloride, and hydrogen sulfide, so with dimethyl ether there is no systematic correlation between x_{Me_2O} (or N_{Me_2O}), on the one hand, and dipole moment or dielectric constant of S, on the other.

Looking at Fig. 113, it can be visualized that at −20°C the whole of each line could be drawn on the existing scale of that diagram. The lines would be less steep and would tend to cluster by the stage where the horizontal at 1 atm was reached. At

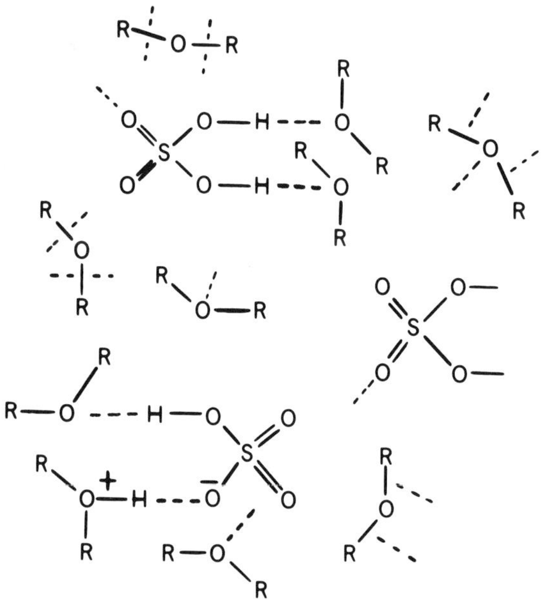

Fig. 116. Postulated modes of reversible interaction between diethyl ether and sulfuric acid at room temperatures. A mole ratio x_{Et_2O} of 2 means that numerically the hydrogen sites in sulfuric acid available for hydrogen bonding are exhausted. Neglecting the possible involvement of the hydrocarbon hydrogen, I conclude that the increase in x_{Et_2O} must entail some other form of intermolecular interaction which I have indicated by the broken lines attached to the other units, isolated in the diagram. Eventually, almost pure ether results when the x_{Et_2O} value is very large, e.g., when the mole fraction N_{Et_2O} is 0.999. In the example of dimethyl ether, the much higher value of $p^o_{Me_2O}$ must be closely approached before that condition prevails (see Fig. 113).

300 mm Hg, for example, the separation of the lines would still be quite distinct. At −30° the systems would be conventionally called liquid–liquid ones (when S was still liquid), but the pattern would be essentially similar.

12

Halogenoalkanes

12.1. Chloromethane, Chloroethane, and Bromomethane and Nonaqueous Liquids S

Data for chloromethane (bp −23.76°C) were observed by Gerrard,[99] who also gave a diagram based on previously reported data and cited references. New data (Table 28) on bromomethane (bp 3.59°C) and chloroethane (bp 13.1°C) now enable diagrams to be drawn, as shown in Figs. 117–122. Although there are some differences in the N_{MeCl}, N_{MeBr}, and N_{EtCl} values for corresponding liquids which must be attributed to factors other than the tendency of the gas to condense, the latter is the main factor. At temperatures below the bp (1 atm), e.g., 0°C for bromomethane and 10°C for chloroethane, the systems are conventionally liquid–liquid systems, but the pattern is essentially the same as for temperatures above the bp. The N_{EtCl} data were calculated from the "cm³ A/cm³ S" data of Kaplan and Romanchuk[198] for the solubilities of chlorethane in carbon tetrachloride and in 1,2-dichloroethane at 0 and 20°C, and these are plotted in Fig. 123. The cross bars indicate the limit of the the pressure measurements. Kaplan et al.[197] gave "cm³A/cm³S" data for kerosine, $d_4^{20} = 0.8429$ (containing 15% aromatic hydrocarbons), at 20°C. Vdovichenko and Kondratenko[364] gave "vol./vol." data for chloromethane in anisole at 20°C and 1 atm, and this leads to a value of $N_{MeCl} = 0.113$, entirely out of step with the N_{MeCl} vs p_{MeCl} data observed by Gerrard.

In connection with refrigerating systems, Zellhoefer (1937)[391] determined the solubilities of low-boiling halogenated hydrocarbons A in a number of liquids S at 32°C and at a pressure p_A which is equal to $p°_A$ at 4.5°C, and expressed these as "g A/cm³ S." For a number of the liquids, the N_{MeCl} and N_{EtCl} values have been calculated, and these are now used in Fig. 124 to illustrate an interesting application of the graphical presentation. On the p_{MeCl} horizontal at 2220 mm Hg, the N_{MeCl} values are registered. Each of these points is on a line which must start from the left corner of a diagram, $p_{MeCl} = 0$, and finish at a point on the right vertical axis where $p_{MeCl} = p°_{MeCl}$ for 32°C. Speculative but probable plots are shown by broken lines. Likewise the N_{EtCl} values for $p_{EtCl} = 557$ mm Hg are registered on that horizontal,

Table 28. Mole Fraction N_A Values for Different Pressures and Temperatures for MeBr, EtCl, MeCl, i-C_4H_{10}, $C(CH_3)_4$, and 2-$CH_3C_4H_9$, and Liquids S

Gas A	t, °C	N_A values at p_A in mm Hg								x_A at 760 mm Hg
		100	200	300	400	500	600	700	760	

(n-$C_8H_{17})_2$O

Gas A	t, °C	100	200	300	400	500	600	700	760	x_A at 760 mm Hg
MeBr	0	0.200	0.370	0.522	0.660	0.790	0.922	(x_A = 11.82)		—
	10	0.135	0.264	0.390	0.502	0.602	0.695	0.784	0.830	4.883
	25	—	—	—	—	0.400	0.472	0.540	0.576	1.361
EtCl	10	0.170	0.324	0.475	0.617	0.746	0.872	(x_A = 6.813)		—
	18	0.118	0.250	0.365	0.476	0.583	0.681	0.796	0.833	4.987
	25	0.108	0.208	0.304	0.392	0.477	0.562	0.644	0.691	2.232
MeCl	0	0.075	0.140	0.200	0.264	0.326	0.384	0.440	0.476	0.907
i-C_4H_{10}	0	0.101	0.195	0.286	0.375	0.457	0.536	0.615	0.664	1.988
$C(CH_3)_4$	20	0.098	0.197	0.292	0.384	0.473	0.562	0.650	0.704	2.379
2-$CH_3C_4H_9$	30	0.150	0.293	0.420	0.548	0.672	0.780	0.886	0.95[a]	—

n-C_8H_{17}Br

Gas A	t, °C	100	200	300	400	500	600	700	760	x_A at 760 mm Hg
MeBr	10	0.120	0.238	0.354	0.465	0.570	0.670	0.768	0.820	4.555
	25	0.080	0.151	0.222	0.289	0.360	0.426	0.492	0.532	1.137
EtCl	18	0.110	0.223	0.330	0.438	0.547	0.655	0.766	0.830	4.866
	25	0.091	0.183	0.271	0.358	0.445	0.530	0.610	0.662	1.958
MeCl	0	0.057	0.117	0.176	0.230	0.284	0.340	0.392	0.425	0.739
i-C_4H_{10}	0	0.070	0.138	0.208	0.282	0.355	0.430	0.516	0.580	1.379
$C(CH_3)_4$	20	0.068	0.138	0.211	0.287	0.365	0.460	0.568	0.634	1.732
2-$CH_3C_4H_9$	30	0.084	0.160	0.264	0.397	0.536	0.684	0.84[a]	0.92[a]	—

C_6H_5Br

Gas A	t, °C	100	200	300	400	500	600	700	760	x_A at 760 mm Hg
MeBr	10	0.104	0.220	0.330	0.437	0.541	0.647	0.751	0.818	4.494
	25	0.070	0.127	0.193	0.261	0.326	0.392	0.454	0.494	0.976

Halogenoalkanes

Compound	t (°C)									
EtCl	18	0.100	0.194	0.295	0.402	0.514	0.626	0.740	0.808	4.208
	25	0.091	0.153	0.236	0.320	0.408	0.497	0.586	0.640	1.774
MeCl	0	0.050	0.100	0.146	0.202	0.254	0.304	0.355	0.388	0.634
i-C_4H_{10}	0	0.033	0.068	0.108	0.154	0.208	0.270	0.360	0.435	0.769
$C(CH_3)_4$	20	0.026	0.052	0.086	0.140	0.207	0.290	0.416	0.537	1.162
2-$CH_3C_4H_9$	30	0.042	0.100	0.180	0.296	0.448	0.604	0.80[a]	—	—
					1,3,5-$C_6H_3(CH_3)_3$					
MeBr	10	0.122	0.223	0.330	0.440	0.544	0.646	0.746	0.805	4.129
	25	0.064	0.150	0.198	0.268	0.335	0.400	0.468	0.500	1.000
EtCl	18	0.104	0.210	0.318	0.422	0.530	0.635	0.748	0.810	4.263
	25	0.086	0.174	0.262	0.348	0.440	0.528	0.615	0.655	1.914
i-C_4H_{10}	0	0.058	0.117	0.180	0.245	0.322	0.404	0.496	0.568	1.346
$C(CH_3)_4$	20	0.065	0.130	0.206	0.280	0.363	0.460	0.570	0.648	1.840
					m-$C_6H_4(CH_3)_2$					
EtCl	18	—	—	—	0.436	0.547	0.655	0.764	0.827	4.834
					n-$C_{10}H_{22}$					
MeBr	10	0.084	0.174	0.265	0.363	0.466	0.581	0.700	0.771	3.374
	25	0.076	0.110	0.164	0.222	0.280	0.340	0.400	0.438	0.779
EtCl	18	0.076	0.160	0.248	0.344	0.460	0.586	0.720	0.775	3.440
i-C_4H_{10}	0	0.093	0.180	0.270	0.345	0.438	0.520	0.600	0.645	1.818
$C(CH_3)_4$	20	nearly on R-line					0.560	0.650	0.714	2.500
2-$CH_3C_4H_9$	30	0.136	0.265	0.396	0.524	0.644	0.760	0.873	0.938[a]	—

continued

Table 28 (contd.)

Gas A	t, °C	\multicolumn{7}{c}{N_A values at p_A in mm Hg}	x_A at 760 mm Hg
		100	200	300	400	500	600	700	760	
\multicolumn{11}{c}{$n\text{-}C_5H_{11}CO_2H$}
MeBr	25	0.054	0.108	0.162	0.216	0.271	0.329	0.388	0.425	0.7375
EtCl	18	0.089	0.175	0.260	0.347	0.442	0.549	0.671	0.746	2.935
$i\text{-}C_4H_{10}$	0	0.040	0.084	0.128	0.180	0.228	0.280	0.360	0.416	0.716
$C(CH_3)_4$	20	0.044	0.089	0.136	0.185	0.247	0.320	0.410	0.484	0.940
$2\text{-}CH_3C_4H_9$	30	0.060	0.126	0.207	0.305	0.420	0.570	0.75[a]	—	—
\multicolumn{11}{c}{$n\text{-}C_8H_{17}OH$}
MeBr	10	0.075	0.151	0.230	0.315	0.403	0.496	0.600	0.676	2.086
	25	0.043	0.088	0.137	0.185	0.232	0.284	0.338	0.372	0.592
EtCl	18	0.059	0.120	0.187	0.268	0.358	0.463	0.584	0.673	2.070
$i\text{-}C_4H_{10}$	25	0.050	0.103	0.160	0.218	0.283	0.350	0.453	0.490	0.961
$C(CH_3)_4$	0	0.045	0.088	0.132	0.178	0.224	0.282	0.344	0.386	0.629
$2\text{-}CH_3C_4H_9$	20	0.040	0.080	0.124	0.173	0.224	0.291	0.367	0.415	0.709
	30	0.060	0.120	0.185	0.260	0.358	0.490	0.676	(x_A = 2.09)	
\multicolumn{11}{c}{Cl_3CCH_2OH}
MeBr	10	0.076	0.152	0.230	0.316	0.412	0.515	0.636	0.716	2.522
	25	0.043	0.090	0.136	0.189	0.240	0.295	0.344	0.387	0.631
EtCl	18	0.068	0.136	0.210	0.302	0.398	0.512	0.642	0.730	2.704

Halogenoalkanes

C₆H₅CH₂OH

	T									
MeBr	10	0.056	0.120	0.195	0.276	0.364	0.457	0.566	0.640	1.778
	25	0.043	0.088	0.130	0.173	0.217	0.264	0.311	0.347	0.531
EtCl	0	0.100	0.205	0.317	0.438	0.590	0.834	—	—	—
	18	0.056	0.112	0.175	0.242	0.317	0.408	0.525	0.627	1.678
	25	0.044	0.088	0.135	0.184	0.236	0.296	0.366	0.416	0.712
MeCl	10	0.075	0.154	0.240	0.340	0.480	0.720	—	—	—
i-C₄H₁₀	0	0.032	0.064	0.096	0.128	0.164	0.200	0.244	0.274	0.378
C(CH₃)₄	0	0.012	0.020	0.028	0.037	0.049	0.064	0.086	0.107	0.120
2-CH₃C₄H₉	20	0.008	0.015	0.026	0.035	0.044	0.053	0.070	0.084	0.092
	30	0.015	0.030	0.048	0.065	0.093	0.125	0.180	—	—

C₆H₅CO₂C₂H₅

	T									
MeBr	10	0.134	0.258	0.370	0.480	0.581	0.678	0.774	0.829	4.841
	25	0.080	0.160	0.236	0.308	0.374	0.441	0.503	0.538	1.165
EtCl	18	—	—	—	—	—	—	—	0.820	4.565
MeCl	0	0.074	0.137	0.200	0.258	0.315	0.369	0.421	—	—
2-CH₃C₄H₉	30	0.050	0.110	0.180	0.255	0.356	0.50	—	0.452	0.825

o-HOC₆H₄CO₂CH₃

	T									
EtCl	18	0.082	0.172	0.265	0.372	0.486	0.606	0.726	0.802	4.0505
	25	0.072	0.147	0.224	0.302	0.382	0.466	0.550	0.612	1.580
C(CH₃)₄	20	0.017	0.040	0.064	0.087	0.116	0.150	0.185	0.211	0.267
2-CH₃C₄H₉	30	0.037	0.075	0.119	0.168	0.246	0.420	0.680	—	—

C₆H₅OC₂H₅

	T									
EtCl	18	0.112	0.222	0.329	0.437	0.548	0.658	0.768	0.831	4.911
MeCl	0	0.060	0.122	0.176	0.234	0.288	0.347	0.406	0.435	0.770

continued

Table 28 (contd.)

Gas A	t, °C	\multicolumn{8}{c}{N_A values at p_A in mm Hg}	x_A at 760 mm Hg
		100	200	300	400	500	600	700	760	

$C_6H_5OCH_3$

Gas A	t, °C	100	200	300	400	500	600	700	760	x_A at 760 mm Hg
EtCl	18	—	—	—	—	0.538	0.637	0.744	0.816	4.435
MeCl	0	0.056	0.105	0.168	0.222	0.279	0.328	0.378	0.416	0.714

$C_6H_5COCH_3$

EtCl	18	0.092	0.187	0.289	0.392	0.501	0.619	0.738	0.812	4.309

$C_6H_5NO_2$

MeBr	10	0.094	0.194	0.294	0.395	0.502	0.608	0.725	0.786	3.673
	25	0.060	0.118	0.176	0.235	0.295	0.355	0.416	0.454	0.832
EtCl	18	0.082	0.165	0.254	0.352	0.460	0.588	0.720	0.806	4.157
	25	0.062	0.131	0.200	0.270	0.342	0.429	0.529	0.596	1.475
i-C_4H_{10}	0	0.012	0.027	0.041	0.053	0.068	0.087	0.111	0.134	0.155
$C(CH_3)_4$	20	0.012	0.016	0.040	0.053	0.067	0.085	0.106	0.123	0.140
2-$CH_3C_4H_9$	30	0.020	0.040	0.066	0.100	0.146	0.240	—	—	—

C_6H_5CN

CH_3Br	10	0.110	0.218	0.330	0.432	0.532	0.642	0.744	0.813	4.336
EtCl	18	0.089	0.185	0.286	0.391	0.502	0.620	0.737	0.806	4.171
2-$CH_3C_4H_9$	30	0.027	0.060	0.090	0.128	0.176	0.270	—	—	—

Halogenoalkanes

				C$_6$H$_5$NH$_2$					
2-CH$_3$C$_4$H$_9$	30	0.009	0.020	0.029	0.041	0.057	0.084	—	—	—
				C$_6$H$_5$N(C$_2$H$_5$)$_2$					
2-CH$_3$C$_4$H$_9$	30	0.066	0.144	0.227	0.320	0.440	—	—	—	—
				HCON(CH$_3$)					
MeBr	10	—	—	—	—	0.522	0.626	0.735	0.794	3.865
	15	0.086	0.170	0.254	0.340	0.432	0.518	0.610	0.675	2.073
	25	0.060	0.120	0.182	0.244	0.306	0.371	0.433	0.470	0.887
EtCl	18	0.080	0.167	0.260	0.362	0.470	0.585	0.716	0.800	3.993
	25	—	—	—	—	—	—	—	0.594	1.464
C(CH$_3$)$_3$	20	0.009	0.016	0.024	0.033	0.040	0.048	0.060	0.071	0.077
2-CH$_3$C$_4$H$_9$	30	0.012	0.022	0.039	0.053	0.072	0.100	—	—	—

[a] Approximate.

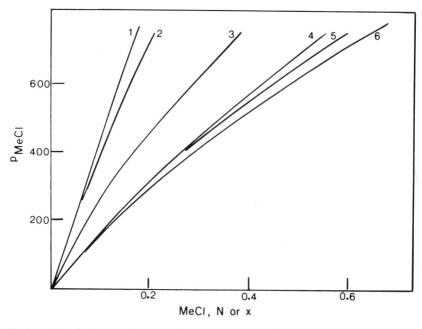

Fig. 117. Solubility of chloromethane MeCl in halobenzenes at different pressures p_{MeCl}. Line 1 is the mole fraction N_{MeCl} plot for C_6H_5Cl from data by Horiuti[177]; line 2 is the corresponding mole ratio x_{MeCl} plot for 25°C. From data by Gerrard,[99] line 3 is for x_{MeCl} in C_6H_5Cl at 10°C; the corresponding plots for 0°C are: (4) C_6H_5I; (5) C_6H_5Br; (6) C_6H_5Cl. The pressures p_{MeCl} are in mm Hg.

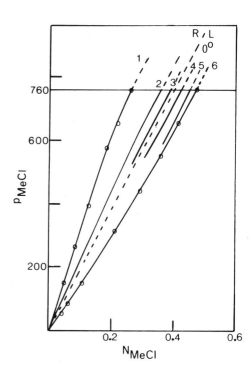

Fig. 118. Solubility of chloromethane MeCl at 0°C and at different pressures p_{MeCl}. The pattern of the mole fraction N_{MeCl} vs p_{MeCl} plots for the following liquids: (1) $C_6H_5CH_2OH$; (2) 1-$C_{10}H_7Br$; (3) C_6H_5Br; (4) n-$C_8H_{17}Br$; (5) $C_6H_5CO_2C_2H_5$; (6) $(n$-$C_8H_{17})_2O$. the disposition of the experimental points is illustrated. The pressures p_{MeCl} are in mm Hg.

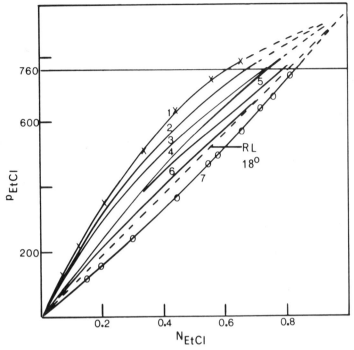

Fig. 119. Solubility of chloroethane EtCl, at 18°C and different pressures p_{EtCl}. Registration of the mole fraction N_{EtCl} vs p_{EtCl} plots on the R-line diagram. The liquids are: (1) $C_6H_5CH_2OH$; (2) n-$C_8H_{17}OH$; (3) Cl_3CCH_2OH; (4) n-$C_5H_{11}CO_2H$; (5) n-$C_{10}H_{22}$ (curve for EtCl in $C_6H_5NO_2$ is similar to 5); (6) C_6H_5Br; (7) $(n$-$C_8H_{17})_2O$. $C_6H_5OC_2H_5$ is just on the right of the R-line. The disposition of experimental points is illustrated. The pressures p_{EtCl} are in mm Hg.

and the probable plots are indicated by broken lines. In this way the approximate values for 760 and 557 mm Hg for the two gases may be compared as far as the duplication of the liquids go. It is worthwhile pointing out briefly that if the chloroethane–carbitol acetate system is deemed to be nearly "ideal" because the N_{EtCl} plot appears nearly "to obey Raoult's law," then we are obliged to conclude that the C_2H_5Cl—$Cl_2CHCHCl_2$ system is decidedly not "ideal." "A/cm^3 S" data for A = $CHFCl_2$, CCl_2F_2, CH_2Cl_2, $C_2Cl_2F_4$, and $CFCl_3$ were also given.

By the same procedure, Zellhoefer, Copley, and Marvel (1938)[392] determined the N_{CHCl_2F} values for 67 oxygen compounds (comprising ethers, carboxylic acids, esters, dihydric alcohols, aldehydes, and ketones), three aromatic amines, quinoline, three amides, nitrobenzene, hydrocarbons, and sulfides. The larger value of N_{CHCl_2F} compared with the "ideal value,"

$$[p_{CHCl_2F}(\text{atm}) \text{ at } 4.5°C]/[p_{CHCl_2F}(\text{atm}) \text{ at } 32.2°C] = N \text{ at } 32.2°C$$

was attributed to hydrogen bonding between the hydrogen in the halomethane and oxygen or nitrogen in the solvent S molecule.

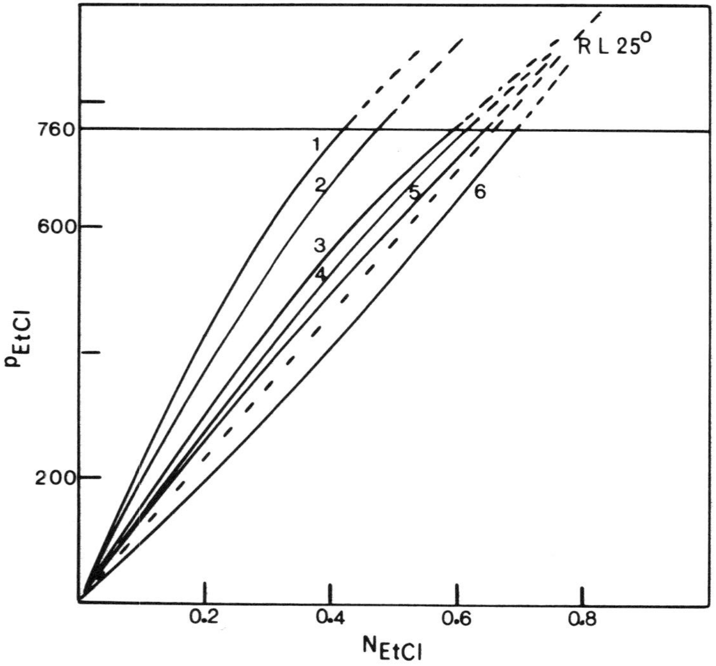

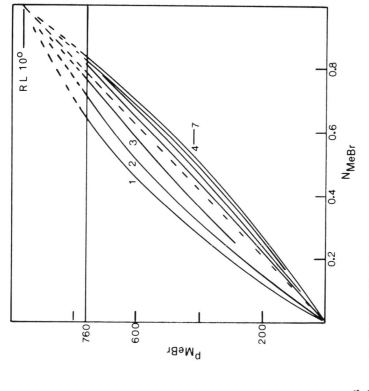

Fig. 120. Solubility of chloroethane EtCl at 25°C and at different pressures p_{EtCl}. Registration of the mole fraction N_{EtCl} plots on the R-line diagram. The liquids are: (1) $C_6H_5CH_2OH$; (2) $n-C_8H_{17}OH$; (3) $C_6H_5NO_2$ [curve for EtCl in $HCON(CH_3)_2$ closely follows 3]; (4) $o-HOC_6H_4CO_2CH_3$; (5) C_6H_5Br; (6) $(n-C_8H_{17})_2O$. The pressures p_{EtCl} are in mm Hg.

Fig. 121. Solubility of bromomethane MeBr at 10°C and at different pressures p_{MeBr}. Registration of the mole fraction N_{MeBr} vs p_{MeBr} plots on the R-line diagram. The liquids are: (1) $C_6H_5CH_2OH$; (2) Cl_3CCH_2OH; (3) $n-C_{10}H_{22}$ (curve for MeBr in $C_6H_5NO_2$ is similar to 3); (4) C_6H_5Br; (5) $n-C_8H_{17}Br$; (6) $C_6H_5CO_2C_2H_5$; (7) $(n-C_8H_{17})_2O$. The line for $HCON(CH_3)_2$ is close to the R-line. The pressures p_{MeBr} are in mm Hg.

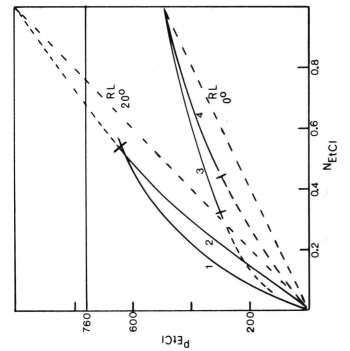

Fig. 123. Solubility of chloroethane EtCl based on the "cm^3 A/cm^3 S" data of Kaplan and Romanchuk.[198] Line 1 is the mole fraction N_{EtCl} vs pressure p_{EtCl} for CCl$_4$ at 20°C; line 2 is for (CH$_2$Cl)$_2$. Line 3 is the N_{EtCl} vs p_{EtCl} plot for CCl$_4$ at 0°C, and line 4 is that for (CH$_2$Cl)$_2$ at 0°C. The crossbars show the *upper* limits of the observed p_{EtCl}. Speculative extrapolation to higher p_{EtCl} shows the fit on the R-line diagram. The pressures p_{EtCl} are in mm Hg.

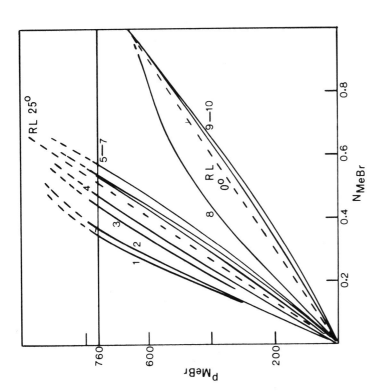

Fig. 122. Solubility of bromomethane MeBr, at different pressures p_{MeBr}. Plots of mole fraction N_{MeBr} vs p_{MeBr} on the R-line diagram. The liquids are, at 25°C: (1) C$_6$H$_5$CH$_2$OH; (2) n-C$_8$H$_{17}$OH; (3) n-C$_{10}$H$_{22}$; (4) HCON(CH$_3$)$_2$; (C$_6$H$_5$Br gives a line just on the left of the R-line); (5) n-C$_8$H$_{17}$Br; (6) C$_6$H$_5$CO$_2$C$_2$H$_5$; (7) (n-C$_8$H$_{17}$)$_2$O; at 0°C: (8) C$_6$H$_5$CH$_2$OH; (9) n-C$_8$H$_{17}$Br; (10) (n-C$_8$H$_{17}$)$_2$O. The pressures p_{MeBr} are in mm Hg.

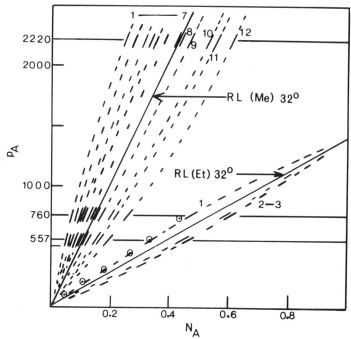

Fig. 124. Graphical presentation of N_{MeCl} and N_{EtCl} data calculated from the "g/cm^3" data of Zellhoefer. For the MeCl group, the liquids S are, from left to right: (1) $(BrCH_2)_2CHOH$; (2) $(ClCH_2)_2CHOH$; (3) decalin; (4) 1-$C_{10}H_7Br$; (5) 1-$C_{10}H_7Cl$; (6) tetralin; (7) $(Cl_2CH)_2$; (8) Cl_2CHCCl_3; (9) $(CH_3COOCH_2CH_2)_2O$; (10) $(C_2H_5OCH_2CH_2)_2O$ [$CH_3(CH_2)_{11}OH$ is just on the right of 10]; (11) $[CH_3O(CH_2)_2OCH_2]_2$; (12) $C_2H_5O(CH_2CH_2O)_4C_2H_5$. For the EtCl group, the liquids are, from left to right: (1) $(Cl_2CH)_2$, $(CH_3COOCH_2CH_2)_2O$, and $[CH_3O(CH_2)_2OCH_2]_2$; (2) $C_2H_5OCH_2CH_2)_2O$; (3) $C_2H_5O(CH_2CH_2O)_4C_2H_5$. The curve for EtCl in $C_2H_5O(CH_2CH_2O)_2OCCH_3$ lies almost on the R-line. The circles are for N_{EtCl} at 32°C for Gerrard's observations on 1-$C_{10}H_7Br$. The pressures p_A (A = MeCl or EtCl) are in mm Hg.

In continuation, these workers (1939) examined the completely halogenated methanes CCl_3F and CCl_2F_2. The solubilities are almost always lower than would be calculated, indicating to those workers that compound formation between solvent and solute does not occur.

Representative plots for N_{CHCl_2F} and the corresponding x_{CHCl_2F} plots (see Gerrard[99]) showed that the N/p plot for dioxane is linear and the x/p plot is curved; whereas another N/p plot is curved, the corresponding x/p plot is essentially linear.

12.2. Solubility in Water

Some interest in the solubility in water has been associated with the mechanisms of hydrolysis and the structure of water, desalination, and biocidal action. The N_{MeX} values (X = F, Cl, Br, I) for 1 atm and 25°C were given by Glew and Moelwyn-Hughes,[130] but there is some discordance with the results of other workers. Contrasted with $N_{MeCl} = 0.0019$, data by Boggs and Buck[28] gave 0.00172; by Svetlanov et al.,[348] 0.00264; by van Arkel and Vles,[361] 0.0074 (?); and by

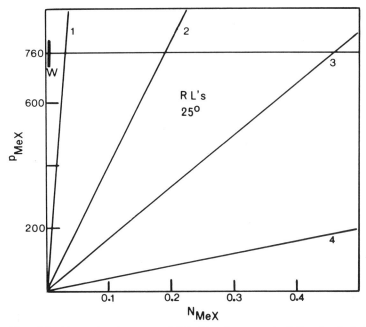

Fig. 125. Position of the N_{MeX} values for water at 25°C with reference to the R-lines for the halomethanes MeX (X = F, Cl, Br, I). The R-lines are: (1) MeF; (2) MeCl; (3) MeBr; (4) MeI. The mole fraction N_{MeX} values for water are registered at W. The pressures p_{MeX} are in mm Hg.

Mamedaliev and Musakhanly,[252] 0.00191. Contrasted with $N_{MeBr} = 0.0029$, data by Haight[145] gave 0.00275 and those by Carey et al.[46] (in connection with demineralizing seawater), 0.0023. According to Swain and Thornton,[351] the solubilities of the four halomethanes are almost the same in heavy as in light water. Whatever might be the accurate N_{MeX} value, however one may depict the condition of water in the immediate vicinity of the halomethane molecule, the N_{MeX} values form a close cluster well on the left of the R-line for fluoromethane (see Fig. 125), just as the hydrocarbons (MeH to BuH) do on the left of the R-line for methane. One molecule of fluoromethane at 25°C and 1 atm requires about 1000 molecules of water (H_2O units) and therefore on a time-average basis the molecule of MeF is *using* 1000 molecules of water. Similarly, one molecule of MeCl is using about 526 molecules of water.

In connection with the use of bromomethane as a fumigant for pest control in fruit growing, Haight measured the number of grams of MeBr absorbed by 100 g of liquid at 25°C and, presumably, ca. 1 atm. Because the value for water was 1.341, whereas the other values were 1.169 (mango juice), 0.921 (papaya juice), 0.911 (pineapple juice), he referred to "the comparatively high solubility in water" and concluded that MeBr is more soluble in water than in the juices. Out of context, this conclusion is misleading; no allowance was made for the effect of molecular weight on the x_{MeBr} value, and the conclusion in terms of mole ratio, or mole fraction, is therefore invalid. Two previous values were cited as "5–10 g/liter of water at 18°C"[234] and "0.1 g/100 g water."[334]

13

Hydrocarbon Gases

13.1. Saturated Hydrocarbons

13.1.1. At and Below 1 atm

In Chap. 2, I used the data for the hydrocarbons CH_4 to n-C_4H_{10} to introduce the R-line diagram. I now give new data for 2-methylpropane and the three pentanes to illustrate still further that for these saturated hydrocarbons the main factor is the condensing tendency of the gas, indicated in a simple way by its bp/1 atm.

A comparison of the data for 2-methylpropane (bp −11.5°C) at 0°C (Fig. 126 and Table 28) with those for n-butane at 10°C shows that the main differences are due to the lower tendency of the i-butane to condense. The positions of the lines for n-butane at 25°C are similar to those for i-butane at 15°C. The plots for 2,2-dimethylpropane (neopentane) (bp 9.5°C), 2-methylbutane (bp 27.8°C), and n-pentane (bp 36.1°C) (Figs. 127–129; cf. Table 28) also show the operation of this factor. The changes in *intra*molecular structure take their toll in terms of a change in the bp, i.e., in the tendency to condense; what further effects on the N_A values result from a change in hydrocarbon will be somewhat difficult to analyze in detail. When comparing the plot for one hydrocarbon at one temperature with that for another at a different temperature, it is necessary to consider the effect of a change of temperature on the intermolecular structure of S and on the complex modes of interaction of the hydrocarbon with S. For data on methane, see Lannung and Gjaldbaek[226]; for data on ethane and propane, see Thomsen and Gjaldbaek.[356]

13.1.2. At Pressures Greater than 1 atm

Equilibrium systems involving gaseous hydrocarbons and relatively low-volatile hydrocarbons have been studied in connection with the petroleum industry and chemical engineering. The statement of Boomer *et al.* (1938)[29] that the solubility of methane in pentane increases rapidly at high pressures (up to 188 atm at 25°C) was based on a plot of the mole ratio form of data. Figure 130 shows that the x_{MeH} plot is

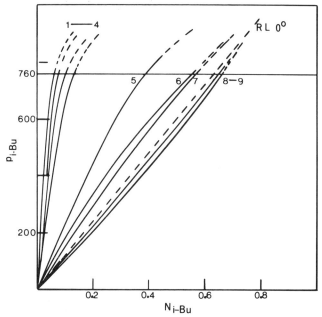

Fig. 126. Solubility of 2-methylpropane (isobutane) i-Bu at 0°C and different pressures $p_{i\text{-Bu}}$. Registration of the mole fraction $N_{i\text{-Bu}}$ vs $p_{i\text{-Bu}}$ plots on the R-line diagram. The liquids are: (1) $C_6H_5NH_2$; (2) $HCON(CH_3)_2$; (3) $C_6H_5CH_2OH$; (4) $C_6H_5NO_2$; (5) $n\text{-}C_8H_{17}OH$; (6) $1,3,5\text{-}C_6H_3(CH_3)_3$; (7) $n\text{-}C_8H_{17}Br$; (8) $n\text{-}C_{10}H_{22}$; (9) $(n\text{-}C_8H_{17})_2O$. The pressures $p_{i\text{-Bu}}$ are in mm Hg.

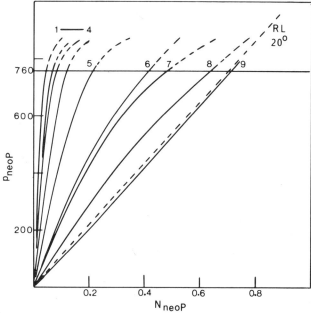

Fig. 127. Solubility of 2,2-dimethylpropane (neopentane, abbreviated as neoP) at 20°C and different pressures p_{neoP}. Registration of the mole fraction N_{neoP} vs p_{neoP} plots on the R-line diagram. The liquids are: (1) $C_6H_5NH_2$; (2) $HCON(CH_3)_2$; (3) $C_6H_5CH_2OH$; (4) $C_6H_5NO_2$; (5) $o\text{-}HOC_6H_4CO_2CH_3$; (6) $n\text{-}C_8H_{17}OH$; (7) $n\text{-}C_5H_{11}CO_2H$; (8) $n\text{-}C_8H_{17}Br$; [the curve for $1,3,5\text{-}C_6H_3(CH_3)_3$ is similar to curve 8]; (9) $(n\text{-}C_8H_{17})_2O$ ($n\text{-}C_{10}H_{22}$ gives a line between the R-line and curve 9). The pressures p_{neoP} are in mm Hg.

Hydrocarbon Gases

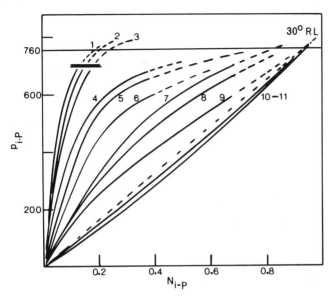

Fig. 128. Solubility of 2-methylbutane (isopentane) i-P at 30°C and different pressures $p_{i\text{-}P}$. Registration of the mole fraction $N_{i\text{-}P}$ plots on the R-line diagram. The liquids are: (1) $C_6H_5NH_2$; (2) $HCON(CH_3)_2$; (3) $C_6H_5CH_2OH$; (the crossbars indicate the separation of two liquid phases above about 700 mm Hg); (4) $C_6H_5NO_2$ (curve for i-P in quinoline is similar to 4); (5) C_6H_5CN; (6) $o\text{-}C_6H_4(OH)CO_2CH_3$; (7) $n\text{-}C_8H_{17}OH$; (8) $n\text{-}C_5H_{11}CO_2H$; (9) $n\text{-}C_8H_{17}Br$; (10) $n\text{-}C_{10}H_{22}$; (11) $(n\text{-}C_8H_{17})_2O$. $C_6H_5CO_2C_2H_5$ gives a line similar to curve 7. The pressures $p_{i\text{-}P}$ are in mm Hg.

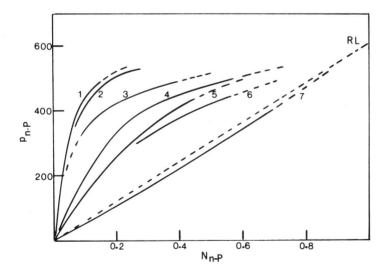

Fig. 129. Solubility of n-pentane n-P at 30°C and different pressures $p_{n\text{-}P}$. Registration of the mole fraction $N_{n\text{-}P}$ vs $p_{n\text{-}P}$ plots on the R-line diagram. The liquids are: (1) $HCON(CH_3)_2$; (2) $C_6H_5CH_2OH$; (3) $C_6H_5NO_2$; (4) $o\text{-}HOC_6H_4CO_2CH_3$; (5) $n\text{-}C_8H_{17}OH$; (6) C_6H_5Br; (7) $(n\text{-}C_8H_{17})_2O$. The pressures $p_{n\text{-}P}$ are in mm Hg.

strongly convex upward, but the N_{MeH} plot is slightly concave upward just on the right of the R-line.

In Fig. 131, I have plotted the $N_{C_2H_6}$ vs $p_{C_2H_6}$ and $x_{C_2H_6}$ vs $p_{C_2H_6}$ from data given by Lee and Kohn (1969)[229] for n-dodecane at 0 and 25°C. The reported $N_{C_2H_6}$ vs $p_{C_2H_6}$ plots showed increasing curvature, concave upward at 25°C, 50°C, 75°C, and 100°C. The work was undertaken to test "some of the simple solution models." The fugacities were tabulated along with the pressures $p_{C_2H_6}$ and were fitted to the Scatchard[325] modification of the regular solution theory and to the Flory–Huggins solution model. The Flory–Huggins model appeared to be accurate enough to represent deviations of this system from Raoult's law for most engineering purposes, even at higher pressures.

Rodrigues, McCaffrey, and Kohn (1968) examined the ethane–n-octane system.[315] For ethene in hexane, cyclohexane, and benzene, see the work of Zhuze and Zhurba (1960).[393]

13.2. Unsaturated Hydrocarbons

Figures 132–134 show plots for ethylene (bp −103°C) based on primary data in refs. 177, 205, 276, 317a, 377, and 393. The acid function sited on hydrogen in acetylene is apparently incisive enough to cause more selective competition with the

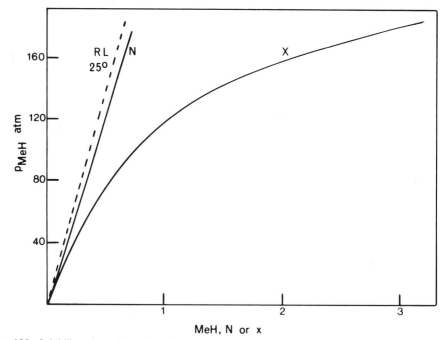

Fig. 130. Solubility of methane MeH in n-pentane at 25°C and at pressures p_{MeH} up to 188 atm. Comparison of the mole fraction N_{MeH} vs p_{MeH} and the mole ratio x_{MeH} vs p_{MeH} plots on the R-line diagram.

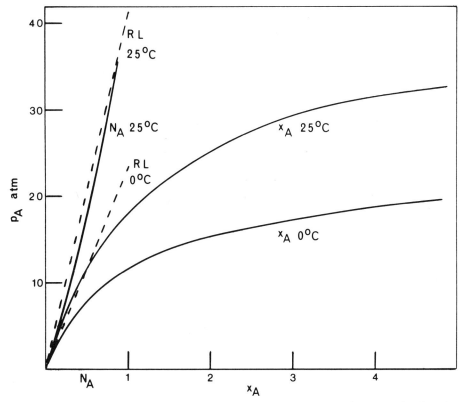

Fig. 131. Solubility of ethane (A) in n-dodecane. Reference lines, mole fraction N_A, and mole ratio x_A plots for pressures p_A at 0 and 25°C. The N_A line for 0°C is close to the R-line.

intermolecular forces of S; the data are almost entirely for one pressure, but the position on the horizontal for 760 mm Hg suggests the form of the $N_{C_2H_2}/p_{C_2H_2}$ plot (see Fig. 135). Data for dimethylformamide by McKinnis (1955)[281] enable the full line to 760 mm Hg to be drawn; it is concave upward, and a speculative extrapolation indicates a long sweep to the right and upward until $p°_{C_2H_2}$ is reached. McKinnis' plot was in the form "ml C_2H_2 per 20 ml (probably should be 1 ml) dimethylformamide vs $p_{C_2H_2}$ in mm Hg," which was a mole ratio plot, and was deemed to be linear up to 760 mm Hg. This has been reproduced in one specialist textbook as an example of a "nonreacting system" that "obeys" Henry's law, which was, however, taken as the mole fraction form. Expressed as g C_3H_6 per mole C_6H_5X at 1 atm at t°C (0 to 80°C), data for C_6H_5X (X = F, Cl, Br, I) were given by Marshtupa et al. (1966).[255]

13.3. Selective Absorption of Gaseous Hydrocarbons

An analysis of a patent specification on this procedure affords an example of the effectiveness of the R-line approach in bringing into focus the essential pattern of

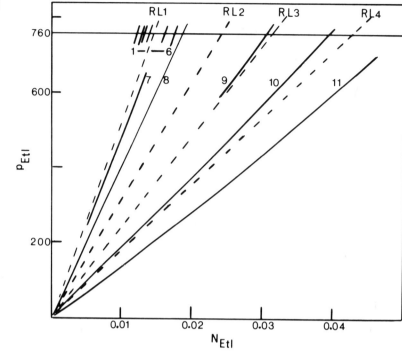

Fig. 132. Solubility of ethene (ethylene) Etl in several liquids. Registration of the mole fraction N_{Etl} vs p_{Etl} plots on the R-line diagram. The liquids are, at 20°C: (1) $(CH_3)_2CO$; (2) C_6H_5Cl; (3) $CH_3CO_2CH_3$; (4) C_6H_6; (5) n-C_6H_{14}; (6) CCl_4; (7) $(ClCH_2)_2$ (at 0°C); (8) $C_6H_4(CH_3)_2$ (at 20°C); (9) the same (at 0°C); (10) the same (at −10°C); (11) the same (at −21°C). The R-lines are for the following temperatures: RL 1 (20°C); RL 2 (0°C); RL 3 (−10°C); RL 4 (−21°C). The pressures p_{Etl} are in mm Hg.

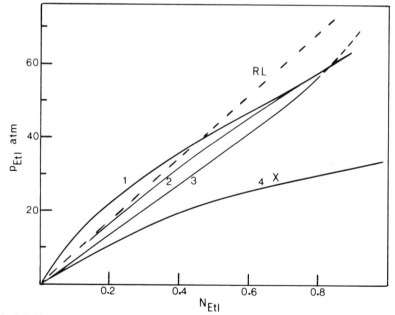

Fig. 133. Solubility of ethene (ethylene) Etl at 30°C and different pressures p_{Etl}. Mole fraction N_{Etl} vs p_{Etl} plots for: (1) C_6H_6; (2) cyclo-C_6H_{12}; (3) n-C_6H_{14}. Line 4 (X) is the mole ratio x_{Etl} vs p_{Etl} plot corresponding to line 3 on the same scale.

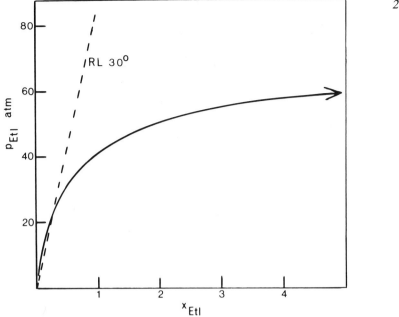

Fig. 134. Solubility of ethene (ethylene) Etl in benzene at 30°C and pressures p_{Etl} up to 60 atm. The mole ratio x_{Etl} vs p_{Etl} plot on the R-line (mole fraction) diagram.

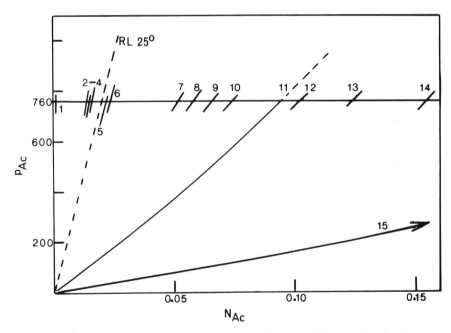

Fig. 135. Solubility of acetylene Ac at 25°C and 1 atm. Registration of the mole fraction N_{Ac} values for 1 atm on the R-line diagram. The liquids are: (1) CCl_4; (2) $CHCl_3$; (3) C_2H_5OH (and CH_3CO_2H); (4) C_6H_5Cl; (5) C_6H_6; (6) $C_6H_5N(CH_3)_2$; (7) C_2H_5Br; (8) $B(OCH_3)_3$; (9) CH_3COCH_3; (10) $CH_3CO_2C_2H_5$; (11) $HCON(CH_3)_2$ (full plot for N_{Ac} vs p_{Ac}); (12) $(CH_3)_2SO$; (13) $(C_2H_5O)_3PO$: (14) $(CH_3OC_2\cdot H_4OC_2H_4)_2O$; (15) $[(CH_3)_2N]_3PO$ (probable plot of N_{Ac} vs p_{Ac} based on the value of N_{Ac} at $p_{Ac} = 1$ atm, which is too large for this scale). The N_{Ac} value for water at 1 atm is 0.00075 and would be registered between line 1 and the left vertical axis. The pressures p_{Ac} are in mm Hg.

Table 29. N_A Values for S_F at 1 atm, 26°C

A = hydrocarbon	bp A/1 atm	N_A
Ethane	−88.6°C	0.0028?
Ethene	−104	
Acetylene	−83.6	0.00923
Propane	−44.5	0.0046
Propene	−47.8	0.014
Propadiene (allene)	−34.5	0.0252
Methylacetylene	−23.2	0.0441
Isobutane (2-methylpropane)	−11.8	0.021
Isobutylene	−6.6	0.0635
1-Butene	−6.3	0.065
2-Butene-*trans*	+0.9	mixture 0.0675
2-Butene-*cis*	+3.7	
(1,3-butadiene	−4.4)	—
n-Butane	−0.5	0.022

data. The purpose of the patent was to protect a procedure for the purification of olefins and acetylenes for use in polymerization and other processes.

Robinson and Sund (1967)[313] described the separation of an unsaturated hydrocarbon from a saturated one, such as ethene from ethane, or propene from propane, by the selective absorption in 1,1,1,3,3,3-hexafluoroisopropanol, S_F, which was deemed a much better liquid for this purpose than others such as dimethylformamide. I have already mentioned the acid–base functions of acetylenes, the acid function sited at a terminal hydrogen, and the basic function sited at the triple bond. The similar acid–base function of the simple olefins is probably of much weaker intensity. Dimethylformamide clearly brings out the acid function of acetylene, as shown by the work of McKinnis; the hexafluoroalcohol S_F should bring out the basic function because there is evidence to show that S_F will have weak basic function sited on oxygen and relatively strong acid function sited on hydrogen. The properties of the corresponding chloro compounds $Cl_3CCHOHCCl_3$ (Gerrard and Howe)[101] and Cl_3CCH_2OH (Gerrard *et al.*), and of chloretone, $Cl_3CC(CH_3)_2OH$ (Gerrard and Wyvill),[122] show a low basic function on oxygen and a relatively strong acidic one on hydrogen. However, hydrogen bonding within the alcohol must also be considered.

The solubility in S_F of 13 hydrocarbons with bp <4°C were reported in the form "ml gas/100 ml S_F" for 26°C at presumably a total pressure of 1 atm (see Table 29). These "vol./vol." values are in the mole ratio form, and since S_F is common to all the systems, the "vol./vol." values are approximately* comparable among themselves but are not so with "vol./vol." values entailing another liquid S because the density and molecular weight of S are disturbing factors. The density of S_F was not stated; nor

---

*Because the gram-mole volume is not necessarily the same for the different gases.

Hydrocarbon Gases

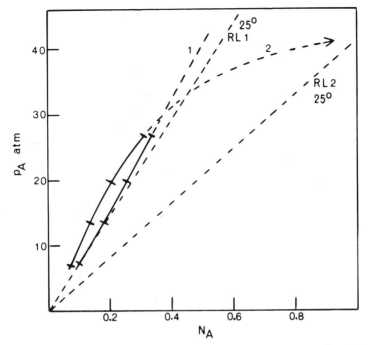

Fig. 136. Use of the R-line for the registration of mole fraction N_A vs pressure p_A data. Analysis of the vol./vol. data from the U.S. Patent.[313] Line 1 is for $A = C_2H_4$ and should be referred to R-line 1, which is for ethene (ethylene). Line 2 is for $C_2H_6 = A$ and is to be referred to R-line 2, which is for ethane. The liquid was $F_3CCHOHCF_3$ at 25°C. The broken parts of each line indicate a speculative extrapolation.

was it by Middleton and Lindsey (1964)[258] who prepared the compound and described it as an "extremely strong hydrogen-bonding donor," and this presumably meant that S_F provides the hydrogen site. To calculate the N_A values from the "vol./vol." data, I have estimated $d^{25} = 1.48$ from a study of other fluorine compounds.

The pressures were recorded as psia, and these have been converted into atmospheres (1 atm = 760 mm Hg). The volumes of gas recorded appear to be those measured at 1 atm; i.e., the mass of gas represented by the volume for 1 atm at 25°C is the mass of gas dissolved by S_F at the stated $p_A > 1$ atm. In Fig. 136, I have plotted the N_A vs p_A values for ethane and ethene. It is seen that the ethene line is just on the left of the R-line for ethene, but the line for ethane is much further to the left of the R-line for ethane. This would appear to indicate that the acid function of S_F has some effect with reference to the basic function of the double bond. However, it is desirable to compare the data with those for an alcohol such as isopropanol.

In Fig. 137, I have indicated how the few data for the other named hydrocarbons fit into the R-line diagram. The propene line is nearer the R-line for propene than the propane line is to its own R-line, and both lines are on the left of each R-line. The lines for the remaining hydrocarbons are decidedly on the left of each R-line.

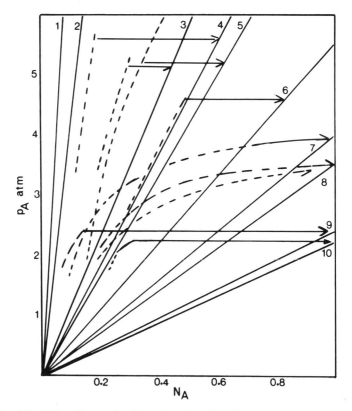

Fig. 137. Use of the R-line diagram for the registration of N_A/p_A values. Analysis of volume/volume data recorded in the U.S. Patent.[313] The full lines are the R-lines (25°C) (bp/1 atm): (1) ethene (−104°C); (2) ethane (−88.6°C) [acetylene (−83.6°C)]; (3) propene (−47.8°C); (4) propane (−44.5°C); (5) allene (−34.5°C); (6) methylacetylene (−23.2°C); (7) 2-methylpropane (−11°C); (8) 1-butene (−6.3°C) [isobutylene (−6.6°C)]; (9) n-butane (−0.5°C); (10) 2-butene, *cis* (3.7°C), *trans* (0.9°C). The broken lines with arrows indicate the particular R-line which each plot must reach at $N_A = 1$, i.e., each broken line shown indicates the name of the particular gas A and the positions of the N_A values with regard to the particular R-line. All the N_A values are on the left of the respective R-line, and there is an indication that the value for an unsaturated hydrocarbon is nearer to its R-line than the value for the parent saturated one is to its R-line. The operation of the basic function of the site of unsaturation is clearly discernible, but the effect is by no means as pronounced as the description of $F_3CCHOHCF_3$ as an "extremely strong hydrogen-bonding donor" would lead us to expect.

"Henry's law constants" were given as mere numbers, e.g., 151,000 for ethane, 87,000 for propene, and 20,900 for butane. These numbers can have no significance whatsoever in the computation of N_A values for the higher pressures. The forms of the lines for pressures <1 atm are obliterated by the scale of operation. Table 30 gives N_A and x_A values for the isomers of butene. For my data on propene and the four butenes, see Chap. 16.

An estimate of the intensity of the basic function of unsaturated hydrocarbons has been given on the basis of the mp curves of binary mixtures with HCl and HBr at low temperatures. Terres and Assemi (1956)[355] described the formation of "more or less stable equimolecular complexes" between HCl or HBr with alkyl benzenes at low temperatures; p-cymene was stated to form a complex with 2 moles of HCl. The "basicity" was seen to be in the following order for the series C_6H_5R: $R = Me < Et < n\text{-}Pr < i\text{-}Pr < t\text{-}Bu$. This order fits the results of Brown et al.[35,36] for the methyl benzenes (see Chap. 9, p. 143). Although cyclohexene was stated to form "a strong 1:1 complex" with HCl, the aliphatic olefins form no other than "weak" complexes, the "basicity" being in the order:

1-decene < 1-nonene < 1-octene < cyclohexene

Basic function appears therefore to be assessed on the basis of the formation of a solid "complex." Data for N_{HCl} vs p_{HCl} for temperatures from, say, 20°C to low temperatures down to the mp of the hydrocarbon would be revealing in this sense.

Table 30. Solubility of 1-Butene (b), 2-methylpropene (isobutylene) (i), trans-2-butene (t), and cis-2-butene (c) in Liquids S at 0°C (b and i), 5°C (t), and 7.8°C (c) at Different Pressures

	Butene isomer	N_A at p_A mm Hg of					x_A at 760 mm Hg
		200	400	600	700	760	
$C_6H_5CH_2OH$	b	0.040	0.087	0.151	0.193	0.226	0.292
	i	—	—	—	—	0.220	0.282
	t	0.055	0.105	0.177	0.234	0.280	0.389
	c	0.057	0.121	0.217	0.296	0.375	0.600
$HCON(CH_3)_2$	b	0.044	0.095	0.160	0.205	0.250	0.333
	i	0.046	0.103	0.175	0.236	0.292	0.411
Cl_3CCH_2OH	b	0.061	0.135	0.236	0.311	0.375	0.600
	t	0.060	0.137	0.266	0.381	0.490	0.961
	c	0.076	0.176	0.335	0.477	0.604	1.525
$n\text{-}C_8H_{17}OH$	b	0.115	0.242	0.390	0.485	0.556	1.252
	i	0.120	0.246	0.396	0.494	0.565	1.297
	c	0.120	0.286	0.480	0.608	0.700	2.333
$n\text{-}C_5H_{11}CO_2H$	b	0.135	0.277	0.461	0.581	0.662	1.960
	i	0.135	0.277	0.461	0.581	0.662	1.960
	c	0.148	0.324	0.533	0.672	0.751	3.310
$C_6H_5CO_2C_2H_5$	b	0.120	0.275	0.495	0.604	0.700	2.333
	i	0.130	0.277	0.470	0.590	0.667	2.000
	t	—	—	—	—	0.80	4.0
	c	—	—	—	—	0.80	4.0
$n\text{-}C_8H_{17}Br$	b	0.210	0.421	0.622	0.724	0.800	4.00
	i	0.193	0.381	0.583	0.688	0.756	3.10
C_6H_5Br	t	—	—	0.623	—	0.840	5.26
	c	—	—	0.632	—	0.836	5.10
$1,3,5\text{-}C_6H_3(CH_3)_3$	b	0.180	0.369	0.590	0.708	0.778	3.504
$n\text{-}C_{10}H_{22}$	i	0.210	0.422	0.628	0.724	0.770	3.348
	t	—	0.480	0.795	—	0.86	6.14
$(n\text{-}C_8H_{17})_2O$	b	0.256	0.480	0.672	0.764	0.816	0.431
$(n\text{–}C_8H_{17})_2O$	i	0.258	0.476	0.666	0.760	0.800	4.00
	t	0.256	0.500	0.726	0.832	0.874	6.924
	c	0.300	0.536	0.740	0.836	0.890	8.054

13.4. Solubility in Alcohols

Kretschmer and Wiebe (1951–52)[217] have determined the N_A values of propane, butane, and isobutane at a total pressure $P = p_A + p_S$ for methanol, ethanol, and propan-2-ol at a few temperatures. Plots of N_A vs approximate p_A are convex upward well on the left of the respective R-line (see Figs. 5, 9, and 126 for approximate position). Table 31 shows that for a fixed alcohol the N_A values are on the order of the bp of the hydrocarbons. For a fixed hydrocarbon, N_A values are $CH_3OH < C_2H_5OH < CH_3CHOHCH_3$. The positions of the N_A/p_A lines show a conflict with the hydrogen bonding of the alcohols. The p_S for the pure alcohol will not be the same as for the solutions, and therefore p_A will not be accurately given by total pressure $-p_S$. It would be misleading to discuss the thermodynamic treatment of the results without consideration of all the details given in the two papers.

Boyer and Bircher (1960)[30] recorded the N_A values for A = N_2, Ar, CH_4, C_2H_4, C_2H_6 for a series of normal alcohols n-ROH (R = CH_3 to C_8H_{17}) for 25°C and 1 atm. It is clear from Table 32 that there are two main trends which illustrate what has been emphasized in this monograph. Apart from the odd items for ethene and ethane and methanol, the N_A values are on the order of the bp of A for a given alcohol, and for a given A they increase as the symmetry of the hydrogen bonding in alcohols probably decreases.

Table 31. N_A at 25°C and Total Pressure (mm Hg)

A	CH_3OH, p_S = 126.9 mm Hg	C_2H_5OH, p_S = 59.0 mm Hg	$CH_3CHOHCH_3$, p_S = 43.5 mm Hg
Propane	0.0096 (759.6)	0.0194 (755.3)	0.0278 (760.5)
Isobutane	0.0210 (764.3)	0.0437 (725.4)	0.0693 (753.1)
n-Butane	0.0317 (762.0)	0.0709 (739.5)	0.1106 (752.7)

Table 32. N_A Values for 25°C, 1 atm

A	CH_3OH	C_2H_5OH	n-C_3H_7OH	n-C_4H_9OH
CH_4	0.000867	0.00128	0.00161	0.00191
C_2H_4	0.00439	0.00597	0.00746	0.00871
C_2H_6	0.00405	0.00686	0.0090	0.0109

A	n-$C_5H_{11}OH$	n-$C_6H_{13}OH$	n-$C_7H_{15}OH$	n-$C_8H_{17}OH$
CH_4	0.00215	0.00234	0.00260	0.00280
C_2H_4	0.00984	0.0108	0.0118	0.01268
C_2H_6	0.0126	0.0143	0.0159	0.0173

13.5. Solubility in Water

The low N_{RH} values have been discussed. In Fig. 138 I show the x_{RH} vs $t°C$ plots for CH_4 to n-C_4H_{10} taken from the vol./g data of Wen and Hung.[371]

Azarnoosh and McKetta (1958)[10] gave data for propane in water for pressures from 1 atm to 500 psia and temperatures from 60 to 280°F. They refer to previous work and mention an extensive bibliography on water–hydrocarbon systems by McKetta and Wehe (1959).[280] For a total pressure of 14.7 psia (1 atm), the mole fraction $\times 10^5$ of propane was given as 5.89, i.e., $N_{C_3H_8} = 0.0000589$ at 60°F. The $N_{C_3H_8}$ vs pressure plots were curved. The same authors (1959)[10] reported data for propylene in water and cited Hiraoka (1954)[170] on acetylene in water for pressures up to 500 psi and temperatures to 86°F, Bradbury et al. (1952)[32] on ethylene in water at pressures up to 8000 psi and temperatures up to 212°F, and Brooks and McKetta (1955)[34] and Brooks, Haughn, and McKetta (1955)[34] on l-butene and water. The propylene data were given as mole fraction $\times 10^4$ for pressures up to 500 psia and temperatures up to 220°F. $N_{C_3H_6} \times 10^4$ was given as 1.66, which is 0.000166 for 1.476 atm at 21°C, probably about 0.00011 at 1 atm. That this value is about twice that for propane indicates some effect of the acid–base function associated with the double bond, but both values are absolutely very small and are entirely out of step with the R-line positions.

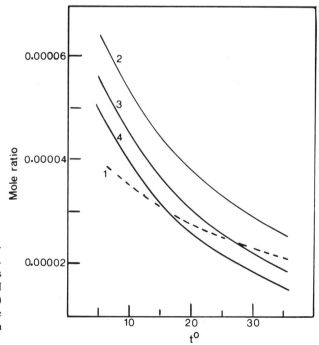

Fig. 138. Solubility of hydrocarbon gases RH in water at 1 atm. Plots of the mole ratio x_{RH} vs temperature $t°C$. The gases RH are: (1) methane; (2) ethane; (3) propane; (4) n-butane. From the volume RH/g water data of Wen and Hung.[371]

Reed and McKetta (1959)[309] dealt with 1,3-butadiene in water, referred to lower hydrocarbons, and gave appropriate citations.

Data for propane in water and aqueous solutions of perchloric, sulfuric, phosphoric, and trifluoroacetic acids have been published by Christopulos (1969)[51] in connection with U.S. Government research.

In his paper on partial molal heats and entropies of solution of gases dissolved in water from the freezing point to near the critical point, Himmelblau (1959)[168] briefly reviewed previous work on O_2, N_2, H_2, He, Xe, and CH_4 for temperatures up to about 300°C. He confined his attention to these "nonpolar gases that do not combine chemically with water nor ionize or dissociate in water." Himmelblau concluded that "even at temperature above 200° it appears as if hydrogen bonds are being stretched and that the rigidity of structure found in water at low temperatures has not completely disappeared."

13.6. Other Examples

A paper by Waters, Mortimer, and Clements (1970)[367] on the "solubility of some light hydrocarbons and hydrogen in some organic solvents" contains data for ethane, ethene, and propane in toluene, ethene in n-hexane, and hydrogen in n-hexane, toluene, and 1,2-dichloroethane, and affords an example of the urgent need to standardize the form of data as mole ratio x_A and mole fraction N_A at specified t°C and p_A. The data were presented in two forms: (a) "mole/liter at 1 atm" ("liter" presumably referring to the volume of solvent S, but one would not know this if quoted out of context), and (b) the Bunsen coefficient α. The "mole/liter" form was obtained from a general expression involving pressures, temperatures, volumes, density of solvent, and the "gram-mole volume" 22.414. To illustrate my point, I present the recorded data for ethylene in Table 33. The inference appears to be that "mole/liter" varies linearly with $p_{C_2H_4}$. To convert "mole/liter" into $x_{C_2H_4}$, I divide 0.199 by the number of moles of toluene in "1 liter," i.e., by $1000 \times d°/M$, where $d°$

Table 33. Solubility Data for Ethene Expressed as "mole/liter" and as the Bunsen Coefficient

Solvent	p of C_2H_4 in mm Hg	Bunsen coefficient α	"Gas solubility, mole/liter at 1 atm"
Toluene at 0°C	243.3	4.450	0.199
	765.0	4.440	0.199
n-Hexane at 0°C	507.0	5.42	0.242
	571.4	5.34	0.239

is the density at 0°C and M is the molecular weight of toluene. This gives $x_{C_2H_4}$ approximately 0.021, $N_{C_2H_4}$ being 0.02. For such low values, the plot would appear approximately linear. For n-hexane, $x_{C_2H_4}$ is approximately 0.0307 and $N_{C_2H_4} = 0.030$, which appears to indicate that the basic function of ethylene is in some conflict with the basic function of toluene. Ethane in toluene for 0°C, 1 atm has $x_{C_2H_6} = 0.0227$; see Fig. 13 for position with regard to the R-line. For propane in toluene at 0°C, Waters et al. gave $\alpha = 26.381$ and "moles/liter" $= 1.177$, which are equivalent to $x_{C_3H_8} = 0.124$ ($N_{C_3H_8} = 0.1095$); see Fig. 9 for position with regard to the R-line.

Turning to the data for hydrogen,[367] I have calculated the x_{H_2} values from the α values (Table 34). Attention was drawn to the variation of the α values for ethylene in toluene [Hannaert et al. (1967),[148] Chien (1959),[50] and Henrici-Olivé and Olive (1967)[155]], the values of these authors falling in between those for benzene [Horiuti (1931)[177]] and xylene [Kireev et al. (1935)[205]]. There was also variation for the ethylene values for n-hexane [McDaniel (1911),[276] Rosenthal (1954)[317a]]. The "solubility coefficients" of hydrogen in 1,2-dichloroethane reported by Kireev and Romanchuk (1936)[207] were stated to be in error, being three times larger than the values by Waters et al. In a paper on the "solubility of propane and carbon dioxide in heptane, dodecane, and hexadecane" at t°C from 25 to 45°C (C_3H_8) and from 10 to 50°C (CO_2), Hayduk, Walter, and Simpson (1972)[151, 152] provide further evidence of the unsuitability of the Ostwald coefficient L for the comparison of solubilities on a molecular basis (see Table 35). Molecular weight and density of the liquid S are disturbing factors.

Table 34. Calculation of x_{H_2} from α^a

S	t, °C	α	x_{H_2}
n-C_6H_{14}	20	0.104	0.00060
$C_6H_5CH_3$	20	0.0607	0.000283
1,2-$C_2H_4Cl_2$	25	0.0502	0.0001745

^{a}See Fig. 22 for positions with regard to the R-line.

Table 35. Ostwald Coefficients and Solubility

S	$L_{C_3H_8}$ at 1 atm, 25°C	$N_{C_3H_8}$	L_{CO_2} at 1 atm, 25°C	N_{CO_2}
Hexane	23.75	0.115 (0.116^a)	—	—
Heptane	21.65	0.117	1.95	0.0117 0.0121^b
Dodecane	14.95	0.123	1.37	0.0127
Hexadecane	12.65	0.133	1.16	0.0138

^{a}Thomsen and Gjaldbaek (1963).
^{b}Gjaldbaek (1953).

From a scrutiny of Fig. 13, it could be predicted that the $N_{C_2H_6}$ vs $p_{C_2H_6}$ plot for 25°C for the whole range of $N_{C_2H_6}$ values would probably be concave upward just on the right of the R-line. Rodriques, McCaffrey, and Kohn (1968)[315] have recorded data for t°C from 0 to 100°C and $p_{C_2H_6}$ up to 22 atm for 0°C and 52 atm for 75°C. Their $N_{C_2H_6}$ vs $p_{C_2H_6}$ plots for octane at 0, 25, 40, 50, 75, and 100°C are concave upward. In Fig. 139, I have given the $x_{C_2H_6}$ plot for 25°C against my R-line.

The vapor pressures of hydrocarbons have special significance in the petroleum industry, and it is relevant to effect a link with this aspect by the brief citation of the following.

Hoffman, Welker, Felt, and Weber (1962)[173] pointed out that satisfactory values for the hypothetical vapor pressure may be obtained by extrapolating the vapor pressure curve into the supercritical region. Methane, ethane, propane, n-pentane, and n-octane were examined.

In a paper by Bradford and Thodos (1966),[33] 4 of the 16 references are to Hildebrand *et al.*, one to Chao and Seader, one to Prausnitz and Edmister, one to Frost and Kalkwarf, and the remainder to Thodos and his colleagues. The solubility parameter δ is a temperature-dependent quantity, and when we are dealing with the difference between the δ_A and δ_S values, we are presented with differential effects. It was pointed out by the authors that the δ value at 25°C has been assumed to be the same for other temperatures for utility purposes and represents a "fictitious" or "empirical" value. Data for n-hydrocarbons from methane to dodecane and for ethylene, propylene, 1,3-butadiene, cyclohexane, and benzene were given.

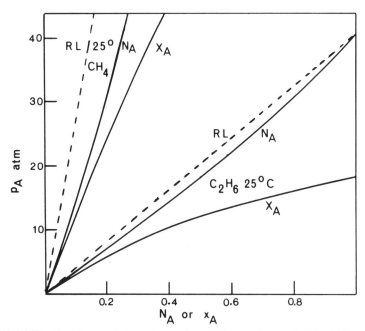

Fig. 139. Solubility of methane and ethane in n-octane. Reference lines and plots of N_A and x_A vs p_A for 25°C on a common scale. For the methane data, see ref. 405 and Fig. 158.

Stepanova (1970)[342] worked at pressures up to 150 kg/cm^2 and at $t°C = 0, 20, 40$ and 60°C. "Henry's law constants" were given, and it was stated that activity coefficients deviated less from unity the higher the bp of the paraffinic hydrocarbon.

Stepanova and Velikovskii (1970)[343] stated that deviation from "ideal behavior" is connected with fugacity and activity. A paper on the "prediction of vapor–liquid equilibrium for polar–nonpolar binary systems" by Finch and Van Winkle is mentioned here because of the terminology and the principles involved.[81] The 12 systems examined comprised systems such as ethylbenzene–hexylene glycol, n-octane–cellosolve, toluene–phenol, and n-heptane–toluene. It was pointed out that whereas the Scatchard–Hildebrand theory has had some success in predicting the vapor–liquid equilibria for nonpolar binary systems, it has proved to be unsatisfactory in the quantitative prediction of such equilibria for polar–polar systems and for polar–nonpolar systems.

14
Examples of Certain Other Gases

14.1. List of Formulas and bp °C/1 atm

NO	−151.8	C_2F_6	−79	H_2Se	−41.5
OF_2	−144.8	PF_5	−75	$ClCHF_2$	−39.8
NF_3	−128.8	F_3CNF_2	−75	$ClCF_2CF_3$	−38
CF_4	−128	$SiClF_3$	−70	C_2H_5F	−37.7
O_3	−111.9	SF_6	−63.8	$F_3CCF_2NF_2$	−35
SiH_4	−112	NO_2F	−63.5	$ClCF_2NO$	−35
PF_3	−101.5	$BrCF_3$	−60	SeF_6	−34.5
ClF	−100.8	NOF	−56	$SiCl_2F_2$	−31.7
BF_3	−100	AsH_3	−55	Cl_2CF_2	−29
$SiHF_3$	−97.5	CH_2F_2	−51.6	CH_3CHF_2	−24.7
PH_3	−87.7	COS	−50.2	CH_2N_2	−23
F_3CNO	−84	F_3CCH_2F	−47.3	COSe	−21.7
COF_2	−83.1	ClO_3F	−46.8	C_2N_2	−21.2
CHF_3	−82.2	NO_3F	−45.9	$F_3C(SF_5)$	−20
$ClCF_3$	−81.1	F_3CCF_2NO	−42	SbH_3	−17.2
H_2Te	−2.2	Cl_2CFCF_3	+3.8	ClO_4F	−15.9
NOBr	−2	SiH_2Cl_2	+8.3	$BrCF_2NO$	ca. −12
F_3CNO	−84	F_3CNO_2	−33.6	$ClCH_2F$	−9.1
	(−93)		(−20)	ClCN	+12.7
SiH_3F	−98.6	CH_3SiHF_2	−35.6	$H_3SiOSiH_3$	−15.2
SiH_2F_2	−77.5	SiH_3Cl	−30.4	Si_2H_6	−14.5
CH_3SiH_3	−57.5	$CH_2=CHSiH_3$	−22.8	$C_2H_5SiH_3$	−13.7
$SiHClF_2$	−50	$(CH_3)_2SiH_2$	−19.6	$(CH_3)_2SiHF$	−9
CH_3SiH_2F	−44	$SiHCl_2F$	−18.4		

14.2. R-Lines

Solubility data for these gases are sparse. The positions of the R-lines for temperatures in the range 0–25°C would require a base scale corresponding to N_A from 0 to 1 to plot them on a common diagram. In Fig. 140 I have given representative plots and indicate the positions of certain observed N_A values on the horizontal at 760 mm Hg. Even when the vapor pressure data are not available, the

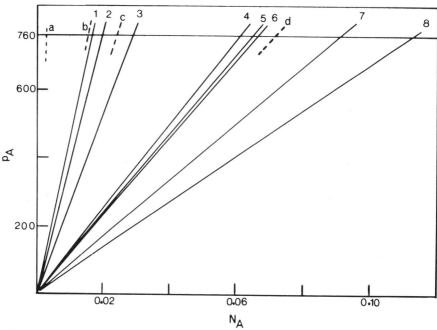

Fig. 140. Reference lines for certain gases A compared with those for Xe, C_2H_2, and Rn. The R-lines are: (1) Xe (25°C); (2) C_2H_2 (25°C); (3) PH_3 (20°C); (4) Rn (25°C); (5) SF_6 (20°C); (6) AsH_3 (20°C); (7) COS (20°C); (8) H_2Se (20°C). The following mole fraction N_A values for 1 atm are registered on the R-line diagram at: (a) N_{SF_6} in $C_6H_5CH_3$ (24.9°C); (b) N_{CF_4} in i-C_8H_{18} (25°C); (c) N_{COS} in C_2H_5OH (22°C); (d) N_{COS} in $C_6H_5CH_3$ (22°C). The N_{SF_6} value for $(CCl_2F)_2$ is 0.0271 (26.4°C), just on the left of the R-line for PH_3. The pressures p_A are in mm Hg.

position of an R-line for a particular gas may be approximately fixed by comparison with the bp of a gas of a known position. From a knowledge of the properties of the gas, and a scrutiny of the background of the solubility spectrum already mapped, a rough idea of the probable N_A values can be predicted.

The possible occurrence of these gases, which may be in small amounts, in chemical, technological, and biotechnological systems and the concern over pollution justify an indication of the positions of the R-lines in the spectrum. Apart from nonreversible interaction with water, in every example the N_A value for water is less than that for any other examined liquid and is far removed from the respective R-line; the absorption coefficient for CF_4 (bp -128°C)[265,266] gives N_{CF_4} = 0.00000362 at 25°, 1 atm, and that for SF_6 (bp -63.8°C) gives 0.00000405,[90] the corresponding value for He being 0.000007. There is no record of any other gas having an N_A value smaller than that for He at 25°, 1 atm for *any* liquid S. Hildebrand and co-workers[5, 171, 209, 242, 310, 311] have given data for CF_4, SF_6, and C_2F_6 for different t°C and 1 atm for six hydrocarbons or CCl_4, CS_2, or $CCl_2F \cdot CClF_2$. For nitromethane at 25°C, 1 atm, $N_{SF_6} = 0.00070$, compared with $N_{Xe} = 0.0020$.[90] For water, N_{H_2Se} is 0.0015 for 25°C, 1 atm,[273] the R-line value being 0.112 for 20°C. Five N_{COS} values for carbonyl sulfide (bp -50.2°C) are obtained from

Examples of Certain Other Gases

published absorption coefficients.[154, 347, 377] The N_{COS} for water at 13.5°C, 1 atm is 0.00051, but that for pyridine, 0.000158, is suspect. In the textbooks carbonyl sulfide is stated to be "moderately soluble in water." For an equimolecular mixture of n-decanol and n-dodecanol (labeled *alcohol*), $N_{COS} = 0.0793$; for *paraffin*, it is 0.1845 at 20°C, 1 atm; the corresponding values for N_{CO_2} were given as 0.0107 and 0.0175, the R-line value for CO_2 being 0.0178 at 20°C.[247]

R-lines for SiH_4 (bp -112°C), PF_3 (bp -101°C), and BF_3 (bp -100°C) would be drawn at the extreme left of the diagram on the scale used, whereas those for C_2N_2 (bp -21.1°C) and SbH_3 (bp -17°C) would require a smaller scale and would appear near the R-line for dimethyl ether at 20°C; that for H_2Te would require a still further reduction in scale and would appear just on the left of the R-line for n-butane for 20°C.

A plot of "cm^3 BF_3(STP) per g n-C_5H_{12}" vs total pressure was nonlinear and was taken to represent a "positive deviation from Henry's law."[41] The N_{BF_3} plot is shown in Fig. 141; the x_{BF_3} plot is clearly distinguishable. The curvature of the N_{BF_3} line shows its trend to reach the R-line at much higher pressures. The N_{BF_3} value for benzene (0.00226 at 47°C, 1 atm) and for toluene (0.00254 at 49°C, 1 atm[373]) are registered in the bottom left corner of the diagram; corresponding values are 0.00247 (22°C) and 0.00279 (27°C).

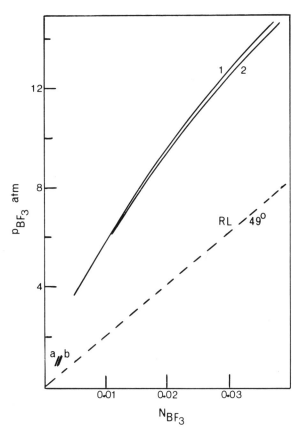

Fig. 141. Solubility of boron trifluoride at different pressures. N_{BF_3} and x_{BF_3} plots for p_{BF_3} up to 14 atm, S = n-pentane. Marks at 1 atm are N_{BF_3} for C_6H_6 (a) (47°C) and for $C_6H_5CH_3$ (b) (49°C).

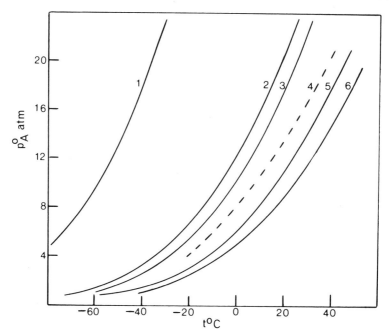

Fig. 142. Plots of vapor pressure p°_A and temperature $t^\circ C$ for the following gases to show where the line for methylsilane (bp/1 atm = -57°C) fits. The lines are for: (1) SiH_4 (bp/1 atm, -112°C); (2) HBr (bp/1 atm, -67°C); (3) H_2S (bp/1 atm, -59.5°C); (4) CH_3SiH_3; (5) COS (bp/1 atm, -50.2°C); (6) H_2Se (bp/1 atm, -41.2°C).

It must again be emphasized that these R-lines represent a property of each gas, *exclusive* of the properties of any liquid S used to absorb the gas.

14.3. Data for Certain Gases Containing Fluorine

Wilhelm and Battino (1971)[379] have reported solubility data for the gases CF_3Cl (bp -81.1°C, 1 atm) and CCl_2F_2 (bp -29°C, 1 atm) in n-heptane, n-octane, cyclohexane, benzene, and carbon tetrachloride at temperatures from about 297.5 to 308.6 K. Data for CF_4 (bp -128°C, 1 atm) and SF_6 (bp -63.8°C, 1 atm) in n-octane for the same temperature range were also given (see Table 36). The primary data were given as the Ostwald coefficient defined as volume of gas absorbed by *one* volume of liquid S. To adjust to $P_A = 1$ atm, they "assumed that Henry's law ($K_H = P_A/N_A$) holds even though it is strictly applicable only for infinitely dilute solutions." They also assumed that P_A is identical with fugacity f_A. For $L < 0.5$, the amount of S was determined from the volume of the "saturated solution," a correction being applied for the change in volume due to the dissolved gas. Finally, the mole fraction N_A was calculated. Values for one temperature are given in Table 37. The R-line for CF_3Cl will be near that for acetylene, R2 in Fig. 140, about 0.02 on the 1-atm horizontal. All the N_{CF_3Cl} values are on the left of the R-line. The position

Table 36. Boiling Points/1 atm and Examples of N_A Values at 1 atm for Gases A (CF_4, SF_6, C_2F_6, C_2H_6) in Certain Liquids S[a]

Liquids	N_A values for 1 atm ($t°C$)			
	CF_4	SF_6	C_2F_6	C_2H_6
CCl_2FCClF_2	0.0048 (25)	0.0271 (26.4)	0.0149 (25)	0.0286 (25)
n-C_7F_{16}	—	0.0224 (25)	—	—
i-C_8H_{18}	0.0029 (24.5)	0.0153 (25)	—	—
n-C_7H_{16}	0.0021 (27)	0.0101 (25)	—	—
CCl_4	0.0012 (24.5)	0.0066 (24.7)	—	—
Cyclohexane	0.0010 (24.7)	0.0054 (26.8)	—	—
C_6H_6	0.00057 (24.8)	0.0026 (25)	0.00111 (25)	0.0151 (25)
CS_2	0.00021 (24.4)	0.00093 (25)	—	—
Bp/1 atm	−129°C	−63.8°C	−79°C	−88.6°C

[a] For full details see refs. 5, 171, 209, 242, 310, and 311.

of the R-line for CF_2Cl_2 will be about 0.18 on the 1-atm horizontal and would require half the scale of Fig. 140 to accommodate it. All the $N_{CF_2Cl_2}$ values are on the left of the R-line and, with reference to the particular R-line, are in positions corresponding to those of the N_{CF_3Cl} values. The *main* effect would seem to be the condensing tendency of the gas.

These authors considered their results from the aspect of the *scaled particle theory* and the evaluation of thermodynamic properties of the systems. The Lennard-Jones (6–12) pair-potential parameters for the solvent were also considered. The agreement between the experimental and calculated values of $\Delta G°$ and N_A at 25°C was deemed to be quite satisfactory for "the less soluble gases, becoming gradually poorer with increasing solubility." "This is to be expected," they state, "since a solution with $N_A > 0.1$ cannot be considered any longer as being very dilute." For details and literature citations, the original paper must be read.

Table 37. Solubility Data for Some Fluorine-Containing Gases

S	t, K	N_A, 1 atm
n-Heptane	297.55	0.0169 (CF_3Cl); 0.126 (CF_2Cl_2)
n-Octane	297.5	0.0163 (CF_3Cl); 0.131 (CF_2CL_2)
n-Octane	297.5	0.00196 (CF_4); 0.097 (SF_6)
Cyclohexane	297.2	0.0100 (CF_3Cl); 0.095 (CF_2Cl_2)
Benzene	297.2	0.00614 (CF_3Cl); 0.072 (CF_2Cl_2)
Carbon tetrachloride	297.6	0.0109 (CF_3Cl); 0.102 (CF_2Cl_2)

14.4. Other R-Line Examples

The use of the R-line diagram for the prediction and checking of data may be illustrated by the example of the methylsilane–methyltrichlorosilane system examined by Shade, Cooper, and Gilbert (1959)[333] in connection with reactions in alkylchlorosilane systems. These workers did not report the bp's of the two compounds, which are $-57°C/1$ atm for methylsilane, CH_3SiH_3, and $+65°C/1$ atm for the trichloro compound, CH_3SiCl_3. In Fig. 142 I give plots of $p°_A$ vs $t°C$ for SiH_4 (bp $-112°C/1$ atm), HBr (bp $-67°C/1$ atm), H_2S (bp $-59.5°C/1$ atm), COS (bp $-50.2°C/1$ atm), and H_2Se (bp $-41.2°C/1$ atm) to show the form of the curves and present a basis for the R-lines. The broken line in Fig. 142 is from the solubility plots of Shade *et al.*, and I am of the opinion that their $p°_{CH_3SiH_3}$ values for $t°C$ from -20 to $+50°C$ are rather low; the slope of the curve does not quite fit the $p°_A$ data for pressures below 1 atm. I compromise a little and estimate $p°_{CH_3SiH_3}$ as 9.0 atm for $0°C$ by comparison with the bp's of H_2S and COS. In Fig. 143 I have drawn the R-lines for $0°C$. Relying on chemical intuition and the available data for H_2S and COS, I would expect the $N_{CH_3SiH_3}$ values for $S = CH_3SiCl_3$ to be near the R-line. At 1 atm, $N_{CH_3SiH_3}$ would be about 0.11 at $0°C$. In the technique used, the two compounds were charged to six stainless steel cylinders, each fitted with a calibrated

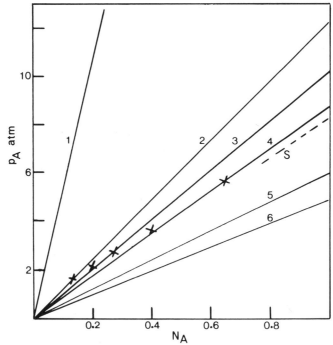

Fig. 143. R-lines at $0°C$ for the following gases given to show the position of the one for methylsilane: (1) SiH_4; (2) HBr; (3) H_2S; (4) CH_3SiH_3; (5) COS; (6) H_2Se. The crosses near line 4 are from data on CH_3SiH_3 by Shade *et al.* and register the mole fraction N_A at different pressures p_A, A being CH_3SiH_3 in CH_3SiCl_3 at $0°C$. Line S is the R-line from data by Shade *et al.*

Examples of Certain Other Gases

Bourdon gauge. Each cylinder was about 100 cm³ capacity and contained moles of CH_3SiH_3 varying from 0.029 to 0.197, and of CH_3SiCl_3 from 0.176 to 0.0421.

The pressures were registered as psia, but I give the equivalents in atm. It is pointless to detail the mode of calculation of the solubility data; suffice it to state that the total pressure was observed, and the partial pressure of methyltrichlorosilane was calculated from Raoult's law and subtracted from the total pressure to give the partial pressure of methylsilane. I have translated their data and registered these as crosses near my R-line in Fig. 143.

In connection with the preparation of diborane, the solubility pattern for "ethane type solvents" was needed. Elliott *et al.* (1952)[76] expressed their results for diethyl ether and tetrahydrofuran at several $t°C$ and pressures $p_{B_2H_6}$ as "moles B_2H_6/100 g solvent." This is a mole *ratio* form. The solubility of diborane in diethyl ether was taken to be slightly greater than that predicted by means of Raoult's law, and it was stated to be proportional to the pressure. The solubility of diborane in tetrahydrofuran was seen to be much greater, and in this example the solubility was stated to increase as the square root of pressure of diborane.

McCarty and Guyon (1954)[275] expressed the solubility in *n*-pentane as mole percent, i.e., the mole fraction form, and declared that "Raoult's law applies very well." In Fig. 144, I have plotted the $N_{B_2H_6}$ vs $p_{B_2H_6}$ and $x_{B_2H_6}$ vs $p_{B_2H_6}$ values, and I

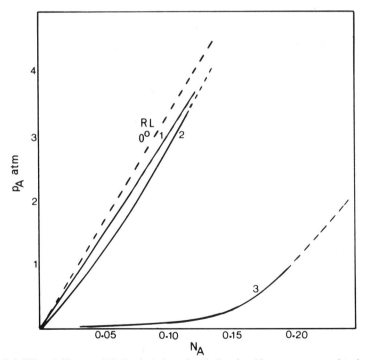

Fig. 144. Solubility of diborane (A). Registration of mole fraction N_A vs pressure p_A (atm) on the R-line diagram. The liquids S are: (1) *n*-pentane (at 0°C); (2) diethyl ether (at 0°C); (3) tetrahydrofuran (at 7°C).

have drawn the R-line for 0°C simply from a consideration of the bp, −92.5°C/1 atm. A value of about 0.03 for $N_{B_2H_6}$ in n-pentane at 0°C and $p_{B_2H_6} = 1$ atm is thereby indicated. The acid function of the diborane is especially effective with tetrahydrofuran.

Two further examples of the usefulness of the R-line diagram in the estimation of N_A or x_A values in difficult systems are afforded by the following papers. Pearson and Robinson (1932)[300] found a molecular weight of 103.0 for carbonyl selenide (bp −22.9°C) from the gas density at 0°C, whereas its actual molecular weight is 107.2. The discrepancy was attributed to the presence of either carbon monoxide or carbon dioxide, which were stated to be appreciably soluble in the liquid. Under the conditions stated, x_{CO_2} would not be more than 0.05, and x_{CO} not more than 0.01. The solubility pattern of COSe itself can be approximately assessed by reference to Fig. 140. The R-line values for COSe at 0°C may be assessed at $N_{COSe} = 0.41$, $x_{COSe} = 0.7$ for 1 atm, based on the vapor-pressure data.

Roozeboom[316] referred to the preparation of chromium oxychloride, the liquid obtained containing "a considerable proportion of free chlorine." The proportion of chlorine was stated to vary "continuously with the temperature and pressure," the phenomenon being deemed to be "one of solution." The solubility pattern may be approximately assessed by reference to Fig. 68; the R-line value for chlorine at 10°C is N_{Cl_2} about 0.20, x_{Cl_2} about 0.25. I need only mention the paper on "liquid–vapor equilibria of the hydrogen–carbon dioxide system" by Spano, Heck, and Barrick (1968)[338] as a different type of example for study against the background of the R-line diagram.

15

Effect of Temperature

15.1. Earlier Aspects

A *positive* temperature coefficient is the term which has been used to indicate that an *increase* in solubility occurs as the temperature is raised, whereas a negative coefficient indicates a *decrease* in solubility with rise in temperature.

Körösy (1937)[213] pointed out that it was then commonly supposed that all gases show a decrease in solubility with rise in temperature, but he affirmed that this was not so. According to Körösy, a large number of gases have a positive coefficient. He attempted to explain this phenomenon by stating that the coefficient depended entirely on the gas, affirming that all gases having a low critical temperature (low cohesional forces) and low solubility show a positive coefficient. There appears to be significant confusion here because "low" solubility was presumably put forward as a *second* requirement, whereas for $p_A = 1$ atm and $t°C$ from 0 to 30°C, say, gases having the low critical temperatures referrred to have, *ipso facto*, low solubilities. Körösy affirmed that gases having a critical temperature above 180 K showed a negative coefficient, as had been expected of gases. He agreed, however, that when the liquid was water, then all gases, even helium, showed a negative coefficient. Neglecting isotope effects, I recognize some 140 elements and compounds as gases; of all these, only seven gases (He, Ne, Ar, H_2, N_2, CO, O_2) have critical temperatures below 180 K. Körösy concluded that Hildebrand's rule about the solubility varying with the inverse of the vapor pressure of the gas does not hold at a "great distance" above the critical temperature.

Burrows and Preece (1953)[39] believed it was then difficult to draw any general conclusions about the solubility of gases. They referred especially to the difficulty over the negative and positive temperature coefficient of solubility. They postulated a similarity between the evolution of gas from a liquid and the evaporation of vapor from a pure liquid. These workers looked upon ΔE_C as the work done on the liquid to form a cavity for the reception of the gas molecule. The quantity ΔE_A was taken to be the energy liberated in putting the gas molecule into the cavity, i.e., the interaction energy. If $\Delta E_A > \Delta E_C$, the temperature coefficient would be negative. If $\Delta E_A < \Delta E_C$, then the temperature coefficient of solubility would be positive.

The quantity ΔH was taken to be the heat given to the liquid when it expels 1 mole of gas isothermally from the saturated solution at a constant pressure. It was assumed that the quantity of liquid is sufficiently large to maintain "a constant concentration of moles of gas per unit volume of solution" during the transfer of gas molecules from the liquid into the space above it. I believe it is now time to analyze this approach in the light of actual solubility data now available for gases in the upper half of the boiling-point range. I should like to see it "applied" to such systems as

$$SO_2 + C_6H_5N(CH_3)_2, \quad HCl + n\text{-}C_8H_{17}OH, \quad NH_3 + H_2O$$

New data given by Burrows and Preece were for helium and three liquids (apiezon G.W. oil, silicone DC 702, and silicone DC 200) and were given as the Ostwald coefficients; data for oxygen and nitrogen with respect to liquids such as kerosene, lubrication oils, and petroleum oils, from the work of others, were cited. For arguments on a molecular basis, such data have no significance.

Jolley and Hildebrand (1958)[192] also referred to the long-puzzling question of the temperature dependence of gas solubility. They seemed to fix a line of demarcation in terms of entropy. The entropy of solution of nonpolar gases in, for example, benzene is positive for any gas for which N_A is <0.00089, and they linked this with a positive coefficient. The effect was deemed to be the reverse when $N_A > 0.00089$.

15.2. Lannung's Data[225]

Lannung (1930) gave solubility results for He, Ne, and Ar in water, methanol, ethanol, acetone, benzene, cyclohexane, and cyclohexanol for different temperatures from 5 to 45°C. He gave his results as the Bunsen absorption coefficient α. He gave the Ostwald coefficient L, where $L = \alpha T/273$ and T is the t K of the measurement, but he defined L as the equilibrium ratio of the "volume concentrations of the gas in the solution and in the vapor phase."

His arguments relate to L and not to mole fraction N_A or to mole ratio x_A. For water, the L values decrease with rise in temperature, but they increase for the other liquids; argon, however, shows some uncertainties. The paper stands out as a type, but it is necessary to convert solubilities to mole ratio (and mole fraction) and thus to allow for the different molecular weights of the liquids S and the change in density of a particular S with change in temperature.

15.3. Data by Gerrard and Colleagues

Gerrard[95] has given x_{HCl} vs $t°C$ plots for a number of organic liquids S. The increase in x_{HCl} as the temperature is lowered to as low a value as $-70°C$ is spectacular. Table 38 shows this effect. For tri-n-butyl borate, x_{HCl} varies from 0.49 at 0°C to 7.80 at $-74°C$. For tetra-n-butyl silicate, x_{HCl} varies from 2.50 at 0°C to 12.30 at $-75°C$.

Effect of Temperature

Table 38. Examples of Change in x_{HCl} with Temperature

$CH_3(CH_2)_3CH_2OH$		$ClCH_2CH_2OH$		$CH_3(CH_2)_4CH_2Cl$	
t, °C	x_{HCl}	t, °C	x_{HCl}	t, °C	x_{HCl}
−72	3.42	−74	2.86	−76	2.64
−42	1.98	−62	2.11	−65	1.19
−27	1.54	−40	1.25	−43	0.46
+2	1.09	−20	0.88	−22	0.18
21	0.87	+6.5	0.56	+0.3	0.09
				+6.0	0.08

n-Decane		$(C_2H_5O)_2CO$		$(n\text{-}C_4H_9O)_3B$	
t, °C	x_{HCl}	t, °C	x_{HCl}	t, °C	x_{HCl}
−9.0	0.062	−75	7.46	−74	7.80
−6.0	0.057	−60.5	3.92	−33	1.33
0	0.050	−42.0	1.94	−18	0.85
+2	0.049	0	0.68	0	0.49
10	0.043	+6.5	0.56		

$(n\text{-}C_4H_9O)_3P{=}O$		$(n\text{-}C_4H_9O)_4Si$		$CH_3CO_2C_2H_5$	
t, °C	x_{HCl}	t, °C	x_{HCl}	t, °C	x_{HCl}
−76	8.74	−75	12.30	−72.5	5.35
−62	6.67	−44	6.30	−54	3.54
−40	2.79	−30	4.65	−30	1.86
+6	2.54	0	2.50	−8.5	1.00
				0	0.821

A number of x_A vs t°C plots are given herein. The effect of a rise in temperature is to move the R-line to the left, and there are no exceptions to this. With the exception of He, Ne, Ar, H_2, and possibly, N_2, for nonaqueous liquids S the observed N_A vs p_A plot is always shifted toward the left as the temperature is increased. In those examples for which the N_A value is given for one pressure, usually referred to 1 atm, then the position on the 1-atm horizontal moves to the left as the temperature is raised.

For several gases, e.g., n-butane, ethyl chloride, methyl bromide, and methylamine, lowering the temperature from 20 to −10°C puts it below the bp of the gas A at 1 atm, and therefore the system becomes, in conventional terms, a "liquid–liquid" one.

A rise in temperature lowers the tendency of the gas to condense at the operational temperature (say, 0 to 25°C), and this will lower the x_A value. However, a rise in t°C will change the weighting of the second and third factors relating to the liquid S and the interaction of the gas A with S. At these operational temperatures and at $p_A = 1$ atm, the x_A values for He, Ne, Ar, H_2, N_2, CO, and O_2 are absolutely very small, and it is understandable that the change in the second factor, relating to the intermolecular structure of S, with rise in $t°C$ may actually make S more receptive on this scale and allow an absolutely small, but relatively more distinctive, increase in the x_A value. Water is the one liquid which does not compensate in this way, and so the x_A values *decrease* with rise in temperature. In converting the Ostwald coefficients for different temperatures into the x_A values, the change in density of liquid S must at least be considered.

16

Prediction

16.1. Approach to the Problem

Hildebrand's parameter formula relates to so-called nonpolar gases and nonpolar liquids; I know of no mathematical relationship which purports to lead to the prediction of N_A over the whole range of gases A and liquids S. Certain trends, such as "like dissolving like," have been recognized; but, as Hildebrand pointed out, much depends on how one defines "likeness." The "like dissolving like" slogan can be grossly misleading. One only needs to lay, side by side, N_A values for pairs of A and S, bearing in mind that p_A and $t°C$ must be specified, to see how ridiculous this slogan can be. A scrutiny of the solubility spectrum now revealed, however, can lead to a useful estimate of the N_A value for a particular gas and particular liquid, despite the many gaps which hinder the process. Chemical knowledge and intuition enable one to appreciate the essential pattern of data and to see how estimates of missing N_A values can be made on a rational basis.

It is only after a detailed scrutiny of the essential pattern of existing data presented in the graphical form as shown herein that one can begin to see the possibilities in terms of prediction. There is a need for more data, and the bubbler-tube manometer procedure enables data to be gathered quickly for those gases in the upper range of bp's. Data obtained thereby can give a rational basis for estimating the N_A values for gases in the lower range of bp's. For example, the disposition of the N_{BuH} vs p_{BuH} lines with respect to the R-line for n-butane at, say, 25°C for a number of liquids S will most probably indicate the disposition of the corresponding N_A values for 1 atm and 25°C for a gas such as carbon monoxide with respect to the R-line for carbon monoxide.

There will, of course, be grounds for doubts, and controversy will emerge. The approach is new, and it will take time to demonstrate its efficacy.

Data for hydrogen selenide are very sparse, and I doubt if there are any on hydrogen telluride. The R-lines for these for 20°C, for example, can be drawn with useful precision, and it is very probable that the N_{H_2Se} and N_{H_2Te} values for 1 atm for a number of liquids S will be disposed about the particular R-line in a way similar to the disposition of the corresponding N_{H_2S} values about the R-line for hydrogen sulfide. The items below are meant to indicate certain aspects of the process of prediction.

16.2. Lachowicz's Review

Lachowicz (1954)[221] (see also Lachowicz and Weale[222]) has reviewed the work on the "solubility of gases in liquids at high pressures" and has cited 66 references. The approach was geared to Henry's law. As other reviewers have remarked, he found difficulties over the critical assessment of experimental data. Such matters as criteria of purity, form of allowances for (or decision to neglect) the difference between the observed pressure and 1 atm, units of pressure, and presentation of data in the form of small-scale graphs engendered uncertainty. Lachowicz dealt only with the "permanent gases" and for these he gave helpful details in the form of tables. It is pertinent to notice the terms in which he refers to Henry's law. Henry's work was described as the first systematic study which led to the publication of a series of results on the solubilities of some gases ("oxygenous and azotic gases," among others) in water. It must again be emphasized that Henry examined only five gases with respect to pressure, and only for water at essentially one temperature.

16.3. Other Papers

Prausnitz (1962)[305] pointed out that chemical engineers are interested in the correlation and prediction of solubilities in connection with the design of technical chemical operations. He deemed Hildebrand's approach to be a highly suitable one for the chemical engineer. The emphasis, however, was on nonpolar systems and on the gases in the lower half of the bp/1 atm range. It was appreciated that although there were many data on the solubility of gases, there was a tendency for these data to be for about 25°C, and even then some were of uncertain accuracy. It was therefore often necessary to estimate gas solubilities from little or essentially no information.

Chang, Gokcen, and Poston (1969)[47] referred to the usefulness of a knowledge of the solubilities of gases in liquid propellants. Liquid propellants are pressurized with gases to provide a protective blanket and to eject the propellants from their containers in the space vehicles. In this way the use of heavy pumps is avoided; but certain undesirable effects have to be contended with, and solubility data are helpful. The mole fraction form of Henry's law was invoked; but here again the gases were He, N_2, O_2, and Ar, gases at the lower end of the bp/1 atm scale, for which the x_A ($\approx N_A$) values are absolutely small at $p_A = 1$ atm and at a temperature range from 0 to 25°C. The operational area is, therefore, a tiny patch at the bottom right corner of the complete R-line diagram. The authors expressed the solubility as parts per million (ppm) of concentration in weight. For He in N_2O_4, this was 3.0 for 0°C and 4.3 for

Prediction

Table 39. Solubility at 1 atm

Gas	Propellant	ppm at 0°C	ppm at 25°C
He	N_2O_4	3.0	4.3
N_2	N_2O_4	182	203
He	N_2H_4	0.5	0.6
N_2	N_2H_4	4.4	6.3
He	Unsymmetrical dimethylhydrazine (UDMH)	4.5	6.2
Ar	UDMH	440	456

25°C, thus showing an increase in solubility (ppm) with rise in t°C. Twenty-three solubility values were given for 0°C, and the same number for 25°C, with respect to the gases named. In Table 39 I indicate the smallness of the values.

The remaining gases mentioned were CO_2 and N_2O_3, one pair of values for each gas being given. For CO_2 in inhibited red fuming nitric acid (IRFNA), the solubilities (ppm) were 1.33×10^4 (0.0133 parts of CO_2 per 1 part of IRFNA) at 0°C and 7.60×10^3 (0.00760 parts of CO_2 per 1 part IRFNA) at 25°C, thus now showing a decrease in ppm with rise in t°C. The expression "dilute solutions" used with reference to *Henry's law* is misleading; at 0°C and 1 atm the values for He, N_2, O_2, and Ar are all absolutely small because they cannot be larger; nor can they be smaller for a given S at equilibrium.

Schröder (1969)[329] examined the gases H_2, CH_4, N_2, and O_2 for water (for pressures up to 80 atm) and H_2 for three alcohols (for pressures varying from 19 to 45 atm). Amirkhanov (1965)[2] referred to the solubility of gases in liquids with "predominant mechanical and molecular physical motions" and gave an equation for calculating the "solubility of gases in different liquids at different pressures."

Two further examples are relevant here because they show the inadequacy of the form of presentation of data not infrequent in the literature. I have already mentioned these two examples under sulfur dioxide, and I now give further essential details from the aspect of prediction.

Lloyd (1918)[244] reported his data to prevent unnecessary work on the part of those who need such data. He expressed his data as "grams of sulfur dioxide per liter of solution," and, as he did not report the density of solution for four of the five liquids S examined, the data are useless for comparison purposes on a molecular basis. The fifth liquid was acetic anhydride, and although he gave data for several temperatures, he gave the density of the *solution* (1.22) for only 0°C, the total pressure being the barometric one (756–760 mm Hg). He reported that 148 g SO_2 were absorbed to give 1 liter of *solution*, i.e., 1220 g of solution. Therefore, 148 g SO_2 were absorbed by 1072 g acetic anhydride, and the mole ratio x_{SO_2} for 0°C is

$$(148/64)/(1072/102) = 0.220 \equiv N_{SO_2} = 0.180$$

Glances at the R-line diagram (Fig. 83) and the plots of Horiuti's data for acetone and methyl acetate for 25°C clearly show that his x_{SO_2} value is right out of the pattern and is most likely grossly in error. The x_{SO_2} value for 10°C, 1 atm for acetone, calculated from Horiuti's data, is 1.70. My own values for acetic anhydride for 747 mm Hg (total pressure) are $x_{SO_2} = 2.90$ (0°C), 1.57 (10°C), 0.945 (20°C), and 0.76 (25°C).

In their review, Markham and Kobe[253] gave Lloyd's results the maximum grading of 4 for quantitative reliability; the same grading was given to Horiuti's detailed results.

Weissenberger and Hadwiger (1927)[370] reported data on "the absorption of sulfur dioxide in organic liquids." They were concerned with the purification of gases by selective absorption, the particular gas being recovered by "mere heating." The volume of sulfur dioxide absorbed by a liquid S (one of six) was measured at 20°C and presumably the prevailing barometric pressure, which was not stated. The observed data were expressed as "cm^3 of SO_2" absorbed by the molar volume of S (i.e., one mole of S). In Table 40 I now record their data and my calculated x_{SO_2} values. I use 24,040 cm^3 as the gram-molecular volume of SO_2 for 20°C.

Against the background of the R-line diagram all these x_{SO_2} values appeared outrageously small (see Figs. 75–83) and were, therefore, in need of checking. I now give my own data in Table 41.

In their paper on the solubility and entropy of solution of argon in selected "nonpolar" solvents ($C_6F_{11}CF_3$, $C_6H_{11}CH_3$, cyclohexane, CCl_4, $C_6H_5CH_3$, C_6H_6, CS_2), Reeves and Hildebrand (1957)[310] expressed their belief that their predictions of isothermal solubility were more reliable than those relating to its dependence on temperature. It was deemed advantageous to work with gas–liquid systems because one could use "very dilute solutions" where solute–solute interaction was deemed to be practically absent and Henry's law closely obeyed.

Whereas no words of mine are needed to commend such detailed studies, I must keep a clear picture of the position of this tiny area in the whole area of operational

Table 40. Absorption of SO_2 in Organic Liquids[a]

S, 1 mole	SO_2, cm^3	x_{SO_2}
Tetrahydronaphthalene	[b]	
Decahydronaphthalene	1181.25	0.049
Cyclohexanone	2353.87	0.098
m-Methylcyclohexanol	2082.82	0.087
"Hydroterpin"[c]	2140[d]	0.089
"Turpentine"	1314[e]	0.055

[a] See also Table 41.
[b] Too small to be of consequence.
[c] See footnote a of Table 41.
[d] 1 cm^3 absorbed 14.5 cm^3 SO_2 (average of two experiments).
[e] 1 cm^3 absorbed 8.3 cm^3 SO_2.

Table 41. Solubility of SO_2 at 760 mm Hg (New Data Observed in Connection with the Analysis of the Results by Lloyd and by Weissenberger and Hadwiger)[a]

S	Weights, g		t, °C	x_{SO_2}	N_{SO_2}
	S	SO_2 (absorbed)			
Tetrahydronaphthalene (tetralin)	7.5100	1.0732	20	0.295 ([b])	0.228
Decahydronaphthalene (decalin)	7.2121	0.2090	20	0.062 (0.0491)	0.060
n-Decane	3.8122	0.3040	0	0.179	0.151
			20	0.105	0.095
Pinene (for turpentine)	3.4166	0.4009	0	0.250	0.200
		0.2201	20	0.124 (0.055)	0.110
Mesitylene 1,3,5-$(CH_3)_3C_6H_3$	3.8690	3.5828	0	1.740	0.635
		1.9450	10	0.944	0.485
		1.2539	20	0.610	0.380
Cyclohexanone	2.0601	2.1639	10	1.611	0.617
		1.4730	20	1.100 (0.098)	0.521
2-Methylcyclohexanone	2.5852	1.9010	10	1.290	0.563
		1.3808	20	0.937	0.484
Cyclohexanol	2.5570	0.4489	20	0.275	0.216
Benzyl benzoate	7.2841	1.5875	20	0.723	0.420
Benzyl acetate	6.1630	2.4370	20	0.928	0.481

[a] Results of Weissenberger and Hadwiger were expressed as vol. SO_2/vol. S; my calculated x_{SO_2} values are given in parentheses. In the original paper (W and H) 1 g of "Hydroterpin" was stated to occupy a volume of "1.133 cm^3." In *Chem. Abstr.*, 1927, *21*, 3052, this was translated as 1 g "hydroterpinol" "absorbs 1.133 cm^3 SO_2," although it was then stated that "1 cm^3 hydroterpinol absorbs 14.308 cm^3 SO_2." The original statement (W and H) was that "in two parallel experiments, 1 cm^3 hydroterpin absorbed 14.3 and 14.8 cm^3 sulfur dioxide, respectively."
[b] "Too small to be of significance." *British Chem. Abstr.*, **B**, 1927, 617 gave "nil."

chemistry. The properties of a molecule in almost complete isolation, "solute–solute interaction being practically absent," may be quite different when the concentration of solute is much greater. The mole fraction, N_{Ar}, for argon in toluene at 1 atm and 288.23 K was given as 10.882×10^{-4}, which is equal to 0.0010882 ($\approx x_{Ar}$). This means that about 984 molecules of toluene are required, on the average, to hold *one* molecule of argon. In terms of moles per liter of solution, the concentration of argon in the gas phase is rather more than four times that in the liquid solution. As the temperature rises from 288.23 to 298.20 K, the N_{Ar} value increases to 0.0010950, and then, at 303.31 K, to 0.0010982 (or 0.0010995). How does the value of the moles of argon per liter of solution change with temperature? How does this change relate to the change in density of argon in the gas phase?

For equilibrium, and within the range of experimental competency, the N_{Ar} value can be no other than ≈0.001088 for toluene at 288.23 K and 1 atm. This system should not be considered as an example of a "dilute solution" in general terms but as the argon–toluene system under the specified conditions of temperature and pressure. Under these specified conditions the N_{Ar} values are absolutely small for all examined liquids.

The relatively large changes in N_{Ar} as the liquid S is changed may then be attributed to the change in intermolecular force pattern and the way this changes on the addition of argon and as the temperature changes. Values of N_{Ar} for $C_6F_{11} \cdot CF_3$ for 1 atm were given as:

$$N_{Ar} \times 10^4 = 48.709 \, (278.09 \text{ K}), \qquad N_{Ar} = 0.0048709$$

$$N_{Ar} \times 10^4 = 45.470 \, (303.16 \text{ K}), \qquad N_{Ar} = 0.004547$$

Therefore, for this liquid S, N_{Ar} decreases with increase in temperature. To get the ratio moles per liter in the gas phase/mole per liter of liquid solution, the density of the fluorocarbon would have to be used, as distinct from the density of toluene. For $C_6H_{11} \cdot CH_3$ and CCl_4, the N_{Ar} value for 1 atm decreases with rise in temperature; but for CS_2, it increases, from 0.0004272 at 253.13 K to 0.0004866 at 298.14 K.

16.4. New Data on Propene and the Four Butenes

New data for these alkenes are discussed here to illustrate the aspect of prediction based on the R-line. The first factor to consider is the bp/1 atm of the gas since this indicates its tendency to condense at the operational temperature. The relevant bp/1 atm data are shown in Table 42.

The bp factor alone will lead to x values for 1-butene and 2-methylpropene being similar for corresponding liquids S at a fixed $t°C$ and 1 atm, whereas they will be smaller than the corresponding ones for the two 2-butenes and n-butane. However, the operation of a more specific acid function probably sited on hydrogen of the olefin might be encouraged by a strong-enough basic function of the liquid S, and the basic function sited at the double bond might be brought out by a strong-enough acid function of S, so that the observed pattern can be complicated. It

Table 42. Boiling points/1 atm

Hydrocarbon	bp/1 atm
Propane	−44.5°C
Propene	−47.8°C
n-Butane	−0.5°C
i-Butane (2-methylpropane)	−11.5°C
1-Butene	−6.3°C
i-Butylene (2-methylpropene)	−6.8°C.
2-Butene (*trans*)	0.9°C
2-Butene (*cis*)	3.5°C

Prediction

may be inferred, however, from the discussion on the work of Robinson and Sund[313] on the use of hexafluoroisopropanol $(F_3C)_2CHOH$ (S_F) for separating an alkane from an alkene that the special acid–base effect will not be pronounced. Although S_F was deemed to be an "extremely strong hydrogen-bonding donor" and therefore one to bring out the basic function of an alkene as distinct from the corresponding alkane, a glance at the actual data recorded in Table 29 and Fig. 137 clearly shows that there is no outstanding effect. Whereas it is true that the $N_{C_3H_6}$ value is nearer to the R-line for propene at 26°C than the $N_{C_3H_8}$ value is to the R-line for propane at the same p_A, each value is definitely on the left (less soluble side) of its R-line. Nor is the basic function of acetylene especially evident from the $N_{C_3H_6}$ value for S_F at 1 atm and 26°C, as in Table 29. On the other hand, the acid function of acetylene *is* clearly revealed by such liquids as dimethylformamide and hexamethylphosphoramide in the work of McKinnis[281] (see Fig. 135):

	$x_{C_2H_2}$	$N_{C_2H_2}$ (at 25°C, 1 atm)
$HCON(CH_3)_2$	0.124	0.110
$[N(CH_3)_2]_3P=O$	0.338	0.252

Both of these $N_{C_2H_2}$ values are well over on the right of the R-line.

There are slight variations in the literature values for the bp and vapor pressures of each alkene named in Table 29. In Fig. 145 I have drawn the two R-lines, one for 1-butene and one for 2-methylpropene, close together. With the exception of the lines for dimethylformamide, the N_A pattern for 1-butene closely overlaps that for 2-methylpropene.

In Fig. 146 I have drawn a common R-line for *trans*-2-butene for 5°C and for the *cis* isomer for 7.8°C, $p°_A$ being taken as the same for these two alkenes at the respective temperatures. Although the patterns tend to correspond, there is a decided tendency for the *cis* isomer line for a given S to be on the right (more soluble side) of the line for the *trans* isomer. The differential change of intermolecular structure of liquids S with rise in temperature must be taken into account. Numerical data are given in Table 30.

For 0°C, the R-line for propene strikes the horizontal at 760 mm Hg at $N_{C_3H_6} = 0.156$. The $N_{C_3H_6}$ values for dimethylformamide and hexamethylphosphoramide are each well on the left of the R-line, but each is definitely larger than the corresponding $N_{C_3H_8}$ value: At 0°C, 1 atm we have

	$N_{C_3H_6}$	$N_{C_3H_8}$
$HCON(CH_3)_2$	0.049	0.028
$[(CH_3)_2N]_3P=O$	0.127	0.091

Thus, the acid function of propene seems to be gently brought out by the decided basic function of these two liquids S. The $N_{C_3H_8}$ value for aniline at 0°C, 1 atm is relatively very low, being 0.0165; the $N_{C_3H_6}$ value is about twice this, being 0.0315, and although it is still small, it does show a slight effect of the basic function of aniline. Whereas mesitylene reveals its basic function with hydrogen chloride, it is not

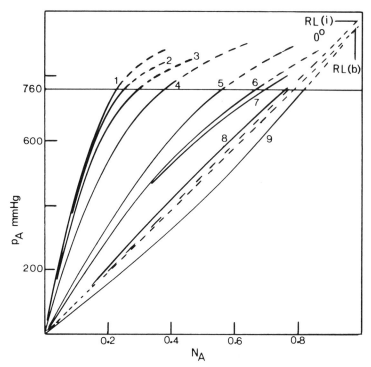

Fig. 145. Solubility of 1-butene (A) and 2-methylpropene (A) in liquids S at 0°C. Plots of mole fraction N_A vs pressure p_A (in mm Hg) on the R-line diagram. For 1-butene the lines are: (1) $C_6H_5CH_2OH$; (2) $HCON(CH_3)_2$; (4) Cl_3CCH_2OH; (5) $n\text{-}C_8H_{17}OH$; (6) $n\text{-}C_5H_{11}CO_2H$; (7) $C_6H_5CO_2C_2H_5$; (9) $(n\text{-}C_8H_{17})_2O$; the line for $n\text{-}C_8H_{17}Br$ closely follows the R-line (b) for 1-butene. For 2-methylpropene (i, isobutylene), the lines are: (3) $HCON(CH_3)_2$; (8) $n\text{-}C_8H_{17}Br$. For corresponding liquids, the following lines are closely followed: 1, 5, 6, 7; the line for $n\text{-}C_{10}H_{22}$ is just on the right of the R-line (i). For $1,3,5\text{-}C_6H_3(CH_3)_3$, the line for 1-butene is just on the left of line 8.

incisive enough to show this specific effect with propene. The values $N_{C_3H_6} = 0.144$ and $N_{C_3H_8} = 0.150$ are in accord with the difference in tendency of these gases to condense, and the intermolecular structure of mesitylene decides the position of the $N_{C_3H_8}$ value with reference to the R-line. We have to recognize a differential effect; the $N_{C_3H_6}$ for mesitylene is about the same as that for the phosphoramide, but the $N_{C_3H_8}$ value for the latter S is much less than for the former S.

The acid–base function of propene appears to be revealed by n-hexanoic acid ($N_{C_3H_6} = 0.138$ and $N_{C_3H_8} = 0.128$ at 0°C, 1 atm), but again each value is on the left of the respective R-line. The corresponding values for trichloroethanol, Cl_3CCH_2OH, are $N_{C_3H_6} = 0.116$ and $N_{C_3H_8} = 0.038$. The values for dichloroacetic acid at 20°C are $N_{C_3H_6} = 0.033$ and $N_{C_3H_8} = 0.021$, each well on the left of the respective R-line for 20°C. Values for n-decane and di-n-octyl ether are decidedly on the *right* of the respective R-lines for propene and propane at 0°C. Values for a selection of liquids are given in Table 43.

Prediction

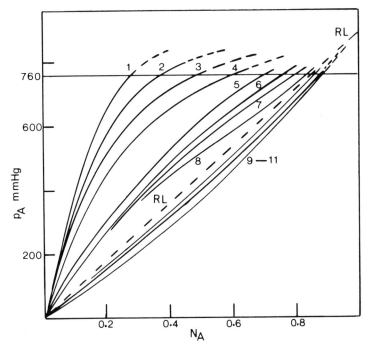

Fig. 146. Solubility of *trans*-2-butene at 5°C and *cis*-2-butene at 7.8°C, the R-line being common for these two temperatures. Plots of the N_A vs pressure p_A values on the R-line diagram. The liquids S are: (1) $C_6H_5CH_2OH$ (*trans*); (2) $C_6H_5CH_2OH$ (*cis*); (3) Cl_3CCH_2OH (*trans*); (4) Cl_3CCH_2OH (*cis*); (5) n-$C_8H_{17}OH$ (*cis*); (6) n-$C_5H_{11}CO_2H$ (*cis*); (7) $C_6H_5CO_2C_2H_5$ (*trans* and *cis*, close fit); (8) C_6H_5Br (*trans* and *cis*, close fit); (9) n-$C_{10}H_{22}$ (*trans*); (10) $(n$-$C_8H_{17})_2O$ (*trans*); (11) $(n$-$C_8H_{17})_2O$ (*cis*).

Table 43. x_A and N_A Data for Propene Compared with Those for Propane at 0°C and 1 atm

S	Propene		Propane	
	x_A	N_A	x_A	N_A
R-line	0.186	0.156	—	—
n-$C_{10}H_{22}$	0.230	0.190	0.322	0.234
$(n$-$C_8H_{17})_2O$	0.305	0.232	0.340	0.260
1,3,5-$C_6H_3(CH_3)_3$	0.168	0.144	0.177	0.150
n-$C_5H_{11}CO_2H$	0.159	0.138	0.146	0.128
Cl_3CCH_2OH	0.127	0.116	0.039	0.0375
$HCON(CH_3)_2$	0.0513	0.0489	0.0283	0.0275
$[(CH_3)_2N]_3P=O$	0.146	0.127	0.100	0.091
$C_6H_5CH_2OH$	0.0435	0.0417	0.0335	0.0324
$C_6H_5NH_2$	0.0325	0.0315	0.0168	0.0165
Cl_2CHCO_2H (20°C)	0.034˙	0.033	0.022	0.021

16.5. The R-Line Approach to the Acetylene Systems[(402)]

Hölemann and Hasselmann gave their data as milligrams of acetylene (Ac) per gram of liquid S, i.e., mg Ac/g S, and with one exception obtained linear plots for mg Ac/g S vs pressure p_{Ac} for p_{Ac} up to 800 mm Hg. Based on what would be the mole ratio form of Henry's law, they gave a table of "solubility coefficients" expressed as mg Ac/g S for 760 mm Hg. In Table 44 I give also the corresponding mole ratio, x_{Ac}, and mole fraction, N_{Ac}, values. It is clear that the weight ratio does not place the different liquids in the same order as the mole ratio does. In Fig. 147 I have registered the N_{Ac} values for 1 atm and 25°C on the R-line diagram. The order left to right in the figure is the order of ascending value of N_{Ac} shown in Table 44. The N_{Ac} values for the alcohols are on the left of the R-line; this is probably due to the hydrogen-bonding structure of the alcohols and the resistance of this to the acid–base function of acetylene. In this respect, methanol is not in its usual position compared with butanol, and I suspect the accuracy of the data; the position of the line for methanol should be just on the left of that for n-butanol. I predict that the line for glycerol will be about halfway between that for ethylene glycol and the left vertical axis. The line for water is indicated by the letter W right against the left axis on the scale used. For the liquids 5 to 13, the increase in N_{Ac} values reflects the increase in intensity of the basic function with respect to the acid function of acetylene.

Figure 148 shows the N_{Ac} vs p_{Ac} and x_{Ac} vs p_{Ac} plots for dimethylformamide at 25°C and for higher pressures. The form of the x_{Ac} line shows why the plot of mg Ac/g S vs p_{Ac} for $p_{Ac} < 1$ atm is approximately linear, since the weight ratio is

Table 44. Solubility of Acetylene in Different Liquids. Comparison of the Observed mg Ac/g S Data with the Mole Ratio x_{Ac} and Mole Fraction N_{Ac} Values for 25°C and a Pressure p_{Ac} of 1 atm (Ac = Acetylene)

Number in Fig. 147	Liquid S	mg Ac/g S	x_{Ac}	N_{Ac}
1	Ethylene glycol	3.5	0.00835	0.00830
2	Butanol	6.3	0.0179	0.0176
—	Allyl alcohol	7.3	0.0163	0.0161
—	Dicyclopentadiene	3.7	0.0188	0.0185
3	Methanol	16.4	0.0202	0.0198
4	1,2-Dichloroethane	5.6	0.0213	0.0209
5	Butyrolactone[a]	14.9	0.0493	0.0470
6	Dioxane	16.6	0.0562	0.0532
7	Ethyl acetoacetate	12.2	0.0610	0.0575
7	Diethyl oxalate	10.9	0.0612	0.0577
8	Ethyl acetate	19.3	0.0653	0.0613
9	Tetrahydrofuran	25.3	0.0702	0.0656
10	Dimethylformamide	37.7	0.106	0.096
11	Tetramethylurea	27.9	0.125	0.111
12	Dimethylacetamide	41.9	0.140	0.123
13	N-Methylpyrrolidone[b]	41.6	0.159	0.137

[a] Incorrectly called butyl 2-hydroxypropionate.
[b] Incorrectly called N-methylpyrrolidine by Stephen and Stephen.[(344)]

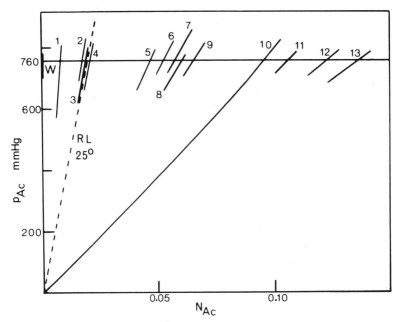

Fig. 147. Solubility of acetylene (Ac) in different liquids at 25°C. Registration of the N_{Ac} values for 1 atm on the R-line diagram. The numbers refer to the compounds listed in Table 44.

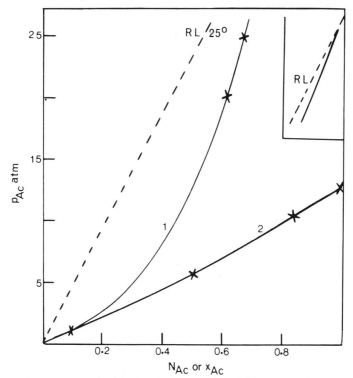

Fig. 148. Solubility of acetylene (Ac) in dimethylformamide at 25°C. Plots of the mole fraction N_{Ac} vs pressure p_{Ac} and mole ratio x_{Ac} vs p_{Ac} on the R-line diagram. The R-line reaches the right vertical axis at 48 atm, as shown in the inset. Line 1 is the N_{Ac} plot up to the observed value at 25 atm; the extrapolation is shown approaching the R-line in the inset. Line 2 is the corresponding x_{Ac} plot. See Fig. 149 for the plot on a smaller scale.

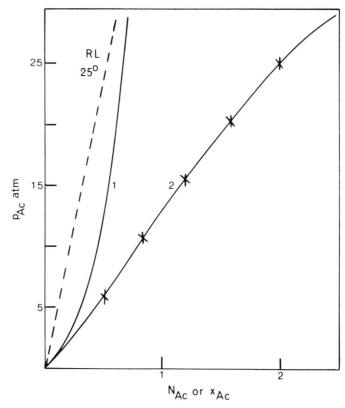

Fig. 149. Solubility of acetylene (Ac) in dimethylformamide at 25°C. (See Fig. 148.) The mole ratio x_{Ac} vs p_{Ac} plot is shown on a smaller scale to show the shape at higher pressures.

equivalent to the mole ratio for a selected liquid S. At higher pressures, the line becomes concave upward at first and then changes into a convex-upward form before leveling in a horizontal direction (Fig. 149). All this fits the concave-upward form of the N_{Ac} plot.

In Fig. 150 the N_{Ac} and x_{Ac} plots are shown for butyrolactone and N-methylpyrrolidone at 25°C and higher pressures, p_{Ac}, in atmospheres. The corresponding lines for N,N-dimethylacetamide are close to those for N-methylpyrrolidone.

16.6. The Carbon Dioxide Systems

The work of Sander (1912)[403] provides an example of the tortuous mode of presentation of solubility data not infrequently encountered in the literature. It affords an opportunity to demonstrate the use of the R-line procedure in sorting out such forms of data and in the correction of misleading statements which emerge from such forms. Sander's data for the solubility of carbon dioxide in water and in other

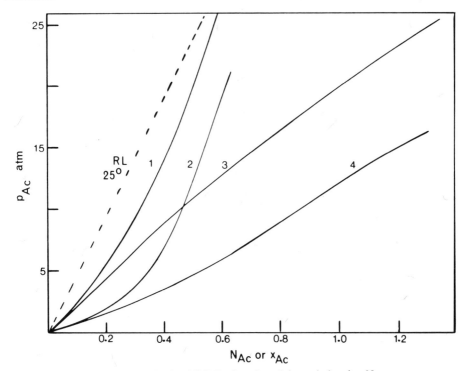

Fig. 150. Solubility of acetylene (Ac) at 25°C. Registration of the mole fraction N_{Ac} vs pressure p_{Ac} and mole ratio x_{Ac} vs p_{Ac} plots on the R-line diagram. Line 1 is the N_{Ac} plot for butyrolactone; line 3 is the corresponding x_{Ac} plot. Line 2 is the N_{Ac} plot for N-methylpyrrolidone; line 4 is the corresponding x_{Ac} plot. The lines for N,N-dimethylacetamide are very close to lines 2 and 4.

liquids at higher pressures were in the form "volume of CO_2/volume of solution," and in that form they cannot be used to calculate the mole ratio x_{CO_2} and mole fraction N_{CO_2}. Stephen and Stephen[344] gave these data in that form, and the data are therefore useless for comparison on a molecular basis. However, Sander did give the volume of the liquid taken, although he did not use it; but it has enabled me to calculate the x_{CO_2} and N_{CO_2} values from his data. The pressure p_{CO_2} was given in kg/cm² (taken herein to be related as 1 atm = 1.033 kg/cm²).

At 25°C and $p_{CO_2} = 20$ kg/cm², benzene (0.080 cm³) absorbed 0.2422 cm³ of CO_2 and gave 0.061 cm³ of solution. At 30 kg/cm³, the volume of solution was 0.0768, and at 50 kg/cm², it was 0.114. According to Lannung and Gjaldbaek (1960) the absorption of a gas by a volume of liquid S always results in an increase in volume of the liquid phase.

Sander gave the expression "(vol. CO_2/vol. solution)/p_{CO_2}" as the "Henry's law constant," but he showed that this "constant" changed with p_{CO_2}. For benzene the "constant" changed as follows:

p_{CO_2}, kg/cm²	15	20	30	40	50
$(V_{CO_2}/V_{solution})/p_{CO_2}$	3.126	3.558	4.177	4.811	5.285

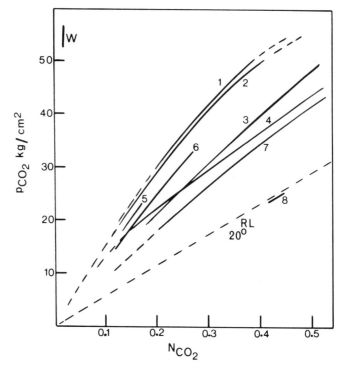

Fig. 151. Solubility of carbon dioxide at 20°C and at various high pressures p_{CO_2}, in kg/cm^2. From the vol./vol. data of Sander.[403] Registration of the mole fraction N_{CO_2} vs p_{CO_2} plots on the R-line diagram. The liquids are: (1) C_2H_5OH; (2) C_3H_7OH; (3) C_6H_5Cl; (4) C_6H_6; (5) C_6H_5Br; (6) $C_6H_5NO_2$; (7) $C_6H_5CH_3$; (8) $CH_3CO_2C_2H_5$.

For a number of liquids S at 20°C, I have shown the positions of the N_{CO_2} vs p_{CO_2} plots on the R-line diagram. With the exception of ethyl acetate, the liquids named give lines on the left of the R-line; the line for water is again near the left vertical axis (see Fig. 151). In Fig. 152, I have plotted the corresponding lines for ethanol, propanol, and benzene for 35°C. In Fig. 153 the N_{CO_2} and x_{CO_2} plots are compared for benzene at 20°C, and in Fig. 154 the corresponding pair for n-propanol is given.

Bearing in mind that the "Henry's law constant" was given as "(vol. CO_2/vol. solution)/p_{CO_2}," the summary given in the abstract[403] indicates that at low temperatures Henry's law was more nearly followed when the volume of gas was related to the volume of the solution rather than to the volume of the solvent. In Fig. 155 I have compared the vol. CO_2/vol. solution plot with that for vol. CO_2/vol. S for propanol and benzene at different p_{CO_2} and at 20°C. To convert vol. A/vol. S (Ostwald coefficient) data into x_A data, the density and molecular weight of the original liquid S are material factors. The statement that the solubility (vol./vol.) decreases with the increase in molecular weight of S, even if the liquids S are chemically related, is, on a molecular basis, invalid. See Table 45.

Prediction

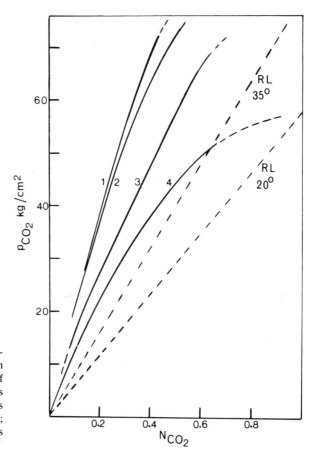

Fig. 152. Solubility of carbon dioxide at several high pressures p_{CO_2}, in kg/cm², and at 35°C. Registration of the mole fraction N_{CO_2} vs p_{CO_2} plots on the R-line diagram. The liquids are: (1) C_2H_5OH; (2) C_3H_8OH; (3) C_6H_6. For comparison, line 4 is for C_6H_6 at 20°C.

In Fig. 156 I have given the R-lines for $t°C = 20, 35, 60, 100°C$. The vapor pressure $p°_{CO_2}$ over liquid carbon dioxide at 20°C is 58.2 kg/cm², and therefore the maximum pressure to be observed is not more than this. At 20°C Sander's maximum observed pressure for benzene was $p_{CO_2} = 50$ kg/cm², and this corresponds to $x_{CO_2} = 1.400$, $N_{CO_2} = 0.582$; therefore the latter must effectively increase to unity as the p_{CO_2} closely approaches $p°_{CO_2} = 58.2$ kg/cm² (see Fig. 152). In Fig. 152 I have shown the N_{CO_2} plot for 35°C for benzene, for which the highest p_{CO_2} observed as 68 kg/cm², $p°_{CO_2}$ being approximately 80 kg/cm². The N_{CO_2} line for 35°C must be compared with the R-line for that temperature. A study of Fig. 152 should show that deviation from linearity must be judged with reference to the N_{CO_2}, the x_{CO_2}, the range of p_{CO_2}, and the position of the R-line. The highest p_{CO_2} recorded by Sander for the particular temperature, $t°C$, is shown as follows:

20°C: 50 kg/cm², except for water at 55 kg/cm²
35°C: 80 kg/cm²;
60°C: 110 kg/cm², except for water at 120 kg/cm²
100°C: 140 kg/cm², except for water at 170 kg/cm²

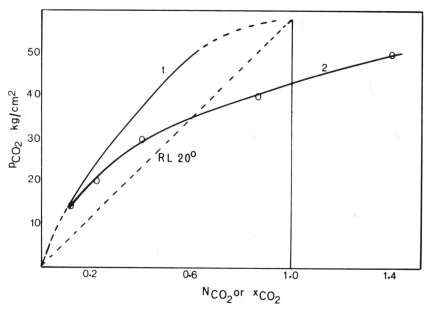

Fig. 153. Solubility of carbon dioxide in benzene at 20°C and at several pressures p_{CO_2}, in km/cm² from the vol./vol. data of Sander. Comparison of the mole fraction N_{CO_2} vs p_{CO_2} plot with the corresponding mole ratio x_{CO_2} plot.

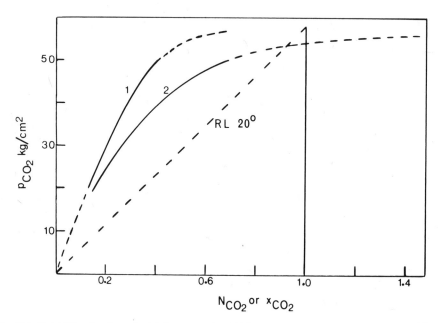

Fig. 154. Solubility of carbon dioxide in propanol at 20°C and different pressures p_{CO_2}, in kg/cm², from the vol./vol. data of Sander. Comparison of the mole fraction N_{CO_2} vs p_{CO_2} plot with the corresponding mole ratio x_{CO_2} plot.

Prediction

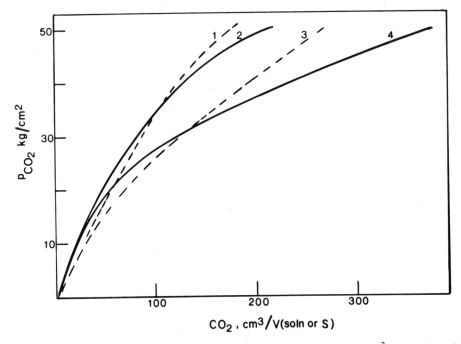

Fig. 155. Solubility of carbon dioxide at 20°C and different pressures p_{CO_2}, in kg/cm². Comparison of the plot of volume of CO_2/volume of solution vs p_{CO_2} with that of volume of CO_2/volume of the liquid S vs p_{CO_2}. For propanol, line 1 is the plot for the solution, and line 2 is that for the liquid S = propanol. For benzene, line 3 is the plot for the solution, and line 4 is that for the liquid S = benzene.

It is seen, therefore, that Sander's end pressures for each temperature tended to approach the $p^\circ_{CO_2}$ value for that temperature.

In Fig. 157 I have given the N_{CO_2} and x_{CO_2} plots for methanol at 0°C and pressures up to about 22 atm from the vol. CO_2/g S data of Krichevskii and Lebedeva (1947).[404]

16.7. Solubility of Methane, Hydrogen, and Nitrogen in Liquids at High Pressures

This work was due to Frolich, Tauch, Hogan, and Peer,[405] and their data provide another example of the use of the R-line procedure in sorting out vol./vol. data. Their data were presented only as small-size diagrams showing plots of "volume A/volume of liquid S" at 25°C vs p_A in atm. The systems were as follows:

water–CH_4, H_2, N_2
ethanol–CH_4, N_2
isopropanol–CH_4, H_2, N_2
propane–CH_4, H_2
butane–CH_4, H_2, N_2

pentane–CH_4, H_2
hexane–CH_4, H_2
octane–CH_4, H_2
cyclohexane–CH_4, H_2

benzene–CH_4
heavy naphtha–CH_4, H_2, N_2
gas oil–CH_4, N_2, H_2
carbon tetrachloride–N_2

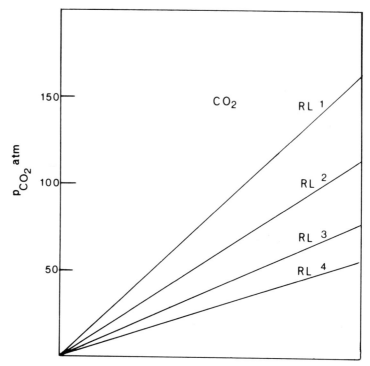

Fig. 156. R-lines for carbon dioxide at different temperatures $t°C$. Each line strikes the right vertical axis at $p°_{CO_2}$, the vapor pressure of CO_2 over its liquid at that temperature: (1) 100°C; (2) 60°C; (3) 35°C; (4) 20°C. To register actual N_{CO_2} values, the base line is to be divided from 0 to 1.

Table 45. Solubility of Carbon Dioxide in Liquids S at 20°C

Liquids S	Vol. CO_2/Vol. solution (Sander)	Vol. CO_2/Vol. S (calc. by Gerrard)	x_{CO_2}	N_{CO_2}
	$p_{CO_2} = 20$ kg/cm²			
C_6H_6	71.16	60.54	0.225	0.183
$C_6H_5CH_3$	57.91	65.10	0.288	0.223
C_6H_5Cl	62.61	54.81	0.232	0.188
C_6H_5Br	50.83	40.10	0.173	0.147
$C_6H_5NO_2$	41.60	33.30	0.141	0.124
	$p_{CO_2} = 30$ kg/cm²			
C_2H_5OH	104.8	101.7	0.247	0.198
n-C_3H_7OH	86.6	82.26	0.264	0.208
	$p_{CO_2} = 25$ kg/cm²			
Water	17.77	15.09	0.0112	0.0110

Prediction

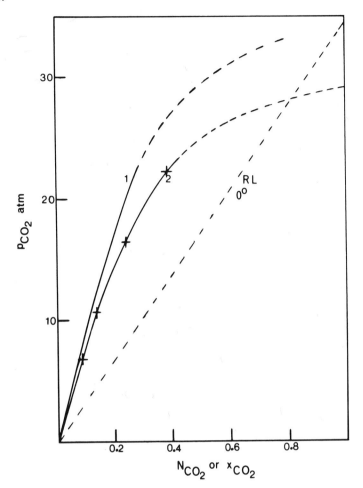

Fig. 157. Solubility of carbon dioxide in methanol at 0°C and several pressures p_{CO_2}, in atm. Plots of the mole fraction N_{CO_2} vs p_{CO_2} and the corresponding mole ratio x_{CO_2} vs p_{CO_2} on the R-line diagram. From the cm^3/g S data of Krichevskii and Lebedeva.[404]

Table 46 shows the invalidity of comparing the vol. A/vol. S values for different liquids S in terms of molecular proportions. See also Tables 47 and 48 for the solubilities of nitrogen and hydrogen at 25°C and 120 atm.

In Fig. 158 I have drawn the N_{MeH} vs p_{MeH} and x_{MeH} vs p_{MeH} plots for methane (MeH) in octane and isopropanol at 25°C. The x_{MeH} plot for octane is slightly convex upward over the observational range; but the N_{MeH} plot is decidedly concave upward in its expected position on the right of the R-line. Using the R-line as a guide, it is rational to extrapolate the N_{MeH} line, as I have shown by the broken line; an extrapolation of the corresponding x_{MeH} plot leads to increased convexity at higher pressures, as indicated in the inset (E), which I have also shown for the example of methane in n-pentane (Fig. 130). The N_{MeH} line for isopropanol is also in its expected position well on the left of the R-line.

Table 46. Solubility of Methane (MeH) in Liquids S at 25°C

Liquid S	cm³ of MeH/cm³ S	p_{MeH}, atm	x_{MeH}	N_{MeH}
Methanol	24.8	50	0.0411	0.0395
Ethanol	20.0	50	0.0478	0.0456
Isopropanol	24.9	50	0.0779	0.0723
Butane	163	89.5	0.657	0.396
Pentane	150	89.5	0.714	0.417
Hexane	140	89.5	0.758	0.431
Octane	125.7	89.5	0.839	0.456

Table 47. Solubility of Nitrogen at 25°C and 120 atm

	x_{N_2}	N_{N_2}
Isopropanol, 15 cm³ N_2/1 cm³ S	0.047	0.0448
Benzene, 15 cm³ N_2/1 cm³ S	0.0551	0.0523

Table 48. Solubility of Hydrogen at 25°C and 120 atm

	x_{H_2}	N_{H_2}
Isopropanol, 9.0 cm³ H_2/1 cm³ S	0.0283	0.0274
Benzene, 9.0 cm³ H_2/1 cm³ S	0.0331	0.0320

It was believed by Frolich *et al.* that Henry's law cannot be applied when there is a tendency for the solute to combine chemically with the solvent, an example being carbon dioxide dissolved in water. "On the other hand," they declared, "carbon dioxide behaves normally in many organic solvents." Referring to the data given by Sander for carbon dioxide and nonaqueous liquids, Frolich *et al.* stated that Henry's law was "obeyed" because the solubility curves became straight "when a pressure correction was applied."

16.8. Solubility of Gases in Nitrate Melts

The paper by Frame, Rhodes, and Ubbelohde (1961)[406] affords an excellent example of the use of the R-line procedure for a type of system which, at first sight, might not appear to be promising. These workers determined the solubilities of oxygen, nitrogen, and water in molten inorganic nitrates. They calculated the mole

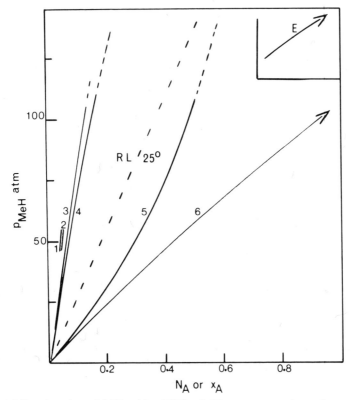

Fig. 158. Solubility of methane (MeH or A) at 25°C and different pressures p_{MeH}, in atm, from the vol. MeH/vol. S data of Frolich et al.[405] Registration of the mole fraction N_A vs p_{MeH} and the mole ratio x_A vs p_{MeH} plots on the R-line diagram. The lines are: (1) CH_3OH (N_A); (2) C_2H_5OH (N_A); (3) i-C_3H_7OH (N_A); (4) i-C_3H_7OH (x_A); (5) n-C_8H_{18} (N_A); (6) n-C_8H_{18} (x_A). At about 150 atm, line 6 begins to flatten, as shown by E in the inset.

ratio x_A from the depression of the freezing point of the particular nitrate. It was stated that oxygen and nitrogen did not affect the freezing points of the melts appreciably, from which it was concluded that the mole ratio x_A must be less than 0.0001. Now in the sodium nitrate system, taken herein as an example, the x_A values for oxygen and nitrogen at 306.8°C and $p_A = 1$ atm are less than 0.0001, and the x_{H_2O} value at 306.8°C and $p_{H_2O} = 22$ mm Hg is 0.00141.

The much greater x_A for water, as compared with those for oxygen and nitrogen, was attributed to the effect of the dipole for water. In Fig. 159 I show the R-line for water at 306°C. Assuming a linear plot for x_{H_2O} vs p_{H_2O}, I estimate the x_{H_2O} value for 760 mm Hg to be 0.053, and I have shown the corresponding N_{H_2O} vs p_{H_2O} plot as being slightly concave upward, well on the right of the R-line. This position does indeed indicate incisive-enough acid–base function involving the water and the nitrate, but this conclusion is quite independent of the N_A values for oxygen and nitrogen. In Fig. 160 I show the R-line for nitrogen at 306°C and register the N_{N_2} value of 0.0001, showing this to be well on the left of the R-line. The actual value is

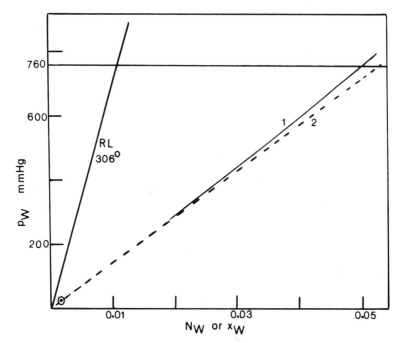

Fig. 159. Solubility of water in molten sodium nitrate (306.8°C) at pressure $p_W = 20$ mm Hg. The position of this value of the mole ratio x_W is registered on the R-line diagram for water. Line 2 is the speculative plot of the x_W vs p_W plot and line 1 is the corresponding N_W vs p_W plot. It is seen that this is well on the right of the R-line.

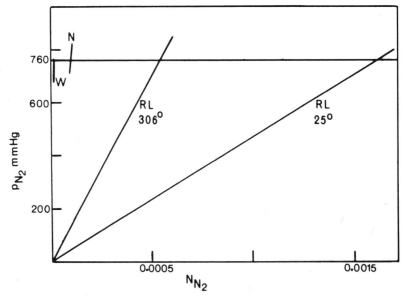

Fig. 160. Solubility of nitrogen in molten sodium nitrate at 306.8°C and 1 atm pressure, p_{N_2}. The upper limit of the mole fraction N_{N_2} value (0.0001) is registered (line N) with reference to the R-line for nitrogen at 306°C. Line W shows the position of the N_{N_2} value for water at 25°C with reference to the R-line for nitrogen at 25°C.

Prediction

finite but beyond the technique used and will be still further on the left of the R-line. This is probably due to the structure of the nitrate melt, and the effect may be related to the absolutely small value of the N_{N_2} for water at 25°C, as indicated in Fig. 160.

At 25°C and $p_{RH} = 1$ atm, the x_{MeH} for methane in n-decane is about 0.005, whereas the corresponding $x_{n\text{-}BuH}$ value for n-butane in n-decane is 1.04. It is not within reason to attribute this great difference in x_{RH} to causes other than the difference in condensing tendency of methane (bp 112 K/1 atm) and n-butane (bp 272.5 K/1 atm). The solubility of each gas in a given liquid must be considered against the particular R-line for the gas.

16.9. Solubility of Noble Gases in Molten Fluorides

Grimes, Smith, and Watson (1958)[407] determined the solubility of He, Ne, Ar, and Xe in molten fluorides, specifically the $NaF + ZrF_4$ system, in connection with the study of nuclear-reactor fuel containing mixtures of metal fluorides and UF_4. They expressed their data as "moles of gas $\times 10^8/cm^3$ melt." They gave the "Henry's law constant" for He at 600°C as 21.6×10^{-8} (i.e., 0.000000216) moles He/cm^3 melt·atm. King[204] gives the "constant" as 4.6×10^6 (i.e., 4,600,000) atm cm^3 melt/mole He. I see no virtue in such expressions; why not simply state the solubility as moles He/cm^3 of melt (specified) at 760 mm Hg (or 1 atm) and 600°C, and that the solubility/pressure plot is deemed to be linear over the range of 0.5 to 2 atm?

In Table 49, I illustrate the essential pattern of data for helium at 1 atm.

Therefore, the melt becomes more absorbing as the temperature is raised, despite the lowering of the concentration of He in the gas phase. To get a rough idea of the position of the N_{He} values on the R-line diagrams for the three high temperatures, I multiply the cm^3 values in Table 49 by a "molecular weight/density" factor of 50. The N_{He} (x_{He}) values then become 600°C, 0.000011; 700°C, 0.000015; 800°C, 0.000022. As I show in Fig. 16, the R-line for He at 25°C strikes the horizontal at 1 atm at $N_{He} = 0.00025$. The R-line for 600°C would be well on the left

Table 49. Solubility of He in Molten $NaF + ZrF_4$, $p_{He} = 1$ atm

t°C	Moles He per cm^3 melt	Concentration of He as mol/cm^3 in the gas phase[a]	Ratio of concentrations in melt and gas phase
600	$22.4 \cdot 10^{-8} = 0.000000224$	0.0000140	60
700	$30.9 \cdot 10^{-8} = 0.000000309$	0.0000125	40
800	$43.9 \cdot 10^{-8} = 0.000000439$	0.0000114	25

[a] Calculated by Gerrard.

of this, and a tenfold increase in scale would be required to depict the line clearly. For 700°C the R-line would be on the left of that for 600°C, and for 800°C it would be still more on the left. It is seen, therefore, that the N_{He} values move differentially to the right as the R-line moves to the left as the temperature is increased. The main factor appears to be the change in intermolecular structure of the melt as the temperature is raised, making it more accommodating.

17

Textbook Statements

17.1. Introduction

Most, if not all, chemists, chemical engineers, biochemical technologists, and others using chemical concepts have been brought up in the conventional belief in the inherent usefulness of the ideal model. The concepts of the ideal solution, Raoult's law, and Henry's law have dominated the treatment of the solubility of gases and liquids in liquids. In my opinion, this has greatly hindered the approach to an understanding of the mechanisms of the dissolution processes in real systems, for such treatment has fostered the search for systems which conform, or nearly conform, to these idealistic notions. This has tended to exclude many real systems even from the barest consideration. An outstanding defect in the textbook treatment is the lack of data on real systems. There are instances of examples being given to show that Henry's law in the mole fraction form is "obeyed," whereas the real data were of the mole ratio form. This area of treatment is in urgent need of revision to bring it in line with reality.

17.2. Illustrative Examples of Textbook Statements

Terms such as *concentration, concentrated, dilute,* and *saturated* are used much too vaguely, and in an unnecessarily misleading way. *Concentration* is understood by certain writers as *molarity,* moles of A per liter of solution, mole fraction not being deemed a concentration. It appears to be understood that the suitable expression of composition is the mole fraction; although it is conceded that in certain examples "Raoult's law is inapplicable, as in the solubility of gases," it may be necessary to consider the *concentration.* I have to confess that I do not understand what is meant by this contention. Furthermore, there appears to be no explanation as to why the mole fraction is more suitable than the mole ratio, which after all is the form of the usual chemical equations.

The concentration C of a component A in a solution is taken to be the amount of A in a unit volume of solution; but in dilute solution, the concentration becomes proportional to the mole fraction. There is the overemphasis on water as the liquid S, and this throws the whole area out of focus, even for water systems. In dilute aqueous solutions, it is contended, concentration becomes approximately equal to molality. All this is far better presented in real terms rather than in mere words: A molality of 1 means 1 mole A in 1000 g S, i.e., in (1000/density of S) cm^3. If the density is 0.800, this is equal to 1250 cm^3. Therefore, there are 0.8 mole A in 1000 cm^3 of S. If one assumes the volume of *solution* to be 1000 cm^3, then the molarity would be 0.8. As the molality is lowered, to 0.1, 0.001, etc., then the more nearly will the volume of solution be that of the original liquid S, and the more nearly will the molarity be 0.8 of the molality; but never will it equal the molality. If the density of S is 1.2, there would be 1 mole in 833 cm^3, i.e., 1.2 moles/1000 cm^3 of S. Again if one assumes this to be 1000 cm^3 of solution, then the molarity would be 1.2. As the molality is made smaller and smaller, the molarity will approach more and more closely 1.2 of the molality. It is only when the density of S is 1 (as is approximately so with water) that the molality and molarity approach coincidence as the molality becomes very small.

Still more mystifying is the belief that whereas molarity changes with temperature, molality m_A and mole fraction N_A are independent of temperature. Certainly for gases and liquids in liquids under conditions of equilibrium both the temperature and pressure are material factors.

For the system A + S, Raoult's law is given as $(p°_A - p_A)/p°_A = N_S$. The alternative form $p_A/p°_A = N_A$ was stated to be essentially Henry's law. Raoult's law and Henry's law are stated to be *merely* alternative ways of expressing the same idea that the vapor pressure of a constituent A of a solution is proportional to the relative number of molecules in solution. The two alternative forms, however, relate to only *one* law, since $p_A/p°_A = 1 - N_S = N_A$. What is meant by the relative number of molecules? Relative to what? Moles of A per mole of S, or moles of A compared with the total number of moles A + S, i.e., mole ratio or mole fraction? In principle, if $p_A \propto N_A$, p_A cannot be proportional to x_A, although for N_A values less than about 0.015 the N_A and x_A values tend toward numerical coincidence. However, Raoult's law is supposed to have universal applicability when N_A is large.

I gather from certain texts that $p_A = KN_A$ is not necessarily Raoult's law but a more general form called Henry's law. But it appears that "we cannot expect Henry's law to be valid unless the concentration of A is small." We can, of course, expect this if we wish to; but why we should not, was not disclosed. The idea is prevalent that the equation $p_A = KN_A$, was *established* and *extensively tested* by Henry.

Henry's law is stated to apply to any *solute*, and Raoult's law to the *solvent*, "and there is a close relation between these two." Whereas Raoult's law is deemed to "conceive" of the ideal solvent ($p_A \propto N_A$), Henry's law relates to the ideal solute ($p_S \propto N_S$). If the solution is ideal, then K in $p_A = KN_A$ is equal to $p°_A$, "and both Henry's law and Raoult's law would convey the same information."

Textbook Statements

It is contended in the books that no matter how much the system A+S deviates from ideality at the *higher* concentrations (as mole fraction N_A), all is well when the N_A value is low, for then A is said to be in a uniform environment and $p_A = KN_A$, but K is not $p°_A$.

References

1. Ahmed, W., Gerrard, W., and Maladkar, V. K., *J. Appl. Chem.*, 1970, **20**, 109, and papers cited therein.
2. Amirkhanov, A. Kh., *Tr. Ufim. Aviats. Inst.*, 1965, No. 6, 135; *Chem. Abstr.*, **67**, 94449 m.
3. Andrews, L. J., *Chem. Rev.*, 1954, **54**, 713.
4. Antropoff, A., *Proc. Roy. Soc.*, 1910, **A83**, 474; *Z. Electrochem.*, 1919, **25**, 269.
5. Archer, G., and Hildebrand, J. H., *J. Phys. Chem.*, 1963, **67**, 1830.
6. Arnol'dov, M. N., Ivanovskii, M. N., Morozov, V. A., Pletenets, S. S., and Subbotin, V. I., *Teplofiz. Svoistva Zhidk., Mater, Vses. Teplofiz. Konf. Svoistvam Veshchestv. Vys. Temp.*, 3rd edn., **1968**, 110.
7. Aksnes, G., and Gramstad, T., *Acta Chem. Scand.*, 1960, **14**, 1485.
8. Atkinson, R. H., Heycock, C. T., and Pope, W. J., *J. Chem. Soc.*, 1920, **117**, 1410.
9. Auwers, K., and Meyer, V., *Chem. Ber.*, 1888, **21**, 1068
10. Azarnoosh, A., and McKetta, J. J., *Pet. Refiner*, 1958, **37**, No. 11, 275; Azarnoosh, A., and McKetta, J. J., *J. Chem. Eng. Data*, 1959, **4**, 211.
11. Bancroft, W. D., and Belden, B. C., *J. Phys. Chem.*, 1930, **34**, 2123.
12. Barth, K., *Z. Phys. Chem.*, 1892, **9**, 176.
13. Bartlett, P. D., and Dauben, H. J., *J. Am. Chem. Soc.*, 1940, **62**, 1339.
14. Basila, M. R., *J. Chem. Phys.*, 1961, **35**, 1151; Basila, M. R., Saier, E. L., and Cousins, L. R., *J. Am. Chem. Soc.*, 1965, **87**, 1665.
15. Baskerville, C., and Cohen, P. W., *Ind. Eng. Chem.*, 1921, **13**, 333.
16. Bates, S. J., and Kirschman, H. D., *J. Am. Chem. Soc.*, 1919, **41**, 1991.
17. Battino, R., and Clever, H. L., *Chem. Rev.*, 1966, **66**, 395.
18. Baud, E., *Bull. Soc. Chim.*, 1915, **17**, 329.
19. Bein, W., *Z. Phys. Chem.*, 1909, **66**, 257.
20. Bell, R. P., *J. Chem. Soc.*, **1931**, 1371.
21. Ben-Naim, A., *J. Phys. Chem.*, 1967, **71**, 1137.
22. Berkeley, P. J., and Hanna, M. W., *J. Phys. Chem.*, 1963, **67**, 846; *J. Am. Chem. Soc.*, 1964, **86**, 2990; *J. Chem. Phys.*, 1964, **41**, 2530.
23. Berlak, M. C., and Gerrard, W., *J. Chem. Soc.*, **1949**, 2309.
24. Beuschlein, W. L., and Simenson, L. O., *J. Chem. Soc.*, 1940, **62**, 610.
25. Blitz, H., *Chem. Ber.*, 1892, **25**, 2542.
26. Blümcke, A., *Wied. Ann.*, 1888, **34**, 10.
27. Boedeker, E. R., and Lynch, C. C., *J. Am. Chem. Soc.*, 1950, **72**, 3234.
28. Boggs, J. E., and Buck, A. E., *J. Phys. Chem.*, 1958, **62**, 1459.
29. Boomer, E. H., Johnson, C. A., and Piercey, A. G. A., *Can. J. Res.*, 1938, **16B**, 319.
30. Boyer, F. L., and Bircher, L. J., *J. Phys. Chem.*, 1960, **64**, 1330.
31. Boyle, R. W., *Phil. Mag.*, 1911, **22**, 840.
32. Bradbury, E. J., McNulty, D., Savage, R. L., and McSweeney, E. E., *Ind. Eng. Chem.*, 1952, **44**, 211.
33. Bradford, M. L., and Thodos, G., *Can. J. Chem. Eng.*, **1966**, 345.

34. Brooks, W. B., and McKetta, J. J., *Pet. Refiner*, 1955, **34**, No. 2, 143; No. 4, 138; Brooks, W. B., Haughn, J. E., and McKetta, J. J., *Pet. Refiner*, 1955, **34**, No. 8, 129.
35. Brown, H. C., and Brady, J. D., *J. Am. Chem. Soc.*, 1949, **71**, 3573; 1952, **74**, 3570.
36. Brown, H. C., and Wallace, W. J., *J. Am. Chem. Soc.*, 1953, **75**, 6268.
37. Brown, I., and Smith, F., *Aust. J. Chem.*, 1960, **13**, 30.
38. Bullock, E., and Tuck, D. G., *Trans. Faraday Soc.*, 1963, **59**, 1293.
39. Burrows, G., and Preece, F. H., *J. Appl. Chem.*, 1953, **3**, 451.
40. Butler, J. A. V., *Trans. Faraday Soc.*, 1937, **33**, 273.
41. Cade, G. N., Dunn, R. E., and Hepp, H. J., *J. Am. Chem. Soc.*, 1946, **68**, 2454.
42. Cady, H. P., Elsey, H. M., and Berger, E. V., *J. Am. Chem. Soc.*, 1922, **44**, 1456.
43. Cahours, A. A. T., Berthelot, M. P. E., and Debray, J. H., *Compt. Rend.*, 1884, **97**, 825.
44. Campbell, F. H., *Trans. Faraday Soc.*, 1915, **11**, 91.
45. Campbell, W. B., and Maass, O., *Can. J. Res.*, 1930, **2**, 42; *Pulp Pap. Mag. Can.*, **1930**, 699.
46. Carey, W. W., Klausutis, N. A., and Barduhn, A. J., *Desalination*, 1966, **1**, 342.
47. Chang, E. T., Gokcen, N. A., and Poston, T. M., *Spacecraft*, 1969, **6**, 1177.
48. Charalambous, J., Frazer, M. J., and Gerrard, W., *J. Chem. Soc.*, **1964**, 1520.
49. Cheung, H., and Zander, E. H., *Chem. Eng. Progr., Symp. Ser.*, 1968, **64**, 34.
50. Chien, J. C. W., *J. Am. Chem. Soc.*, 1959, **81**, 86.
51. Christopulos, J. A., *U.S. C.F.S.T.I. AD Rep.* 1969, No. 700958.
52. Claussen, W. F., *J. Chem. Phys.*, 1951, **19**, 259, 662, 1425.
53. Claussen, W. F., and Polglase, M. F., *J. Am. Chem. Soc.*, 1952, **74**, 4817.
54. Clever, H. L., Battino, R., Saylor, J. H., and Gross, P. M., *J. Phys. Chem.*, 1957, **61**, 1078.
55. Clever, H. L., *J. Phys. Chem.*, 1958, **62**, 375.
56. Cook, T. M., and Gerrard, W., unpublished work.
57. Cooke, V. F. G., and Gerrard, W., *J. Chem. Soc.*, **1955**, 1978.
58. Copley, M. J., Ginsberg, E., Zellhoefer, G. F., and Marvel, C. S., *J. Am. Chem. Soc.*, 1941, **63**, 254.
59. Copley, M. J., Zellhoefer, G. F., and Marvel, C. S., *J. Am. Chem. Soc.*, 1938, **60**, 2666; 1939, **61**, 3550.
60. Coulson, C. A., *Disc. Faraday Soc.*, 1965, **No. 40**, 285.
61. Davies, M., *Chem. Ind.*, **1970**, 826.
62. Davis, M. M., and Hetzer, H. B., *J. Res. Nat. Bur. Stand.*, 1951, **46**, 496; 1958, **60**, 569.
63. Deaton, W. M., and Frost, E. M., Jr., *Gas Hydrates and Their Relation to the Operation of Natural Gas Pipelines*, U.S. Bur. Mines Monograph, 1949, 8.
64. De Bruyn, C. A. L., *Rec. Trav. Chim. Pays-Bas Belg.*, 1892, **11**, 29, 112; *Z. Phys. Chem.*, 1892, **10**, 782.
65. Dim, A., Gardner, G. R., Ponter, A. B., and Wood, T., *J. Chem. Eng. Jpn.*, 1971, **4**, 92.
66. Dmitrieva, I. B., Konovalov, E. E., and Klents, A. I., *Teplofiz. Svoistva Zhidk., Mater. Vses. Teplofiz. Konf. Svoistvam Veshchestv Vys. Temp.* 3rd edn., **1968**, 117.
67. Dobson, H. J. E., and Masson, I., *J. Chem. Soc.*, 1924, **125**, 668.
68. Dolezalek, F., *Z., Phys. Chem.*, 1908, **64**, 727; 1910, **71**, 191; 1913, **83**, 40.
69. Dolezalek, F., and Schulze, A., *Z. Phys. Chem.*, 1913, **83**, 45.
70. Dorofeeva, N. G., and Kudra, O. K., *Ukr. Khim. Zh.*, 1958, **24**, 592, 706.
71. Dunn., J. S., and Rideal, E. K., *J. Chem. Soc.*, 1924, **125**, 676.
72. duPont de Nemours and Co., *Chem. Eng. News*, 1955, **33**, 2366.
73. Eigen, M., and De Maeyer, L., *Proc. Roy. Soc.*, 1958, **A247**, 505.
74. Eisenberg, D., and Kauzmann, W., *The Structure and Properties of Water*, 1969, Clarendon Press, Oxford.
75. Eley, D. D., *Trans. Faraday Soc.*, 1939, **35**, 1281, 1421.
76. Elliott, J. R., Roth, W. L., Roedel, G. F., and Boldebuck, E. M., *J. Am. Chem. Soc.*, 1952, **74**, 5211.
77. Engler, C., *Annalen*, 1867, **142**, 291.
78. Fauser, G., *Math. Naturwiss. Ber. Ung.*, 1888, **6**, 154.
79. Feakins, D., and Watson, P., *J. Chem. Soc.*, **1963**, 4734, 4686.
80. Fernandes, J. B., and Sharma, M. M., *Indian Chem. Eng. (Trans.)*, 1965, **7**, 38.
81. Finch, R. N., and Van Winkle, M., *Am. Inst. Chem. Eng. J.*, 1962, **8**, No. 4, 455.
82. Fontana, C. M., and Herold, R. J., *J. Am. Chem. Soc.*, 1948, **70**, 2881.

83. Foote, H. W., and Fleischer, J., *J. Am. Chem. Soc.*, 1934, **56**, 870.
84. Fox, J. H. P., and Lambert, J. D., *Proc. Roy. Soc.*, 1951, **A210**, 557.
85. Frank, H. S., *Disc. Faraday Soc.*, 1967, No. 43, 137, 147.
86. Frank, H. S., and Evans, M. W., *J. Chem. Phys.*, 1945, **13**, 507.
87. Frank, H. S., and Wen, W. Y., *Disc. Faraday. Soc.*, 1957, **24**, 133.
88. Franks, F., *Hydrogen-Bonded Solvent Systems, Proceedings of the Symposium*, **1968**, 31.
89. Frazer, M. J., and Gerrard, W., *Nature*, 1964, **204**, 1299.
90. Friedman, H. L., *J. Am. Chem. Soc.*, 1954, **76**, 3294.
91. Gel'perin, N. I., Matveev, I. G., and Bil'shau, K. V., *Zh. Prikl. Khim. Leningr.*, 1958, **31**, 1323.
92. Gautier, A., *Ann. Chim. Phys.*, 1869, **17**, 103.
93. Gerrard, W., *J. Chem. Soc.*, **1939**, 99, and subsequent papers.
94. Gerrard, W., *Organic Chemistry of Boron*, 1961, Academic Press, London.
95. Gerrard, W., *J. Chim. Phys.*, **1964**, 73.
96. Gerrard, W., *Chem. Ind.*, **1966**, 815.
97. Gerrard, W., *Chem. Ind.*, **1969**, 295, 460, 1460.
98. Gerrard, W., *Chem. Ind.*, **1971**, 884.
99. Gerrard, W., *J. Appl. Chem. Biotechnol.*, 1972, **22**, 623.
100. Gerrard, W., *J. Appl. Chem. Biotechnol.*, 1973, **23**, 1.
101. Gerrard, W., and Howe, B. K., *J. Chem. Soc.*, **1955**, 505.
102a. Gerrard, W., and Hudson, H. R., *Chem. Rev.*, 1965, **65**, 697.
102. Gerrard, W., and Hudson, H. R., Organic Derivatives of Phosphorous Acid and Thiophosphorous Acid, in: *Organic Phosphorus Compounds*, Vol. 5, Kosolapoff, G. M., and Maier, L., eds., 1973, 23–329, Wiley, New York.
103. Gerrard, W., and Jones, J. V., *J. Chem. Soc.*, **1952**, 690.
104. Gerrard, W., and Kilburn, K. D., *J. Chem. Soc.*, **1956**, 1536.
105. Gerrard, W., Kulkarni, S. V., and Wyvill, P. L., unpublished work.
106. Gerrard, W., and Lines, C. B.; unpublished work.
107. Gerrard, W., Machell, G., and Tolcher, P., *Research (London)*, 1955, **8**, S7.
108. Gerrard, W., and Macklen, E. D., *J. Appl. Chem.*, 1956, **6**, 241.
109. Gerrard, W., and Macklen, E. D., *Chem. Rev.*, 1959, **59**, 1105.
110. Gerrard, W., and Macklen, E. D., *J. Appl. Chem.*, 1959, **9**, 85.
111. Gerrard, W., and Macklen, E. D., *J. Appl. Chem.*, 1960, **10**, 57.
112. Gerrard, W., and Macklen, E. D., *Chem. Ind.*, **1959**, 1070 (a), 1521 (b), 1549 (c).
113. Gerrard, W., Madden, R. W., and Tolcher, P. H., *J. Appl. Chem.*, 1955, **5**, 28.
114. Gerrard, W., and Maladkar, V. K., *Chem. Ind.*, 1969, 1554.
115. Gerrard, W., and Maladkar, V. K., *Chem. Ind.*, **1970**, 925.
116. Gerrard, W., Mincer, A. M. A., and Wyvill, P. L., *Chem. Ind.*, **1958**, 894.
117. Gerrard, W., Mincer, A. M. A., and Wyvill, P. L., *J. Appl. Chem.*, 1959, **9**, 89.
118. Gerrard, W., Mincer, A. M. A., and Wyvill, P. L., *J. Appl. Chem.*, 1960, **10**, 115.
119. Gerrard, W., and Shepherd, B. D., *J. Chem. Soc.*, **1953**, 2069.
120. Gerrard, W., and Whitbread, E. G. G., *J. Chem. Soc.*, **1952**, 914.
121. Gerrard, W., and Woodhead, A. H., *J. Chem. Soc.*, **1951**, 519.
122. Gerrard, W., and Wyvill, P. L., *Research (London)*, 1949, **2**, 536.
123. Gjaldbaek, J. C., *K. Dan. Vidensk. Selsk. Mat.-Fys. Medd.*, 1948, **24**, No. 13.
124. Gjaldbaek, J. C., *Acta Chem. Scand.*, 1952, **6**, 623; 1953, **7**, 537.
125. Gjaldbaek, J. C., and Andersen, E. K., *Acta Chem. Scand.*, 1954, **8**, 1398.
126. Gjaldbaek, J. C., and Hildebrand, J. H., *J. Am. Chem. Soc.*, 1949, **71**, 3147.
127. Gjaldbaek, J. C., and Hildebrand, J. H., *J. Am. Chem. Soc.*, 1950, **72**, 609.
128. Gjaldbaek, J. C., and Niemann, H., *Acta Chem. Scand.*, 1958, **12**, 611, 1015.
129. Glasstone, S., *Trans. Faraday Soc.*, 1937, **33**, 158.
130. Glew, D. N., and Moelwyn-Hughes, E. A., *Disc. Faraday Soc.*, 1953, **15**, 150.
131. Glockler, G., *Trans. Faraday Soc.*, 1937, **33**, 224.
132. Goldschmidt, H., *Z. Phys. Chem.*, 1915, **89**, 129, 131; 1925, **114**, 1; 1921, **99**, 116; Goldschmidt, H., and Ass, F., *Z. Phys. Chem.*, 1924, **112**, 423.
133. Gordy, W., *J. Am. Chem. Soc.*, 1938, **60**, 605.

134. Gordy, W., *J. Chem. Phys.*, 1939, **7**, 93.
135. Gordy, W., *J. Chem. Phys.*, 1941, **9**, 215.
136. Gordy, W., and Martin, P. C., *J. Chem. Phys.*, 1939, **7**, 99.
137. Gordy, W., and Stanford, S. C., *J. Chem. Phys.*, 1940, **8**, 170.
138. Gordy, W., and Stanford, S. C., *J. Am. Chem. Soc.*, 1940, **62**, 497.
139. Gordy, W., and Stanford, S. C., *J. Chem. Phys.*, 1941, **9**, 204.
140. Gordy, W., and Martin, P. C., *Phys. Revs.*, 1937, **52**, 1075.
141. Gramstad, T., *Acta Chem. Scand.*, 1961, **15**, 1337.
142. Guggenheim, E. A., *Trans. Faraday Soc.*, 1937, **33**, 151.
143. Guss, L. S., and Kolthoff, I. M., *J. Am. Chem. Soc.*, 1940, **62**, 1494.
144. Guthrie, F., *Philos. Mag.*, 1885, **18**, 495.
145. Haight, G. P., *Ind. Eng. Chem.*, 1951, **43**, 1827.
146. Halban, H., *Z. Phys. Chem.*, 1913, **84**, 129.
147. Hamai, S., *Bull. Chem. Soc. Jpn.*, 1935, **10**, 5, 207.
148. Hannaert, H., Haccuria, M., and Mathieu, M. P., *Ind. Chim. Belg.*, 1967, **67**, 156.
149. Hantzsch, A., *Chem. Ber.*, 1931, **64B**, 661.
150. Harned, H. S., and Davis, R., Jr., *J. Am. Chem. Soc.*, 1943, **65**, 2030.
151. Hayduk, W., and Cheng, S. C., *Can. J. Chem. Eng.*, 1970, **48**, 93.
152. Hayduk, W., Walter, E. B., and Simpson, P., *J. Chem. Eng. Data*, 1972, **17**, 59.
153. Heafield, T. G., Hopkins, G., and Hunter, L., *Nature*, 1942, **149**, 218.
154. Hempel, W., *Z. Angew. Chem.*, 1901, **14**, 865.
155. Henrici-Olivé, G., and Olivé, S., *Angew. Chem.*, 1967, **79**, 764.
156. Henry, W., *Philos. Trans.*, 1803, **93**, 29, 274.
157. Herington, E. F. G., *Disc. Faraday Soc.*, 1953, **No. 15**, 256.
158. Hildebrand, J. H., *J. Am. Chem. Soc.*, 1916, **38**, 1452.
159. Hildebrand, J. H., *Phys. Rev.*, 1923, **21**, 46.
160. Hildebrand, J. H., *Trans. Faraday Soc.*, 1937, **33**, 144.
161. Hildebrand, J. H., *Disc. Faraday Soc.*, 1953, **No. 15**, 9.
162. Hildebrand, J. H., *J. Chim. Phys.*, **1964**, 53.
163. Hildebrand, J. H., *Proc. Nat. Acad. Sci. U.S.*, 1967, **57**, 542.
164. Hildebrand, J. H., and Scott, R. L., *Solubility of Nonelectrolytes*, 3rd edn., 1950, Reinhold.
165. Hildebrand, J. H., and Scott, R. L., *Regular Solutions*, 1962, Prentice-Hall.
166. Hill, A. E., *J. Am. Chem. Soc.*, 1931, **53**, 2598.
167. Hill, A. E., and Fitzgerald, T. B., *J. Am. Chem. Soc.*, 1935, **57**, 250.
168. Himmelblau, D. M., *J. Phys. Chem.*, 1959, **63**, 1803.
169. Hinkel, L. E., and Treharne, G. J., *J. Chem. Soc.*, **1945**, 866.
170. Hiraoka, H., *Rev. Phys. Chem. Jpn.*, 1954, **24**, No. 1, 13.
171. Hiraoka, H., and Hildebrand, J. H., *J. Phys. Chem.*, 1964, **68**, 213.
172. Hite, G., Smissman, E. E., and West, R., *J. Am. Chem. Soc.*, 1960, **82**, 1207.
173. Hoffman, D. S., Welker, J. R., Felt, R. E., and Weber, J. H., *Am. Inst. Chem. Eng. J.*, **1962**, 508.
174. Högfeldt, E., *Arkiv. Kemi.*, 1954, **7**, 315; *Rec. Trav. Chim.*, 1956, **75**, 790.
175. Hogfeldt, E., and Bolander, B., *Arkiv. Kemi.*, 1963, **21**, 161.
176. Holtzer, A., and Emerson, M. F., *J. Phys. Chem.*, 1969, **73**, 26.
177. Horiuti, J., *Sci. Pap. Inst. Phys. Chem. Res. Tokyo*, 1931, **17**, 125.
178. Howland, J. J., Miller, D. R., and Willard, J. E., *J. Am. Chem. Soc.*, 1941, **63**, 2807.
179. Hudson, J. C., *J. Chem. Soc.*, **1925**, 1332.
180. Huggins, M. L., *Ann. N.Y. Acad. Sci.*, 1942, **43**, 1.
181. Inglis, K. H., *Proc. Phys. Soc.*, 1906, **20**, 152.
182. International Critical Tables, Vol. 3, 254, McGraw-Hill, New York.
183. Ionin, M. V., Kurina, N. V., and Sudoplatova, A. E., *Tr. Khim. Khim. Tekhnol.*, 1963, **1**, 47.
184. Ionin, M. V., and Shverina, V. G., *Zh. Obshch. Khim.*, 1965, **35**, 209.
185. Ives, D. J. G., *Some Reflections on Water*, 1963, University of London (available from Dillon, Malet St., London, W.C.1).
186. Jain, S. C., and Rivest, R., *Can. J. Chem.*, 1964, **42**, 1079.
187. Jakobsen, R. J., and Brasch, J. W., *Spectrochim. Acta*, 1965, **21**, 1753.

References

188. Janz, G. J., and Danyluk, S. S., *J. Am. Chem. Soc.*, 1959, **81**, 3846, 3850, 3854.
189. Janz, G. J., and Danyluk, S. S., *Chem. Rev.*, 1960, **60**, 209.
190. Johnson, J. R., Christian, S. D., and Affsprung, H. E., *J. Chem. Soc.* (A) **1967**, 1924; see also **1963**, 1896; **1966**, 77.
191. Johnstone, H. F., and Leppla, P. W., *J. Am. Chem. Soc.*, 1934, **56**, 2233.
192. Jolley, J. E., and Hildebrand, J. H., *J. Am. Chem. Soc.*, 1958, **80**, 1050.
193. Jones, W. J., Lapworth, A., and Lingford, H. M., *J. Chem. Soc.*, 1913, **103**, 252.
194. Josien, M-L., and Saumagne, P., *Bull. Soc. Chim. Fr.*, **1956**, 937; Josien, M-L., Castinel, C., and Saumagne, P., *Bull. Soc. Chim. Fr.*, **1957**, 648; Josien, M-L., Dizabo, P., and Saumagne, P., *Bull. Soc. Chim. Fr.*, **1957**, 423.
195. Josien, M-L., and Sourisseau, G., *Bull. Soc. Chim. Fr.*, **1955**, 178; Josien, M-L., Sourisseau, G., and Castinel, C., *Bull. Soc. Chim. Fr.*, **1955**, 1539.
196. Just, G., *Z. Phys. Chem.*, 1901, **37**, 342.
197. Kaplan, S. I., Monakhova, Z. D., Reformatskaya, A. S., and Bessonova, E. I., *Zh. Prikl. Khim.*, 1937, 10, 2022.
198. Kaplan, S. I., and Romanchuk, M. A., *Zh. Obshch. Khim.*, 1936, **6**, 950.
199. Kapoor, K. P., Luckcock, R. G., and Sandbach, J. A., *J. Appl. Chem. Biotechnol.*, 1971, **21**, 97.
200. Katayama, T., Mori, T., and Nitta, T., *Kagaku Kogaku*, 1967, **31**, 559.
201. Kauzmann, W., *Adv. Protein Chem.*, 1959, **14**, 1.
202. Kendall, J., *Trans. Faraday Soc.*, 1937, **33**, 2.
203. Kilpatrick, M., and Luborsky, F. E., *J. Am. Chem. Soc.*, 1953, **75**, 577.
204. King, M. B., *Phase Equilibrium in Mixtures*, 1969, Pergamon, London.
205. Kireev, V. A., Kaplan, S. I., and Romanchuk, M. A., *Zh. Obshch. Khim.*, 1935, **5**, 444.
206. Kireev, V. A., Kaplan, S. I., and Vasneva, K. I., *Zh. Obshch. Khim.*, 1936, **6**, 799.
207. Kireev, V. A., and Romanchuk, M. A., *Zh. Obshch. Khim.*, 1936, **6**, 81.
208. Knight, R. W., and Hinshelwood, C. N., *J. Chem. Soc.*, **1927**, 466.
209. Kobatake, Y., and Hildebrand, J. H., *J. Phys. Chem.*, 1961, **65**, 331.
210. Kohn, G., *Chem. Ber.*, 1932, **65**, 589.
211. Kolthoff, I. M., and Coetzee, J. F., *J. Am. Chem. Soc.*, 1957, **79**, 870, 1852, 6110.
212. Korinek, G., and Schneider, W. G., *Can. J. Chem.*, 1957, **35**, 1157.
213. Körösy, F., *Trans. Faraday Soc.*, 1937, **33**, 416.
214. Kostina, T. K., Baum, B. A., Kurochkin, K. T., and Gel'd, P. V., *Zh. Fiz. Khim.*, 1971, **45**, 813.
215. Kreglewski, A., *Bull. Acad. Polon. Sci. Ser. Sci. Chim.*, 1965, **13**, 723, 729.
216. Krestov, G. A., and Nedel'ko, B. E., *Izv. Vyssh. Ucheb. Zaved., Khim. Khim. Tekhnol.*, 1970, **13**, 158.
217. Kretschmer, C. B., and Wiebe, R., *J. Am. Chem. Soc.*, 1951, **73**, 3778; 1952, **74**, 1276.
218. Krieve, W. F., and Mason D. M., *J. Phys. Chem.*, 1956, **60**, 374.
219. Krivonos, F. F., *J. Appl. Chem., U.S.S.R.*, 1958, **31**, 487.
220. Kunerth, W., *Phys. Rev.*, 1922, **19**, 512.
221. Lachowicz, S. K., *J. Imp. Coll. Chem. Eng. Soc.*, 1954, **8**, 51.
222. Lachowicz, S. K., and Weale, K. E., *Ind. Eng. Chem. Data*, 1958, **3**, 162.
223. Lake, R. F., and Thompson, H. W., *Proc. Roy. Soc.*, 1966, **291**, 469.
224. Landolt-Bornstein, *Physikalisch-Chemische Tabellen* (a series), Springer, Berlin.
225. Lannung, A., *J. Am. Chem. Soc.*, 1930, **52**, 68.
226. Lannung, A., and Gjaldbaek, J. C., *Acta Chem. Scand.*, 1960, **14**, 1124.
227. Lassettre, E. N., *Chem. Rev.*, 1937, **20**, 259.
228. Lawrence, J. H., Loomis, W. F., Tobias, C. A., and Turpin, F. H., *J. Physiol.*, 1946, **105**, 197.
229. Lee, K. H., and Kohn, J. P., *J. Chem. Eng. Data*, 1969, **14**, No. 3, 292.
230. LeFevre, R. J. W., *Trans. Faraday Soc.*, 1937, **33**, 86, etc.
231. Lehfeldt, R. A., *Philos. Mag.*, 1895 (5), **40**, 395; 1898 (5), **46**, 42.
232. Lennard-Jones, J., and Pople, J. A., *Proc. Roy. Soc.*, 1951, **A205**, 155; Pople, J. A., *Proc. Roy. Soc.*, 1951, **A205**, 163.
233. Leonova, L. S., and Ukshe, E. A., *Elektrokhimiya*, 1970, **6**, 892.
234. Lepigre, A. L., *Bull. Soc. Encour. Ind. Natl.*, 1936, **135**, 385.
235. Lewis, G. N., *Z. Phys. Chem.*, 1901, **38**, 205.

236. Lewis, G. N., *Proc. Am. Acad.*, 1907, **43**, 259.
237. Lewis, G. N., *J. Am. Chem. Soc.*, 1908, **30**, 668.
238. Lewis, G. N., *Valence and Structure of Atoms and Molecules*, 1923, Chemical Catalog Co., New York.
239. Lewis, G. N., *J. Franklin Inst.*, **1938**, 293.
240. Lindner, J., *Monatsh. Chem.*, 1912, **33**, 613.
241. Linebarger, C. E., *J. Am. Chem. Soc.*, 1895, **17**, 615.
242. Linford, R. G., and Hildebrand, J. H., *Trans. Faraday Soc.*, 1970, **66**, 577; *J. Phys. Chem.*, 1969, **73**, 4410.
243. Lippincott, E. R., and Schroeder, R., *J. Chem. Phys.*, 1955, **23**, 1099.
244. Lloyd, S. J., *J. Phys. Chem.*, 1918, **22**, 300.
245. Longuet-Higgins, H. C., *Disc. Faraday Soc.*, 1965, No. 40, 7.
246. Luck, W. A. P., *Disc. Faraday Soc.*, 1967, No. 43, 115.
247. Luther, H., and Hiemenz, W., *Chem. Ing. Tech.*, 1957, **29**, 530.
248. Maass, C. E., and Maass, O., *J. Am. Chem. Soc.*, 1928, **50**, 1352.
249. Maass, O., and Morrison, D. M., *J. Am. Chem. Soc.*, 1923, **45**, 1675.
250. Maass, O., and McIntosh, D., *J. Am. Chem. Soc.*, 1912, **34**, 1273; 1913, **35**, 535.
251. Magat, M., *Disc. Faraday Soc.*, 1967, No. 43, 145.
252. Mamedaliev, Yu. G., and Musakhanly, S., *Zh. Prikl. Khim.*, 1940, **13**, 735.
253. Markham, A. E., and Kobe, K. A., *Chem. Rev.*, 1941, **28**, 519.
254. Markham, A. E., and Kobe, K. A., *J. Am. Chem. Soc.*, 1941, **63**, 449.
255. Marshtupa, V. P., Babin, E. P., Kolpakchi, A. A., and Beginina, M. S., *Khim. Prom.*, 1966, **42**, 513.
256. Mellor, J. W., *Comprehensive Treatise on Inorganic and Theoretical Chemistry*, Suppl. II, Pt. I, Longmans, London.
257. Metzger, G., and Sauerwald, F., *Z. Anorg. Chem.*, 1950, **263**, 324.
258. Middleton, W. J., and Lindsey, R. V., *J. Am. Chem. Soc.*, 1964, **86**, 4948.
259. Mills, G. A., and Urey, H. C., *J. Am. Chem. Soc.*, 1940, **62**, 1019.
260. Mitra, S. S., *J. Chem. Phys.*, 1962, **36**, 3286.
261. Möller, H. G., *Z. Phys. Chem.*, 1909, **69**, 449.
262. Morgan, O. M., and Maass, O., *Can. J. Res.*, 1931, **5**, 162.
263. Morris, J. C., *J. Am. Chem. Soc.*, 1946, **68**, 1692.
264. Morrison, T. J., and Billet, F., *J. Chem. Soc.*, **1952**, 3819.
265. Morrison, T. J., and Johnstone, N. B., *J. Chem. Soc.*, **1954**, 3441.
266. Morrison, T. J., and Johnstone, N. B., *J. Chem. Soc.*, **1955**, 3655.
267. Mounajed, T., *Compt. Rend.*, 1933, **197**, 44.
268. Mulliken, R. S., *J. Phys. Chem.*, 1952, **56**, 801 (and other papers in that series).
269. Mulliken, R. S., *J. Chim. Phys.*, **1964**, 20.
270. Murphy, R. A., and Davis, J. C., Jr., *J. Phys. Chem.*, 1968, **72**, 3111.
271. Murray, F. E., and Schneider, W. G., *Can. J. Chem.*, 1955, **33**, 797.
272. Myers, R., and Willet, R. D., *J. Inorg. Nucl. Chem.*, 1967, **29**, 1546.
273. McAmis, A. J., and Felsing, W. A., *J. Am. Chem. Soc.*, 1925, **47**, 2633.
274. McBain, J. W., *J. Chem. Soc.*, 1912, **101**, 814.
275. McCarty, L. V., and Guyon, J., *J. Phys. Chem.*, 1954, **58**, 285.
276. McDaniel, A. S., *J. Phys. Chem.*, 1911, **15**, 587.
277. McGlashan, M. L., *J. Chem. Educ.*, 1963, **40**, 516.
278. McGlashan, M. L., and Rastogi, R. P., *Trans. Faraday Soc.*, 1958, **54**, 496.
279. McGlashan, M. L., and Wingrove, R. J., *Trans. Faraday Soc.*, 1956, **52**, 470.
280. McKetta, J. J., and Wehe, A. H., in *Petroleum Production Handbook*, Vol. 2, T. C. Frick, ed., 1962, McGraw-Hill, New York, pp. 22-1–22-26; gives an extensive bibliography on the various hydrocarbon–water systems.
281. McKinnis, A. C., *Ind. Eng. Chem.*, 1955, **47**, 850.
282. Namiot, A. Yu., *Zh. Strukt. Khim.*, 1961, **2**, 408.
283. Namiot, A. Yu., *Zh. Strukt. Khim.*, 1967, **8**, 408.
284. Naumann, A., *Chem. Ber.*, 1914, **47**, 247.
285. Némethy, G., and Scheraga, H. A., *J. Chem. Phys.*, 1962, **36**, 3382, 3401; *J. Phys. Chem.*, 1962, **66**, 1773.

References

286. Nussbaum, E., and Hursh, J. B., *J. Phys. Chem.*, 1958, **62**, 81.
287. O'Brien, S. J., Kenny, C. L., and Zuercher, R. A., *J. Am. Chem. Soc.*, 1939, **61**, 2504.
288. O'Brien, S. J., and Kenny, C. L., *J. Am. Chem. Soc.*, 1940, **62**, 1189.
289. O'Brien, S. J., and Byrne, J. B., *J. Am. Chem. Soc.*, 1940, **62**, 2063.
290. O'Brien, S. J., and Bobalek, E. G., *J. Am. Chem. Soc.*, 1940, **62**, 3227.
291. O'Brien, S. J., *J. Am. Chem. Soc.*, 1941, **63**, 2709; 1942, **64**, 951.
292. O'Brien, S. J., and King, C. V., *J. Am. Chem. Soc.*, 1949, **71**, 3632.
293. Oda, R., Yoshida, Z., and Osawa, E., *J. Soc. Chem. Ind., Jpn.*, 1960, **63**, 1556.
294. Olson, A. R., and Youle, P. V., *J. Am. Chem. Soc.*, 1940, **62**, 1027.
295. Ostwald, W., *J. Prakt. Chem.*, 1885, **32**, 314.
296. Parsons, L. B., *J. Am. Chem. Soc.*, 1925, **47**, 1820.
297. Partington, J. R., *An Advanced Treatise on Physical Chemistry*, II, 1951, Longmans, Green & Co.
298. Partington, J. R., *General and Inorganic Chemistry*, 1966, Macmillan, London.
299. Peach, M. E., and Waddington, T. C., *J. Chem. Soc.*, **1961**, 1238.
300. Pearson, T. G., and Robinson, P. L., *J. Chem. Soc.*, **1932**, 652.
301. Pearson, R. G., and Vogelsong, D. C., *J. Am. Chem. Soc.*, 1958, **80**, 1048.
302. Planck, M., *Z. Phys. Chem.*, 1888, **2**, 405.
303. Plesch, P. H., *Chem. Ind.*, **1966**, 905.
304. Pleskov, V. A., *J. Phys. Chem. USSR*, 1948, **22**, 3, 351.
305. Prausnitz, J. M., *J. Phys. Chem.*, 1962, **66**, 640.
306. Radley, J. A., Elliott, G., and Cornish, E. H., Brit. Patent 894,456 (1962).
307. Raoult, F. M., (a) For earlier references, see *Chem. Soc. Abstr.*, 1880, 523; 1882, 1260; 1883, 7, 278, 952; (b) Loi Générale de Congélation des dissolvants, *Ann. Chim. Phys.*, 1884, **2** (6), 66–125; (c) Freezing points of salt solutions, etc., *Compt. Rend.*, 1884, **97**, 941; **98**, 509, 1047; **99**, 324; 1885, **100**, 982; 1886, **101**, 1056; **102**, 1307, **114**, 268; *Z. Phys. Chem.*, 1888, **2**, 488; **9**, 343; (d) Vapor tensions of ethereal solutions, etc., *Compt. Rend.*, 1887, **103**, 1125; **104**, 976; *Z. Phys. Chem.*, 1888, **2**, 353; *Ann. Chim. Phys.*, 1890, **20** (6), 297.
308. Rawcliffe, C. T., and Rawson, D. H., *Principles of Inorganic and Theoretical Chemistry*, 1969, Heinemann, London.
308a. Reamer, H. H., Sage, B. H., and Lacey, W. N., *Ind. Eng. Chem.*, 1952, **44**, 609.
309. Reed, C. D., and McKetta, J. J., *J. Chem. Eng. Data*, 1959, **4**, 294.
310. Reeves, L. W., and Hildebrand, J. H., *J. Am. Chem. Soc.*, 1957, **79**, 1313.
311. Reeves, L. W., and Hildebrand, J. H., *J. Phys. Chem.*, 1963, **67**, 1918.
312. Robb, R. A., and Zimmer, M. F., *J. Chem. Eng. Data*, 1968, **13**, 200.
313. Robinson, W. T., and Sund, E. H., U.S. Patent 3,333,399 (1967).
314. Rodebush, W. H., and Ewart, R. H., *J. Am. Chem. Soc.*, 1932, **54**, 419.
315. Rodrigues, A. B. J., McCaffrey, D. S., Jr., and Kohn, J. P., *J. Chem. Eng. Data*, 1968, **13**, 164.
316. Roozeboom, H. W. B., *Rec. Trav. Chim.*, 1884, **3**, 104; 1885, **4**, 107, 379; 1886, **5**, 358; *Z. Phys. Chem.*, 1888, **2**, 454.
317. Roscoe, H. E., and Dittmar, W., *J. Chem. Soc.*, 1860, **12**, 128.
317a. Rosenthal, W., Thesis, Faculty of Science, University of Strasbourg, 1954.
318. Rowland, H. A., *Proc. Am. Acad.*, 1879–1880, **7**, 75.
319. Rowlinson, J. S., *Liquids and Liquid Mixtures*, 1959, Butterworths, London.
320. Russell, J., and Maass, O., *Can. J. Res.*, 1931, **5**, 436.
321. Saum, A. M., *J. Polymer Sci.*, 1960, **42**, 57.
322. Savinov, V. B., Konovalov, E. E., and Peizulaev, Sh. I., *Teplofiz. Svoistva Zhidk., Mater, Vses. Teplofiz. Konf. Svoistvam Veshchestv Vys. Temp.*, 3rd edn., **1968**, 118.
323. Saylor, J., *J. Am. Chem. Soc.*, 1937, **59**, 1712.
324. Saylor, J. H., and Battino, R., *J. Phys. Chem.*, 1958, **62**, 1334.
325. Scatchard, G., *Chem. Rev.*, 1931, **8**, 321.
326. Schleyer, P. R., and West, R., *J. Am. Chem. Soc.*, 1959, **81**, 3164.
327. Schmid, H., Maschka, A., and Sofer, H., *Monatsh. Chem.*, 1964, **95**, 348.
328. Schneider, W. G., *J. Chem. Phys.*, 1955, **23**, 26.
329. Schröder, W., *Z. Naturforscher*, 1969, **24b**, 500.
330. Schulze, A., *Z. Phys. Chem.*, 1920, **95**, 257.
331. Seel, R. M., and Sheppard, N., *Spectrochim. Acta*, 1969, **25A**, 1287.

332. Seidell, A, and Linke, W. F., *Solubilities of Inorganic and Organic Compounds*, 1958, 1965.
333. Shade, R. W., Cooper, G. D., and Gilbert, A. R., *J. Chem. Eng. Data*, 1959, **4**, 213.
334. Shepard, H. H., *Chemistry and Toxicology of Insecticides*, Burgess Publishing Co., Minneapolis.
335. Sims, T. H., *Justus Liebigs Ann. Chem.*, 1861, **118**, 333; *J. Chem. Soc.*, 1862, **14**, 1.
336. Smith, T. L., *J. Phys. Chem.*, 1955, **59**, 188.
337. Smith, W. T., and Parkhurst, R. B., *J. Am. Chem. Soc.*, 1922, **44**, 1918.
338. Spano, J. O., Heck, C. K., and Barrick, P. L., *J. Chem. Eng. Data*, 1968, **13**, 168.
339. Speyers, C. L., *Am. J. Sci.*, 1900, (iv), **9**, 341; *J. Am. Chem. Soc.*, 1899, **21**, 282, 725; *J. Phys. Chem.*, 1898, **2**, 347, 362.
340. Stackelberg, M., and Muller, H. R., *J. Chem. Phys.*, 1951, **19**, 1319; *Naturwissenschaften*, 1951, **38**, 456; 1952, **39**, 20.
341. Steele, B. D., McIntosh, D., and Archibald, E. H., *Philos. Mag.*, 1905, **205**, 99.
342. Stepanova, G. S., *Gazov. Delo*, **1970**, 26.
343. Stepanova, G. S., and Velikovskii, A. S., *Gazov. Delo*, **1970**, 31.
344. Stephen, H., and Stephen, T., *Solubilities of Inorganic and Organic Compounds*, 1963, Pergamon Press, Oxford.
345. Stevenson, D. P., *J. Am. Chem. Soc.*, 1962, **84**, 2849.
346. Stevenson, D. P., and Coppinger, G. M., *J. Am. Chem. Soc.*, 1962, **84**, 149.
347. Stock, A., and Kuss, E., *Chem. Ber.*, 1917, **50**, 159.
348. Strohmeier, W., and Echte, A., *Z. Electrochem.*, 1957, **61**, 549.
349. Stuart, A. V., and Sutherland, G. B. B. M., *J. Chem. Phys.*, 1956, **24**, 559.
350. Svetlanov, E. B., Velichko, S. M., Levinskii, M. I., Treger, Yu. A., and Flid, R. M., *Zh. Fiz. Khim.*, 1971, **45**, 877.
351. Swain, C. G., and Thornton, E. R., *J. Am. Chem. Soc.*, 1962, **84**, 822.
353. Tartar, H. V., and Garretson, H. H., *J. Am. Chem. Soc.*, 1941, **63**, 808.
353. Taylor, A. E., *J. Phys. Chem.*, 1900, **4**, 675.
354. Taylor, N. W., and Hildebrand, J. H., *J. Am. Chem. Soc.*, 1923, **45**, 682.
355. Terres, E., and Assemi, M. T., *Brennst-Chem.*, 1956, **37**, 257.
356. Thomsen, E. S., and Gjaldbaek, J. C., *Acta Chem. Scand.*, 1963, **17**, 127, 134.
357. Timmerman, J., *Physico-Chemical Constants of Pure Organic Compounds*, Vol. I, 1950; Vol. II, 1965, Elsevier, New York.
358. Tsiklis, D. S., and Svetlova, G. M., *Zh. Fiz. Khim.*, 1958, **32**, 1476.
359. Tuck, D. G., and Diamond, R. M., *Proc. Chem. Soc.*, **1958**, 236.
360. Valentiner, S., *Z. Phys. Chem.*, 1930, **61**, 563.
361. Van Arkel, A. E., and Vles, S. E., *Rec. Trav. Chim.*, 1936, **55**, 407.
362. Van Laar, J. J., *Z. Phys. Chem.*, 1910, **72**, 723, 1913, **83**, 599.
363. Van Liempt, J. A. M., and van Wijk, W., *Rec. Trav. Chim.*, 1937, **56**, 632.
364. Vdovichenko, V. T., and Kondratenko, V. I., *Khim. Prom.*, **1967**, 290.
365. Vilcu, R., and Gainar, I., *Rev. Roum. Chim.*, 1967, **12**, 181.
366. Waddington, T. C., and Klanberg, F., *Naturwissenschaften*, 1959, **46**, 578; *J. Chem. Soc.*, **1960**, 2329, 2332, 2339.
367. Waters, J. A., Mortimer, G. A., and Clements, H. E., *J. Chem. Eng. Data*, 1970, **15**, 174.
368. Weaver, J. R., and Parry, R. W., *Inorg. Chem.*, 1966, **5**, 718.
369. Weimer, R. F., and Prausnitz, J. M., *Spectrochim. Acta*, 1966, **22**, 77.
370. Weissenberger, G., and Hadwiger, H., *Z. Angew. Chem.*, 1927, **40**, 734.
371. Wen, W. Y., and Hung, J. H., *J. Phys. Chem.*, 1970, **74**, 170.
372. West, R., and Kraihanzel, C. S., *J. Am. Chem. Soc.*, 1961, **83**, 765.
373. Wheeler, C. M., Jr., and Keating, H. P., *J. Phys. Chem.*, 1954, **58**, 1171.
374. Whitney, R. P., and Vivian, J. E., *Ind. Eng. Chem.*, 1941, **33**, 741.
375. Wicke, E., *Angew. Chem. Int. Ed. Engl.*, 1966, **5**, 106.
376. Winkler, L. W., *Chem. Ber.*, 1901, **34**, 1408.
377. Winkler, L. W., *Z. Phys. Chem.*, 1906, **55**, 350.
378. Wilhelm, E., and Battino, R., *Chem. Rev.*, 1973, **73**, 1.
379. Wilhelm, E., and Battino, R., *J. Chem. Thermodyn.*, 1971, **3**, 379.
380. Williams, D., *Phys. Rev.*, 1936, **50**, 719 (see also Plyler, E. K., and Williams, D., *Phys. Rev.*, 1936, **49**, 215).

381. Wishnia, A., *Proc. Nat. Acad. Sci. U.S.A.*, 1962, **48**, 2200.
382. Witschonke, C. R. and Kraus, C. A., *J. Am. Chem. Soc.*, 1947, **69**, 2472.
383. Wright, R. H., and Maass, O., *Can. J. Res.*, 1932, **6**, 588.
384. Wynne-Jones, W. F. K., *J. Chem. Soc.*, **1930**, 1064.
385. Yen, L. C., and McKetta, J. J., Jr., *Am. Inst. Chem. Eng. J.*, 1962, **8**, 501.
386. Yen, L. C., and McKetta, J. J., Jr., *J. Chem. Eng. Data*, 1962, **7**, 288.
387. Yoneda, H., *Bull. Chem. Soc. Jpn.*, 1958, **31**, 708.
388. Young, S., *J. Chem. Soc.*, 1902, **81**, 768.
389. Young, S., and Fortey, E. C., *J. Chem. Soc.*, 1903, **83**, 45.
390. Zawidski, J., *Z. Phys. Chem.*, 1900, **35**, 129.
391. Zellhoefer, G. F., *Ind. Eng. Chem.*, 1937, **29**, 548.
392. Zellhoefer, G. F., Copley, M. J., and Marvel, C. S., *J. Am. Chem. Soc.*, 1938, **60**, 1337; Zellhoefer, G. F., and Copley, M. J., *J. Am. Chem. Soc.*, 1938, **60**, 1343.
393. Zhuze, T. P., and Zhurba, A. S., *Bull. Acad. Sci. USSR, Div. Chem. Sci.*, **1960**, 335.
394. *Water and Aqueous Solutions: Structure, Thermodynamics, and Transport Processes*, Horne, R. A., ed., 1972, Wiley-Interscience, New York.
395. *Water, A Comprehensive Treatise*, Franks, F., ed., Volumes 1–5, 1972–1975, Plenum Press, New York.
396. Shipman, L. L., and Scheraga, H. A., *J. Phys. Chem.*, 1974, **78**, 909.
397. Hagler, A. T., Scheraga, H. A., and Némethy, G., *J. Phys. Chem.*, 1972, **76**, 3229.
398. Ben-Naim, A., and Stillinger, F. H., in *Water and Aqueous Solutions: Structure, Thermodynamics, and Transport Processes*, Horne, R. A., ed., 1972, Wiley-Interscience, New York.
399. *Physico-Chemical Processes in Mixed Aqueous Solvents*, Franks, F., ed., 1967, Heinemann, London.
400. Kaulgud, M. V., and Patil, K. J., *J. Phys. Chem.*, 1974, **78**, 714.
401. Symons, M. C. R., *Nature*, 1972, **239**, 257.
402. Hölemann, P., and Hasselmann, R., *Forschungsberichte des Wirtschafts und Verkehrsministeriums Nordrein-Westfalen*, Nr. 109, 1954 (Westdeutscher Verlag/Köln und Opladen). Untersuchungen über die Löslichkeit von Azetylen in verschiedenen organischen Lösungsmitteln im Auftrag der Forschungstelle fur Azetylen, Dusseldorf–Dortmund.
403. Sander, W., *Z. Phys. Chem.*, 1911, **78**, 513; *J. Chem. Soc. Abstr.*, (ii), 1912, 251.
404. Krichevskii, I. R., and Lebedeva, E. S., *Zh. Fiz. Khim.*, 1947, **21**, 715.
405. Frolich, P. K., Tauch, E. J., Hogan, J. J., and Peer, A. A., *Ind. Eng. Chem.*, 1931, **23**, 548.
406. Frame, J. P., Rhodes, E., and Ubbelohde, A. R., *Trans. Faraday Soc.*, 1961, **57**, 1075.
407. Grimes, W. R., Smith, N. V., and Watson, G. M., *J. Phys. Chem.*, 1958, **62**, 862.
408. Culberson, O. L., Horn, A. B., and McKetta, J., Jr., *Trans. Am. Inst. Mining Metall. Eng. Pet. Div.*, 1950, **189**, 1.
409. Culberson, O. L., and McKetta, J., Jr., *Trans. Am. Inst. Mining Metall. Eng. Pet. Div.*, 1951, **192**, 223.

Index

The index is intended to be used in conjunction with the Table of Contents, pp. vii–x. Names of individual compounds are given only when it is expedient.

Absorption coefficient, 77, 224
  of Bunsen, 73, 78-80, 93, 99-101, 182-184, 218, 219, 232
  of Ostwald, 72, 78, 83, 88, 90, 182-184, 219, 226, 232, 234, 248
Acetic acid, 45, 157, 158, 162
Acetic anhydride, 237, 238
Acetone, 238
Acetylene, ethyne, 26, 88, 226, 241, 242
Acetylenes, 119, 212
Acid–base function, 26, 59, 73, 88, 94, 95, 109, 115-121, 131, 138, 143, 145, 146, 150, 152, 156, 157, 164, 166, 187, 189, 212-215, 241, 242, 244
Activity, 52, 158, 184, 221
Adsorption on silica, 118
Alcohols, 26, 73, 85, 86, 88, 91, 92, 94, 106, 121-124, 133, 134, 138, 146, 149, 164, 168, 181, 183, 187, 212, 216, 237, 244, 248, 251-254,
Amines, 169, 181, 184
Ammonia, 27, 30, 31, 56, 64
Anilines, 109, 164, 169, 176, 181
Antimony hydride, 225
Argon, 22, 72, 80, 88, 232, 236-240
Aromatic hydrocarbons, 136, 141-143, 145, 148, 151, 152, 169, 176, 181, 215, 219, 225, 240, 242
Associated liquid, 53, 55
Association constant, 117

Benzene, 136, 140, 158, 227, 249, 254
Biocidal action, 202
Boron trihalides, 123, 130, 148, 225
Bovine serum albumen in water, 63
Bromine, 103

1-Bromonaphthalene, 183
n-Butane, 9-11, 26, 225, 257
Butenes, 240-243

Carbon dioxide, 29, 49, 56, 57, 73, 88, 219, 230, 237, 246-251
Carbon monoxide, 25, 29, 56, 73, 86, 230, 235
Carbonyl chloride, 26
Carbonyl selenide, 230
Carbonyl sulfide, 224, 225, 228
Carboxylic acids, 13, 85, 94, 157, 158, 187, 242
Charge transfer theory, 115, 118, 167
Chemical interaction, 26, 53, 55, 73, 75, 115
Chloretone, $CCl_3(CH_3)_2COH$, 212
Chlorine, 70, 74, 230
Chloroform, 53, 155, 164, 167, 169, 176, 187
Chloromethane, 26, 189
Chromium oxychloride, 230
Complex formation, 53, 138, 148
Critical temperature, 231
Cyanogen, 225
Cyclic ethers, 125, 135, 229

Degree of dissociation, 155
Depression of the freezing point, 36
Depression of the vapor pressure, 37
Desalination, 202
Diborane, 229
Di-n-butyl ether, 119
Di-n-butyl sulfide, 119
Dielectric constant, 55, 135, 149, 151, 167, 189
Diethyl ether, 37-41, 50, 53, 150, 151, 167, 187
Dimethoxyethane, 119
N,N-Dimethylacetamide, 119, 246
Dimethyl ether, 26, 151, 225

Dimethylformamide, 119, 169, 176, 187, 209, 212, 241-245
Dinitrogen tetroxide, 237
Dinitrogen trioxide, 237
Diphenyl ether, 119
Diphenyl sulfide, 119
Dipole–dipole interaction, 53, 59, 60, 119, 138, 255
Dipole moment, 53, 55, 91, 166, 167, 189
Dissociation constant, 117, 161
Dolezalek's theory, 47

Electrolytic conductance, 146, 149, 161, 187
Electron donor–acceptor terminology, 56, 57, 109, 115, 118, 120, 147, 152, 155-157, 167
Electronegativity, 119
Enthalpy, 61, 138, 141, 218
Entropy, 61, 138, 141, 218, 232, 238
Eötvös' constant, 135
Esters of carboxylic acids, 124, 136, 169, 238, 248
Esters of inorganic acids, 126, 145, 147, 232
Ethane, 16, 18, 19, 56
Ethers, 125, 135, 138, 142, 144, 148, 168, 176, 187, 189, 229, 242
Ethylene, ethene, 65
Ethylene glycol, 139, 140, 181

Flickering-cluster model, 60
Force constants, 56, 91
Free energy, 61, 119, 153, 227
Fugacity, 52, 98, 99, 140, 145, 208, 221, 226

Gas, definition of, 2
Gas hydrates, 61
Glycerol, 52, 85, 168, 181
Glycols, 134, 136, 139, 168, 181

Halogenoalkanes, 68, 69, 71, 72, 74, 118, 119, 125, 130, 135, 141, 189, 191, 219, 224, 226, 238
Halogenobenzenes, 125, 140, 141, 181, 251, 252
Heat of mixing, 53
Heavy water, 203
Helium, 21, 26, 71, 74, 218, 224, 232, 237, 257
Hemoglobin in water, 63
Henry's law, 1, 51, 70, 75, 78, 79, 91, 93, 94, 97, 98, 102, 104, 105, 109, 112, 139, 141, 142, 144, 145, 182, 183, 209, 225, 226, 236-238, 248, 254, 259, 260
Henry's law constant, 30, 36, 51, 69, 91-94, 100, 139, 143-145, 214, 221, 247, 248
1,1,1,3,3,3-Hexachloroisopropanol, 212
1,1,1,3,3,3-Hexafluoroisopropanol, 212

Hexamethylphosphoramide, 241
n-Hexanoic acid, 13, 38, 242
Homolytic fission, 117
Hydrazines, 237
Hydrocarbon liquids, 125, 141, 144, 145, 151-154, 163, 215, 219, 226, 227, 229, 238, 242, 257
Hydrocarbon gases, 11, 26, 27, 61, 63, 72, 85, 86, 91, 203
Hydrocarbon gas hydrates, 62
Hydrogen, 24, 56, 57, 64, 73, 86, 218, 219, 230, 237, 251, 254
Hydrogen bonding, 95, 96, 116-118, 147, 148, 152, 155, 156, 159, 163-169, 181, 187, 189, 199, 212, 213, 218
Hydrogen fluoride, hydrofluoric acid, 151, 152
Hydrogen halides, 26, 30-34, 64, 161, 187, 189, 212, 228
Hydrogen ion, 60, 117, 158
Hydrogen selenide, 228, 235
Hydrogen sulfide, 26, 29, 73, 189, 228, 236
Hydrogen telluride, 225, 235
Hydrophobic interactions, 59, 63, 185
Hydroxyl ion, 60, 163

Ideal behavior, 1, 88, 93, 199, 221, 260
Ideal gas and deviation factor, 78, 79
Ideality, 53, 55, 78
Ideal solubility, 56, 97, 99, 144, 145, 199
Ideal solutions, 52, 53, 78, 142, 158, 164, 199, 260
Infrared spectroscopy, 95, 134-136, 138, 141, 148, 151, 156, 157
Intermolecular forces, 26, 53, 55, 115, 116, 205, 209
Intermolecular potentials, 61
Internal pressure, 55, 90
Interstitial dissolution, 184, 185
Intramolecular electron density pattern, 25, 26, 205
Iodine in water, 103
Ionization constants of carbonic acid, 92, 152
Ionization potential, 118, 157
Ionizing solvent, 151

Krypton, 22

Lewis acids and bases, 116
Lewis's system of thermodynamics, 51
London dispersion forces, 119
Lowry–Brönsted theory, 117
Lysozyme in water, 63

Mercaptans, 95, 119, 147
Methane, 16, 19, 20, 56, 65, 218, 237, 251, 254, 257

Index

Methanol, 52, 106, 151, 153, 251
Methylamines, 64
Methyl benzenes, 118, 143, 151, 152, 181, 242
Methylchlorosilane, 228
Methylsilane, 228
Molal volume, 56
Mole fraction, 1, 6, 11-13
Mole ratio, 1, 6, 11-13
Molten fluorides, 257
Molten inorganic nitrates, 257
Molten rubidium chloride, 101

Narcotic effect of noble gases, 87
Neon, 21, 64, 74, 232
Nitriles, 153-157, 169
Nitrobenzene, 38, 150, 169, 183, 252
Nitrogen, 23, 29, 48, 56, 72, 73, 80, 86, 218, 232, 236, 237, 251, 254-256
Nitromethane, 150
Nitrosyl chloride, 99
Nitrous oxide, 29
Noble gases, 20, 26, 70, 83, 232, 236, 257
Nonpolar gases, 56, 80, 88, 91, 218
Nonpolar liquids, 56, 61, 88, 91, 238
Normal–abnormal terminology, 53, 55
Nuclear magnetic resonance, 118, 157

n-Octane, 74
Oxonium compounds, 133
Oxygen, 24, 29, 48, 73, 80, 86, 218, 232, 236, 254-256

Pair potentials, 57, 227
Partial molal volumes, 56
Partial pressure, 78
Pauli principle, 161
Perfect solutions, 53
Pest control, 203
Phenols, 125, 134
Phoreograms, 149, 150
Phosphine, 29
Phosphorus halides, 121, 130, 148, 225
Physical interaction, 26, 53, 55, 75, 115
Polarizability, 26, 166
Polarizability of gases, 84
Polar–nonpolar terminology, 53, 55, 56, 88, 91, 143, 157, 163, 221, 235, 236, 238
Propane, 15
Propene, propylene, 240-243
Proton donor–acceptor terminology, 116, 118, 119, 147, 155
Proton magnetic resonance, 167
Proton transfer, 116, 159, 163, 189

Radon, 23
Raoult's law, 1, 90, 98, 100, 110, 140, 144, 145, 199, 299, 259, 260
Raoult's law, deviation from, 53, 109, 140-142, 167, 184, 208
Refrigeration systems, 191
Regular solutions, 142
Rocket propellants, 236

Scaled particle theory, 227
Silane, SiH_4, 225, 228
Silicon tetrachloride, 122, 130, 148
Sodium lauryl sulfate, in water, 63
Solubility parameter, 91, 184, 220, 235
Solubility spectrum, 17, 224, 235
Stannic chloride, 130
Steric hindrance, 121
Substitutional dissolution, 184
Sulfur compounds, 127, 146, 147
Sulfur dioxide, 26, 49, 74, 163, 189, 237-239
Sulfur hexafluoride, 74, 224, 226
Sulfuric acid, 40, 187
Sulfurous acid, 111, 112

Tetraalkylammonium salts in water, 63, 64
Tetra-n-butoxysilane, 145
Tetrahydrofuran, 229, 230, 244
Thionyl halides, 123
Titanium tetrachloride, 130, 135, 148
Total bond energy, 135
Trialkyl phosphites, 145
Tri-n-butyl phosphate, 74, 145
2,2,2-Trichloroethanol, 181, 212
Triphenyl phosphite, 145
Tris(trichloroethyl) borate, 145
Tris(trichloroethyl) phosphite, 145
True solubility, 145

Ultraviolet spectra, 166

Van der Waals forces, 53
Volume change on mixing, 53

Water
 activity in hydrochloric acid, 158
 solubility in inorganic nitrate melts, 254
 solubility in liquids, 75

Xenon, 23, 218

Zawidski's diagrams, 44, 45